Y0-CUQ-349

Static and Dynamic Fracture Mechanics

Thanks are due to Professor A. Carpinteri for the use of Figure 29 on page 360, which appears on the front cover of this book.

International Centre for Mechanical Sciences

Static and Dynamic Fracture Mechanics

Editors:
M.H. Aliabadi
Wessex Institute of Technology, Sothampton, UK
C.A. Brebbia
Wessex Institute of Technology, Southampton, UK
V.Z. Parton
École Centrale Paris, France

Computational Mechanics Publications
Southampton Boston

M.H. Aliabadi
Wessex Institute of Technology
Ashurst Lodge
Ashurst
Southampton
SO4 2AA, UK

C.A. Brebbia
Wessex Institute of Technology
Ashurst Lodge
Ashurst
Southampton
SO4 2AA, UK

V.Z. Parton
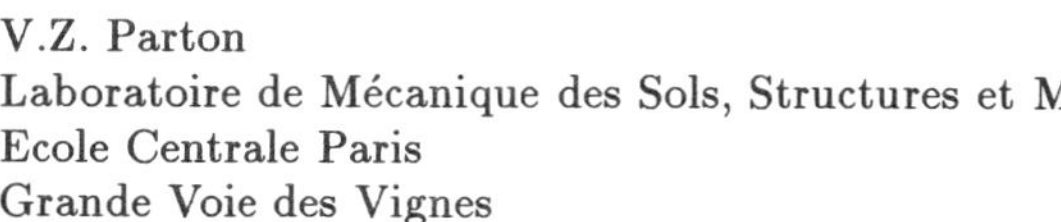
Laboratoire de Mécanique des Sols, Structures et Materiaux
Ecole Centrale Paris
Grande Voie des Vignes
92295 Châtenay-Malabry Cedex
France

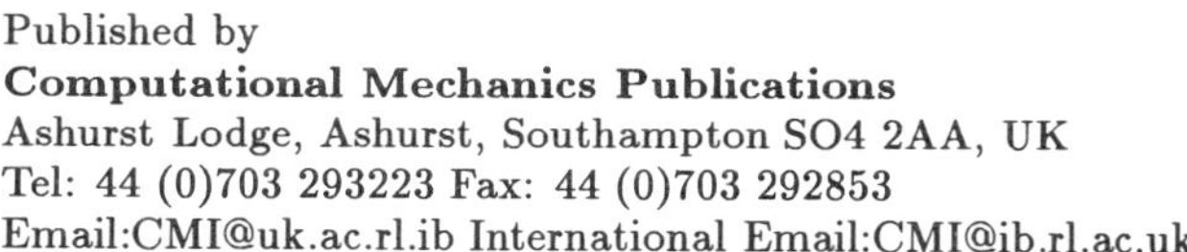
Published by
Computational Mechanics Publications
Ashurst Lodge, Ashurst, Southampton SO4 2AA, UK
Tel: 44 (0)703 293223 Fax: 44 (0)703 292853
Email:CMI@uk.ac.rl.ib International Email:CMI@ib.rl.ac.uk

For USA, Canada and Mexico

Computational Mechanics Publications Inc
25 Bridge Street, Billerica, MA 01821, USA
Tel: 508 667 5841 Fax: 508 667 7582

British Library Cataloguing-in-Publication Data

A catalogue for this book is available from the British Library

ISBN 1-85312-251-3 Computational Mechanics Publications, Southampton
ISBN 1-56252-175-6 Computational Mechanics Publications, Boston

Library of Congress Catalog Card Number 93-72573

Printed and bound in Great Britain by Bookcraft Ltd, Bath, UK

Contents

PREFACE

The science of fracture mechanics has important practical significance in engineering, since it provides a rational basis for the calculation of the strength of cracked structures and the determination of growth rates of cracks in fatigue. Its use enables components, which may contain crack-like flaws, to be designed with greater safety margins. This reduces the risk of costly in-service failures.

This book originated as a result of a series of lectures by the authors given at the International Centre for Mechanical Science (CISM) in Udine, Italy. Following the meeting the editors encouraged the lecturers to write an extended and updated version of their notes, which were then edited into the present book.

Topics covered in this volume include fundamentals of fracture mechanics, the boundary element method and its application to both static and dynamic crack problems in fracture mechanics, fundamentals of dynamic crack propagation, finite element analysis of cracked structures, cracking of strain-softening materials and the modelling of fracture in concrete materials. Each chapter has been written by a leading researcher in the field and is aimed at researchers and postgraduate students interested in the numerical methods used in theoretical and applied fracture mechanics.

The editors would like to express their appreciation to all the authors for their contributions to this book. Special thanks are also due to Mrs Rosamary Reinoso for her careful work on the preparation of the final manuscript.

The Editors

Chapter 1

Development of fracture mechanics

D.P. Rooke

DRA, Farnborough, Hampshire, GU14 6TD, UK

Summary

This chapter outlines the development of fracture mechanics, from the pioneering work of A.A. Griffith in the 1920s to current technological applications of fracture control strategies. The chapter is divided into six major sections. In the first the general concepts of linear-elastic fracture mechanics (LEFM) are described. In particular the stress intensity factor is introduced and its use demonstrated in the calculation of the static strength of cracked bodies and in the calculation of the rate of growth of fatigue cracks.

Knowledge of the basic principles of solid mechanics is essential for the study of fracture mechanics. In the second section those results and equations that will be needed in later sections are reviewed. The basic equations of elasticity are first stated and their physical significance explained, and then some special 'stress-function' solution of these equations are outlined. In particular, solutions which are suitable for solving crack problems in both two and three dimensions are considered; the case of mixed mode fracture is also considered. The principles of fracture mechanics are formulated in terms of the strain-energy release rate and the relationship to stress intensity factors is outlined. Some procedures for extending LEFM theory to problems with plastic deformations are also described.

Many theoretical (analytical and numerical) and experimental methods for evaluating stress intensity factors are now available. Techniques involving Green's functions have been used extensively for determining stress intensity factors of cracks under complex loading conditions. These techniques, which offer scope for further development, are discussed in the third section. The Green's functions developed give an insight into the importance of different types of loading and structures and demonstrate the versatility of the technique. Two generalisations into what are known as 'weight function' techniques are also described in this section.

The fourth section describes a simple but versatile method of obtaining approximate stress intensity factors for complex structural configurations from those for simple configurations. The method is illustrated with several examples for both single and multiple cracks. The sources of error are identified and the errors shown to be acceptable for most engineering applications.

The section on residual static strength is concerned with the calculation of the residual strength of a cracked structure. The basic premise in residual strength theory is that a cracked material will fail by rapid unstable crack growth if the applied stress intensity factor is equal to or greater than the fracture toughness of that material. It is a well established fact that the plane strain fracture toughness K_{Ic}, applicable to thick sections, is a material constant. For materials in thin sections, the fracture toughness K_{c} is not a constant, but is a function of material thickness, crack length and specimen size. This leads to difficulties in developing a universal procedure for calculating residual strengths. Some of the difficulties are described and also some of the attempts to overcome them.

The ultimate aim of fracture mechanics is fracture control. Whatever the conditions, elastic or plastic, static or dynamic and whatever the materials, the aim is to develop procedures for controlling fracture in practical engineering components. Apart from the economic cost of fracture it is also important to realise that in some situations uncontrolled fracture can put human life at risk. The final section discusses the principles of fracture control and procedures for implementation. Many fracture control plans already exist, indeed in some industries there are statutory requirements for such plans.

1 General concepts of fracture mechanics

1.1 Introduction

The first systematic investigation of fracture phenomena was carried out by A.A. Griffith[1] nearly seventy years ago. His experimental studies were carried out on glass, a brittle material. In order to explain the measured strengths of different glass-rods, Griffith postulated the existence of crack-like flaws which may or may not increase in size under the action of external loads. Griffith further postulated that the growth of the cracks was controlled by the balance between the available strain energy and the energy required to form new crack surfaces. The idea of an energy balance criterion was extended by Irwin[2] and Orowan[3] to ductile materials. They suggested that the energy of deformation must also be considered since, for ductile materials, it can be much larger than the surface energy.

Irwin[4] showed that the elastic-stress field in the vicinity of a crack-tip always has the same functional dependence on the spatial coordinates and the stress magnitudes are determined by one parameter, the stress intensity factor. This parameter can be related to the strain energy used by Griffith.

The fundamental postulate of linear elastic fracture mechanics (LEFM) is that the behaviour of a crack, (i.e. whether it grows or not, and how fast it grows) is determined solely by the value of the stress intensity factor. This factor, which is a function of the applied loading and the geometry of the cracked component, has been evaluated for many hundreds of structural configurations. Some of the methods available for obtaining stress intensity factor solutions to crack problems will be described in later sections. Most of the available solutions for the simpler structural configurations, have been collected together in handbooks for the convenience of engineers and designers.

The stress intensity factor is used in three main areas: i) the determination of the residual strength of a cracked structure; ii) the determination of the rate of growth of a crack in a structure subjected to variable loading (fatigue); iii) the determination of the strength of a cracked, loaded structure in a corrosive environment (stress corrosion).

These topics will be discussed briefly in this section. Finally, this section will indicate that LEFM has limitations due to the fact that in real materials some deformation occurs at the crack-tip which is both non-linear and non-elastic.

1.2 Energy balance

The first systematic study of fracture phenomena was carried out by A.A. Griffith[1] who measured the tensile strengths of glass rods. He observed that freshly drawn glass rods fractured at a higher stress than old rods of the same diameter and that thin rods fractured at a higher stress than thick rods. To explain these phenomena he postulated the existence of crack-like flaws which weakened the glass rods; fracture occurred when these flaws spread across the section. He suggested that as the rods aged more flaws would appear at the surfaces, because of mechanical damage or corrosion; these old rods would tend to be weaker than new ones since the old rods were more likely to be cracked. It also follows that thick rods would tend to be weaker than thin ones since the thick rods with larger surface area are more likely to be cracked. Since the depth, and hence severity of these naturally occurring cracks is largely random, it follows that the rods with more cracks are more likely to contain the severe ones.

How failure initiated at sharp cracks could not be explained by extrapolating what was known about notches. It was known: (i) that notches acted as stress concentrators and high stresses occurred at the root of the notches in a loaded body; (ii) failure resulted if the stress at the notch tip exceeded the material strength; (iii) the stress concentration effect was larger the smaller the root-radius of the notch. The stress concentration of a crack (zero root-radius) tends to infinity and thus failure should occur at near-zero stresses for all cracks which was contrary to experimental observation. Griffith suggested that the criterion for failure due to crack growth was determined by the balance of strain energy and surface energy. He postulated that if the strain energy released by the strain field when the crack advanced a small distance was greater than the energy required to form the new surfaces then unstable crack growth leading to failure would follow. Symbolically, unstable crack growth occurs if

$$G \geq 2\Gamma \tag{1}$$

where G is the strain energy release rate (per unit area of crack growth) and Γ is the work required to form unit area of new crack surface (two surfaces). The strain energy release rate is a function of the loading and the crack size.

By considering a crack as the limiting case of a thin ellipsoidal cavity Griffith showed that the strain energy release rate was given by

$$G = \pi\sigma^2 a/E \tag{2}$$

where σ is the tensile stress applied remote from the crack in a direction perpendicular to the crack, $2a$ is the crack length and E is the Young's modulus of the material. From eqn (2) it can be seen that at the onset of instability ($G = 2\Gamma$) the failure stress σ_c is related to the critical crack length a_c as follows:

$$\sigma_c \propto 1/\sqrt{a_c} \tag{3}$$

This functional dependence was verified experimentally.

Griffith's experiments were conducted on glass, a brittle material, which fractures with little or no permanent deformation. Most structural materials, e.g. metals, are ductile so that fracture is accompanied by permanent deformation. Irwin[2] and Orowan[3] independently suggested that Griffith's energy-balance criterion could be extended to ductile materials. They suggested that failure by unstable crack-growth would occur if

$$G \geq 2\Gamma + \Delta \tag{4}$$

where Δ is the non-recoverable work associated with the permanent deformation at the crack-tip. For ductile materials such as metals $\Delta >> \Gamma$, it therefore follows that the energy-balance criterion becomes

$$G \geq \Delta \tag{5}$$

This relationship illustrates why more work must be done to fracture a ductile material than to fracture a brittle material. Materials with a large work of fracture are said to be 'tough' - the parameter used to measure this property is called the 'fracture toughness'.

1.3 Stress intensity factors

Irwin[4] solved several two-dimensional crack problems in linear elasticity and showed that the stress field in the vicinity of the crack-tip was always of the same form. He showed that the stress-field σ_{ij} at the point (r, θ) near the crack-tip is given by

$$\sigma_{ij}(r,\theta) = \frac{K}{\sqrt{2\pi r}} f_{ij} + \quad \text{other terms} \tag{6}$$

where the origin of coordinates is at the tip. As the coordinate r approaches zero the leading term in eqn (6) dominates; the 'other terms' are constant or tend to zero. The constant K in the first term is known as the stress intensity factor. It therefore follows that the stress field in the vicinity of the crack-tip is characterised by the stress intensity factor.

By considering the elastic work to close-up the tip of a crack Irwin[4] derived a relationship between the strain energy release rate and the stress intensity factor; it was

$$K^2 \propto G \tag{7}$$

The constant of proportionality in eqn (7) is a function of the elastic constants of the material. This relationship provides a link between the crack-tip stress field and the energy required for a crack to grow. This means that the energy-balance criterion for crack growth can now be interpreted in terms of a critical K value that is required for crack growth to start.

Since the basic assumption of fracture mechanics is that the growth (stable or unstable) of a crack is controlled by the stress field at the crack-tip, it follows that crack growth will be characterised by the parameter K. This implies that two different cracks which have the same K-value will behave in the same manner. Thus, in order to predict the growth of a crack in a structural component, it is necessary to evaluate the stress intensity factor for that component. In general K will be a function of the crack size and shape, the type of loading and the geometrical configuration of the structure. The stress intensity factor is often written in the following form:

$$K = Y\sigma\sqrt{\pi a} \tag{8}$$

where σ is a stress, a is a measure of the crack-length and Y is a non-dimensional function of the geometry.

In the thirty-odd years since Irwin[4] demonstrated the importance of the stress intensity factor in determining crack-tip stress fields many different methods have been devised for obtaining K and many K-solutions now exist. These solutions can be found in collected works and in many other publications. Some of the actual methods devised will be described in detail in later sections. To give some idea of the scope of the available methods the more common ones are listed below.

Simple theoretical methods[5]
- superposition
- stress concentrations
- stress distributions
- compounding
- Green's functions
- weight functions

Advanced theoretical methods (numerical)
- collocation (mapping)
- integral transforms
- alternating techniques
- force/displacement matching
- boundary integral equations
- finite elements

Experimental methods
- compliance
- photoelastic
- fatigue crack-growth

The method chosen to solve a particular problem will depend on the complexity of the problem, the time and money available and the degree of accuracy to which the solution is required.

1.4 Residual strength

The criterion for failure due to unstable crack-growth can be expressed in the following way: failure occurs if

$$\left.\begin{array}{ll} K \geq K_{\mathrm{Ic}} & \text{(plane strain)} \\ K \geq K_{\mathrm{c}} & \text{(plane stress)} \end{array}\right\} \tag{9}$$

where K_{Ic} or K_{c} are considered to be constants, sometimes called the 'fracture toughness' of the material. Experience has shown that K_{Ic} is indeed a constant for a given material but is applicable to thick sections only. For thinner sections stable crack-growth can occur and K_{c} is found to vary with crack-length and specimen geometry. An alternative, but equivalent in the elastic regime, approach is to use the crack-opening-displacement COD as a criterion for failure. An attempt to incorporate stable crack-growth in the description of the material leads to what is known as the R-curve.

From eqns (8) and (9) the failure criterion can be written as

$$Y\sigma_{\mathrm{c}}\sqrt{\pi a_{\mathrm{c}}} = K_{\mathrm{c}} \tag{10}$$

where σ_{c} and a_{c} are respectively the critical stress and the critical crack length. The functional relationship between σ_{c} and a_{c} given in eqn (10) is a generalisation of that derived by Griffith (see eqn (3)). Two possible practical situations can now be considered with the aid of eqn (10); they are : i) what is the maximum crack-size that is safe at a given stress level? and ii) what is the maximum safe operating stress for a given crack-size?

From eqn (10) it follows that, for a given stress-level σ_c, the critical crack length depends on both the toughness K_c and the stress σ_c . In particular

$$a_c \propto (K_c/\sigma_c)^2 \tag{11}$$

Thus the critical crack-length increases as the toughness increases and decreases as the stress-level increases. In practice cracks must be detectable before they reach the length a_c, thus the minimum detectable size a_{min} sets a lower limit on the allowable value of the ratio (K_c/σ_c).

If a structure is known to be cracked the maximum safe operating stress, (i.e. the stress below which failure by rapid crack-growth will not occur) must be determined in order to compare with the proposed working stress. It follows from eqn (10) that for a given crack-length the critical stress depends on both the toughness and the crack-length. In fact

$$\sigma_c \propto K_c/\sqrt{a_c} \tag{12}$$

Thus the critical stress increases as the toughness increases and decreases as the crack-length increases. It follows that a cracked structure with a critical stress σ_c must be considered unsafe if the proposed operating stresses exceed σ_c at any time during the working life. In fact, the possibility of σ_c decreasing with time, due to fatigue crack-growth, must also be taken into account.

1.5 Fatigue crack growth

Paris[6] suggested that the growth of cracks which results when the applied stress is varied (fatigue) may also be described by the stress intensity factor, even though the maximum stresses may be much less than the critical stress. He postulated that the rate of growth per cycle of stress $(\mathrm{d}a/\mathrm{d}N)$ was a function of the stress intensity range $\Delta K (= K_{max} - K_{min})$, i.e.

$$\frac{\mathrm{d}a}{\mathrm{d}N} = Cf(\Delta K) \tag{13}$$

where C is a constant to be determined experimentally. Many experimental data are now available to confirm that, for many stress conditions, fatigue crack growth is largely controlled by ΔK. The simplest explicit form of eqn (13) suggested was

$$\frac{\mathrm{d}a}{\mathrm{d}N} = C(\Delta K)^m \tag{14}$$

where C and m are constants. Later work suggested that C was not strictly constant but depended on such parameters as $R(= K_{min}/K_{max})$, K_c and the threshold stress intensity factor range ΔK_{th}. The simple power-law must also break down at both high and low values of ΔK. At high values if $K_{max} = K_c$ then static failure $(\mathrm{d}a/\mathrm{d}N = \infty)$

must occur, and at low values if ΔK falls below the threshold value ΔK_{th} then no crack-growth occurs ($\mathrm{d}a/\mathrm{d}N = 0$).

A relationship between crack-growth rate and the stress intensity factor which reflects the loading conditions and the crack-size means that calculations can now be made of the fatigue life-times of cracked structures. In other words given a crack of a certain size, it is possible to compute the number of cycles of stress required before the crack grows to a critical size. If $\Delta N_{i,f}$ is the number of cycles for a crack of initial length a_i to grow to a final length of a_f, then it follows, in general, that

$$\Delta N_{i,f} = \int_{a_i}^{a_f} \left(\frac{\mathrm{d}a}{\mathrm{d}N}\right)^{-1} \mathrm{d}a \tag{15}$$

By substituting for $\mathrm{d}a/\mathrm{d}N$ from eqn (13), $\Delta N_{i,f}$ becomes

$$\Delta N_{i,f} = \frac{1}{C} \int_{a_i}^{a_f} \frac{\mathrm{d}a}{f(\Delta K)} \tag{16}$$

The above integral can be evaluated either analytically or numerically for any given configuration provided that the stress intensity factor for that configuration is known.

The above description of fatigue crack propagation is adequate for many practical applications where the stress undergoes periodic or narrow-band random fluctuations. However load-interaction effects, which are non-linear, can occur. For instance if a large peak in the applied stress occurs within a periodic sequence of small peaks, the crack-growth rate can actually decrease after the large peak; this phenomenon is known as 'retardation'. This effect and others are important in the calculation of fatigue life-times for structures which undergo complex loading sequences. For instance, an aircraft wing experiences the following sequence of loading: taxiing loads, take-off loads, gust loads, landing loads and taxiing loads each flight. These non-linear effects are usually incorporated into practical calculations as modifications to the stress intensity factor.

1.6 Stress corrosion cracking

The toughness of many materials is reduced in the presence of a corrosive environment. This means that a crack can grow under a static stress at values of K less than K_{Ic}. It is often assumed (see for example Ref. 6) that there is a threshold value of K below which corrosion induced crack-growth in a stressed component will not occur. The threshold value for what is called stress corrosion cracking is known as K_{Iscc}. Values of K_{Iscc} can be much less than K_{Ic} ; for some materials in certain corrosive environments $K_{Iscc} \approx 0.1K_{Ic}$.

After the onset of stress corrosion cracking, the crack velocity increases very rapidly with increasing stress intensity factor. This rapid increase in crack velocity (approximately an exponential dependence on K) means that little use is made of such information in design. The parameter K_{Iscc} is used to assess materials and to ensure that there are adequate safety factors in operating stress levels so that corrosion induced cracking does not start.

1.7 Limitations (plasticity)

As stated earlier fracture mechanics as introduced in this section, is based on linear elastostatics. The application of linear elastic theory led to eqn (6) which predicts stresses which approach infinity as the crack tip ($r = 0$) is approached. Of course in real materials infinite stresses cannot exist and plastic (or permanent) deformations will occur in the vicinity of the tip. Provided that these deformations are contained in a region that is small compared to the region dominated by the K-term in the stress field (see eqn (6)) the concepts of LEFM can still be used.

The region in which these non-elastic deformations occur is known as the plastic-zone. Some methods have been suggested for modifying LEFM concepts to take into account the size of this zone and the effect its presence has on the load-carrying capacity of a cracked structure. Where the plastic zone is large compared to the crack-length new measures of toughness need to be considered; the concepts of elastic-plastic fracture mechanics will be dealt with in more detail, later in this course, as will dynamic effects.

2 Fundamental stress analysis for fracture mechanics

2.1 Basic equations of elasticity

The state of stress in an elemental volume of a loaded three-dimensional body shown in Fig. 1, can be defined in terms of six components of stress, which can be expressed in tensor notation as σ_{ij} for $i, j = 1, 2, 3$; they have a symmetry such that $\sigma_{ij} = \sigma_{ji}$. If the 1,2 and 3 directions coincide with the cartesian directions x, y and z, then the stress components are the normal components σ_{xx}, σ_{yy} and σ_{zz} and the shear components σ_{xy}, σ_{yz} and σ_{zx}. The first suffix indicates the directions of the normal to the plane on which the stress acts and the second suffix indicates the direction in which it acts. Similarly the six strain components corresponding to the stress components can be written in tensor notation as $\varepsilon_{ij} = \varepsilon_{ji},\ i, j = 1, 2, 3$.

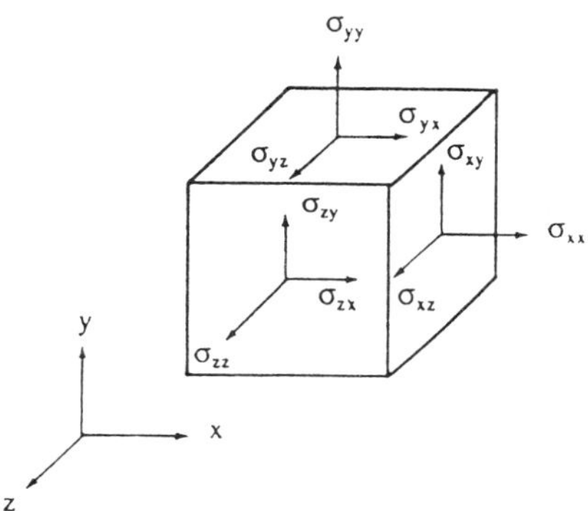

Figure 1: Stress components on a three-dimensional element

The strains ε_{ij} are defined in terms of the displacements u_i (for small deformations only) as follows:

$$\varepsilon_{ij} = \frac{1}{2}\left(\frac{\partial u_i}{\partial x_j} + \frac{\partial u_j}{\partial x_i}\right) \quad i, j = 1, 2, 3 \tag{17}$$

The detailed development of the governing equations of elasticity will not be reproduced here as they are readily available in standard textbooks, e.g. Timoshenko[8] and Mushkelishvili[9] for the general development, and Aliabadi & Rooke[10] for crack problems. The development is based on two important physical requirements, equilibrium of the stress state and compatibility of the strains; these lead to differential relations involving the stresses and strains.

The relationship between stresses and strains for an elastic, isotropic material is known as Hooke's law and is given in cartesian co-ordinates by

$$\varepsilon_{ii} = [\sigma_{ii} - \nu(\sigma_{jj} + \sigma_{kk})]/E \ i \neq j \neq k; \quad \text{and} \quad \varepsilon_{ij} = (1+\nu)\sigma_{ij}/E \ i \neq j \tag{18}$$

where E and ν are Young's modulus and Poisson's ratio respectively. The equations of equilibrium can now be written in terms of the displacements, and are known as Navier's equations:

$$\mu\nabla^2 u_i + (\lambda+\mu)\frac{\partial e}{\partial x_i} = 0 \qquad i = 1,2,3 \tag{19}$$

where μ is the shear modulus, $\lambda = 2\mu\nu/(1-2\nu)$ is known as the Lame constant and $e = \varepsilon_{xx} + \varepsilon_{yy} + \varepsilon_{zz}$. The traction t_i, which is required in the boundary element formulation of elasticity is a vector force per unit area; it is defined in terms of the stresses as $t_i = \sigma_{ij}n_j$, where n_j are the cartesian components of the outward normal to a surface.

The equations of elasticity can be greatly simplified by the two-dimensional specialisations of plane strain and plane stress conditions. Plane strain conditions are generally assumed applicable to thick plates, that is, those bodies in which geometry and loading do not vary significantly in one of the co-ordinate (usually z) directions. In these problems the dependent variables are assumed to be functions of the (x, y) co-ordinates only. The displacement component in the z-direction, (i.e. u_z) is zero at every cross section and the strain components $\varepsilon_{zz}, \varepsilon_{yz}$ and ε_{zx} will therefore vanish. Since ε_{zz} is zero, the stress σ_{zz} is now given, from Hooke's law, by $\sigma_{zz} = \nu(\sigma_{xx} + \sigma_{yy})$.

For thin plates, if no loadings are applied to the surface of the plate, the stress components σ_{zz}, σ_{xz} and σ_{yz} are zero on both sides of the plate and they are assumed to be zero within the plate. The non-zero components σ_{xx}, σ_{yy} and σ_{xy} are averaged over the thickness and assumed to be independent of z. This state of stress is commonly referred to as generalised plane stress. The strain components ε_{yz} and ε_{zx} also vanish on the surfaces and the component ε_{zz} is given by $\varepsilon_{zz} = -\nu(\varepsilon_{xx} + \varepsilon_{yy})/(1-\nu)$.

2.2 Airy stress functions

The equations of equilibrium in two dimensions are automatically satisfied if

$$\sigma_{xx} = \frac{\partial^2\psi}{\partial y^2}, \qquad \sigma_{yy} = \frac{\partial^2\psi}{\partial x^2}, \qquad \sigma_{xy} = -\frac{\partial^2\psi}{\partial x \partial y} \tag{20}$$

where the function ψ is known as an Airy stress function. The compatibility equation for plane stress can now be expressed[1] (via Hooke's law) in terms of ψ as follows:

$$\frac{\partial^4\psi}{\partial x^4} + 2\frac{\partial^4\psi}{\partial x^2\partial y^2} + \frac{\partial^4\psi}{\partial y^4} = 0 \tag{21}$$

The same equation can be obtained for plane strain; it may be written as $\nabla^4\psi = \nabla^2(\nabla^2\psi) = 0$ and is called the biharmonic equation since ∇^2 is the usual harmonic operator.

2.3 Muskhelishvili's complex functions

The Airy stress functions defined above are limited to bodies with smooth boundaries and represent a special case of the more general complex stress functions developed by Muskhelishvili.[9] Following Muskhelishvili's work on complex functions, stresses for the two-dimensional equations of elasticity may be defined in terms of two functions, the Airy stress function ψ and some other function φ, such that

$$\sigma_{xx} = \frac{\partial^2\psi}{\partial x^2} - 2\frac{\partial^2\varphi}{\partial y\partial x}, \quad \sigma_{yy} = \frac{\partial^2\psi}{\partial y^2} + 2\frac{\partial^2\varphi}{\partial y\partial x} \text{ and } \sigma_{xy} = \frac{\partial^2\psi}{\partial x\partial y} - \frac{\partial^2\varphi}{\partial y^2} + \frac{\partial^2\varphi}{\partial x^2} \tag{22}$$

The functions ψ and φ are given in terms of two analytical functions $\phi(z)$ and $\chi(z)$ of the complex variable $z(= x + iy)$, such that

$$\psi = \text{Re}\{\bar{z}\phi(z) + \chi(z)\} \text{ and } \varphi = \text{Im}\{\bar{z}\phi(z) + \chi(z)\} \tag{23}$$

where Re and Im denote the real and imaginary parts respectively ($\bar{z}$ is the complex conjugate of z). The stresses can thus be expressed in terms of ϕ and χ, as can the displacements, by use of Hooke's law and the strain/displacement relationships.

2.4 Westergaard's stress function

Westergaard devised stress functions[11] suitable for solving two-dimensional crack problems; they will be examined here, because they illustrate the principal properties of crack stress fields. For instance the Westergaard stress function $Z_{\text{I}}(z)$ for an infinite sheet, containing a crack of length $2a$, subjected to a remote tensile stress σ (see Fig. 2) is given by

$$Z_{\text{I}}(z) = \frac{\sigma z}{\sqrt{z^2 - a^2}} \tag{24}$$

if the crack faces are traction-free. The stress fields which describe mode I deformation are obtained[11] from

$$\sigma_{xx} = \text{Re}\{Z_{\text{I}}\} - y\text{Im}\{Z'_{\text{I}}\}, \qquad \sigma_{yy} = \text{Re}\{Z_{\text{I}}\} + y\text{Im}\{Z'_{\text{I}}\}$$

and

$$\sigma_{xy} = -y\text{Re}\{Z'_{\text{I}}\} \tag{25}$$

and the displacements from

$$2\mu u_x = \frac{1}{2}(\kappa - 1)\text{Re}\{\tilde{Z}_{\text{I}}\} - y\text{Im}\{Z_{\text{I}}\}$$

and

$$2\mu u_y = \frac{1}{2}(\kappa - 1)\text{Im}\{\tilde{Z}_{\text{I}}\} - y\text{Re}\{Z_{\text{I}}\} \tag{26}$$

where $\kappa = (3 - 4\nu)$ for plane strain and $(3 - \nu)/(1 + \nu)$ for plane stress, $Z' = \text{d}Z/\text{d}z$ and $\tilde{Z} = \int Z(z)\text{d}z$. Equations (25) and (26) are the same as the equations derived from Mushkelishvili's analysis if $2\phi'(z) = Z(z)$ where a dash denotes differentiation with respect to z; for details see Eftis, Subramonian & Liebowitz,[12] Sih[13] and Eftis & Liebowitz.[14]

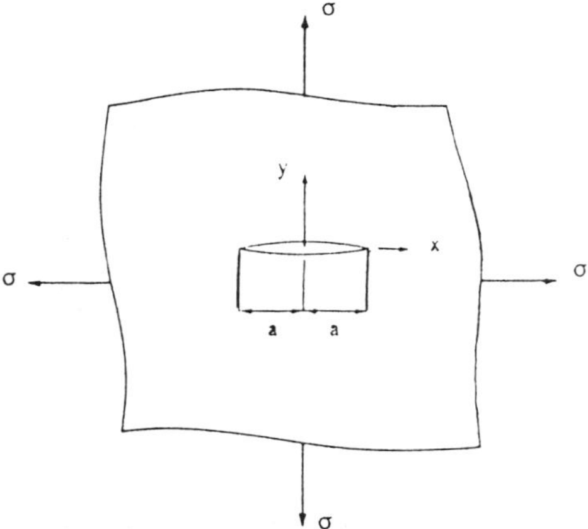

Figure 2: A crack of length $2a$ in a remotely loaded finite sheet

In order to obtain the stress field near the crack tip it is more convenient to translate the origin of the co-ordinate system to the tip, that is to put $z = a + \xi$, so that the stress function in eqn (24) can now be written as

$$Z_{\mathrm{I}}(z) = \frac{\sigma(a+\xi)}{\sqrt{(a+\xi)^2 - a^2}} \tag{27}$$

The first order approximation to eqn (27) for $\xi \ll a$ is

$$Z_{\mathrm{I}}(z) = \frac{\sigma a}{\sqrt{2a\xi}} = \frac{\sigma\sqrt{\pi a}}{\sqrt{2\pi r}}e^{-i\theta/2} = \frac{K_I}{\sqrt{2\pi r}}e^{-1\theta/2} \tag{28}$$

where $\xi = re^{i\theta}$ and $K_{\mathrm{I}} = \sigma\sqrt{(\pi a)}$ for an isolated crack in an infinite sheet.

Substitution of eqn (28) into eqn (25) gives the near-tip stress fields as

$$\sigma_{xx} = \frac{K_{\mathrm{I}}}{\sqrt{2\pi r}}\cos\frac{\theta}{2}\left[1 - \sin\frac{\theta}{2}\sin\frac{3}{2}\theta\right], \quad \sigma_{yy} = \frac{K_{\mathrm{I}}}{\sqrt{2\pi r}}\cos\frac{\theta}{2}\left[1 + \sin\frac{\theta}{2}\sin\frac{3}{2}\theta\right]$$

and

$$\sigma_{xy} = \frac{K_{\mathrm{I}}}{\sqrt{2\pi r}}\sin\frac{\theta}{2}\cos\frac{\theta}{2}\cos\frac{3}{2}\theta \tag{29}$$

Thus the opening-mode stress intensity factor K_{I} may, in general, be defined by

$$K_{\mathrm{I}} = \lim_{r\to 0}\left\{\sqrt{2\pi r}\sigma_{yy}(r, \theta = 0)\right\} \tag{30}$$

Similarly substituting eqn (28) into eqn (26), the near-tip displacement fields are obtained as

$$u_x = \frac{K_{\mathrm{I}}}{\mu}\sqrt{\frac{r}{2\pi}}\cos\frac{\theta}{2}\left[\frac{(\kappa-1)}{2} + \sin^2\frac{\theta}{2}\right], \quad u_y = \frac{K_{\mathrm{I}}}{\mu}\sqrt{\frac{r}{2\pi}}\sin\frac{\theta}{2}\left[\frac{(\kappa+1)}{2} - \cos^2\frac{\theta}{2}\right] \tag{31}$$

In a similar manner the near-tip stress field for mode II can be obtained from

$$\sigma_{xx} = 2\mathrm{Im}\{Z_{\mathrm{II}}\} + y\mathrm{Re}\{Z'_{\mathrm{II}}\}, \quad \sigma_{yy} = -y\mathrm{Re}\{Z'_{\mathrm{II}}\}$$

and

$$\sigma_{xy} = \mathrm{Re}\{Z_{\mathrm{II}}\} - y\mathrm{Im}\{Z'_{\mathrm{II}}\} \tag{32}$$

and the displacements from

$$2\mu u_x = \frac{\kappa+1}{2}\mathrm{Im}\{\tilde{Z}_{\mathrm{II}}\} + y\mathrm{Re}\{Z_{\mathrm{II}}\} \text{ and } 2\mu u_y = -\frac{\kappa-1}{2}\mathrm{Re}\{\tilde{Z}_{\mathrm{II}}\} - y\mathrm{Im}\{Z_{\mathrm{II}}\} \tag{33}$$

For an infinite sheet, containing a crack of length $2a$, subjected to a remote shear stress τ, the stress function Z_{II} is given by

$$Z_{\mathrm{II}}(z) = \frac{\tau z}{\sqrt{z^2 - a^2}} \tag{34}$$

The crack-tip stress fields are given by

$$\sigma_{xx} = -\frac{K_{\mathrm{II}}}{\sqrt{2\pi r}}\sin\frac{\theta}{2}\left[2 + \cos\frac{\theta}{2}\cos\frac{3\theta}{2}\right], \quad \sigma_{yy} = \frac{K_{\mathrm{II}}}{\sqrt{2\pi r}}\cos\frac{\theta}{2}\sin\frac{\theta}{2}\cos\frac{3\theta}{2}$$

and

$$\sigma_{xy} = -\frac{K_{\mathrm{II}}}{\sqrt{2\pi r}}\cos\frac{\theta}{2}\left[1 - \sin\frac{\theta}{2}\sin\frac{3\theta}{2}\right] \tag{35}$$

and the displacement fields by

$$u_x = \frac{K_{\mathrm{II}}}{\mu}\sqrt{\frac{r}{2\pi}}\sin\frac{\theta}{2}\left[\frac{\kappa+1}{2} + \cos^2\frac{\theta}{2}\right] \text{ and } u_y = \frac{K_{\mathrm{II}}}{\mu}\sqrt{\frac{r}{2\pi}}\cos\frac{\theta}{2}\left[\frac{\kappa-1}{2} + \sin^2\frac{\theta}{2}\right] \tag{36}$$

The sliding-mode stress intensity factor K_{II} may, in general, be defined by

$$K_{\mathrm{II}} = \lim_{r\to 0}\left\{\sqrt{2\pi r}\sigma_{yy}(r, \theta = 0)\right\} \tag{37}$$

It is thus seen that both the stresses and the displacements at the crack-tip are proportional to the stress intensity factor and that, in general, the stresses (or tractions) are singular $0(r^{-1/2})$.

2.5 William's eigenfunction series expansion

The stress and displacement fields near a V-notch (Fig. 3) can be presented in the form of a series expansion. These fields are due to Williams[15] and are applicable to traction-free boundary conditions over the notch surfaces which are separated by an angle 2γ. As γ tends to zero the solution for a mathematical crack ($\gamma = 0$) is obtained. For the details of the derivation see the original paper[15] or Ref. 10. The results for the stresses are expressed in a series form as follows:

$$\sigma_{xx} = \sum \mathrm{Re}[a_n f_{11}(\lambda_n; r, \theta) + b_n f_{21}(\xi_n; r, \theta)]$$

$$\sigma_{yy} = \sum \mathrm{Re}[a_n f_{12}(\lambda_n; r, \theta) + b_n f_{22}(\xi_n; r, \theta)]$$

and

$$\sigma_{xy} = \sum \mathrm{Re}[a_n f_{13}(\lambda_n; r, \theta) + b_n f_{23}(\xi_n; r, \theta)] \tag{38}$$

and the displacement fields as

$$u_x = \frac{1}{2\mu} \sum \mathrm{Re}[a_n g_{11}(\lambda_n; r, \theta) + b_n g_{21}(\xi_n; r, \theta)]$$

and

$$u_y = \frac{1}{2\mu} \sum \mathrm{Re}[a_n g_{12}(\lambda_n; r, \theta) + b_n g_{22}(\xi_n; r, \theta)] \tag{39}$$

where

$$\begin{aligned}
f_{11} &= \lambda_n r^{\lambda_n - 1}[(2 + \lambda_n \cos 2\alpha + \cos 2\lambda_n \alpha) \cos(\lambda_n - 1)\theta - (\lambda_n - 1) \cos(\lambda_n - 3)\theta] \\
f_{12} &= \lambda_n r^{\lambda_n - 1}[(2 - \lambda_n \cos 2\alpha - \cos 2\lambda_n \alpha) \cos(\lambda_n - 1)\theta + (\lambda_n - 1) \cos(\lambda_n - 3)\theta] \\
f_{21} &= \xi_n r^{\xi_n - 1}[-(2 + \xi_n \cos 2\alpha - \cos 2\xi_n \alpha) \sin(\xi_n - 1)\theta + (\xi_n - 1) \sin(\xi_n - 3)\theta] \\
f_{22} &= \xi_n r^{\xi_n - 1}[(-2 + \xi_n \cos 2\alpha - \cos 2\xi_n \alpha) \sin(\xi_n - 1)\theta - (\xi_n - 1) \sin(\xi_n - 3)\theta] \\
f_{13} &= \lambda_n r^{\lambda_n - 1}[-(\lambda_n \cos 2\alpha + \cos 2\lambda_n \alpha) \sin(\lambda_n - 1)\theta - (\lambda_n - 1) \sin(\lambda_n - 3)\theta] \\
f_{23} &= \xi_n r^{\xi_n - 1}[-(\xi_n \cos 2\alpha - \cos 2\xi_n \alpha) \cos(\xi_n - 1)\theta - (\xi_n - 1) \cos(\xi_n - 3)\theta] \\
g_{11} &= r^{\lambda_n}[(\kappa + \lambda_n \cos 2\alpha + \cos 2\lambda_n \alpha) \cos \lambda_n \theta - \lambda_n \cos(\lambda_n - 2)\theta] \\
g_{12} &= r^{\lambda_n}[(\kappa - \lambda_n \cos 2\alpha - \cos 2\lambda_n \alpha) \sin \lambda_n \theta + \lambda_n \sin(\lambda_n - 2)\theta] \\
g_{21} &= r^{\xi_n}[(\kappa + \xi_n \cos 2\alpha - \cos 2\xi_n \alpha) \sin \xi_n \theta - \xi_n \sin(\xi_n - 2)\theta] \\
g_{22} &= r^{\xi_n}[(\kappa - \xi_n \cos 2\alpha + \cos 2\xi_n \alpha) \cos \xi_n \theta + \xi_n \cos(\xi_n - 2)\theta]
\end{aligned} \tag{40}$$

with $\alpha = \pi - \gamma$. For symmetrical loading the eigenvalues λ_n are solutions of

$$\sin 2\lambda_n \alpha + \lambda_n \sin 2\alpha = 0 \tag{41}$$

and for anti-symmetrical loading the eigenvalues ξ_n are solutions of

$$\sin 2\xi_n \alpha - \xi_n \sin 2\alpha = 0 \tag{42}$$

In general λ_n and ξ_n are complex with $\mathrm{Re}(\lambda_n), \mathrm{Re}(\xi_n) \geq 1/2$. For a crack ($\gamma = 0$) the first eigenvalues $\lambda_1, \xi_1 = 1/2$, giving the familiar $r^{-1/2}$ dependence of the stress near the crack-tip.

It can be seen from eqns (38) and (40) that the stress σ_{yy} ahead of the crack ($\theta = 0$) is given by

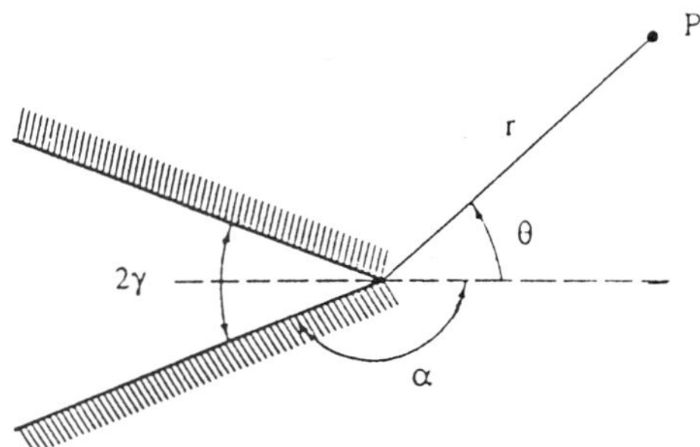

Figure 3: Geometry of a V-notch

$$\sigma_{yy} = a_1/\sqrt{r} \tag{43}$$

Now comparing eqn (43) with Westergaard's solution (29) it is seen that

$$a_1 = K_{\rm I}/\sqrt{2\pi} \tag{44}$$

and similarly, consideration of σ_{xy} for mode II deformation leads to

$$b_1 = -K_{\rm II}/\sqrt{2\pi} \tag{45}$$

In general the infinite series solution derived for a V-notch in a semi-infinite plane leads to more complicated stress and displacement fields than for a crack. However the coefficients of $r^{(\lambda_1-1)}$ and $r^{(\xi_1-1)}$ in the leading terms of the solution can be used to define generalised stress intensity factors K^λ, as follows:

$$K_{\rm I}^\lambda = \sqrt{2\pi}\lambda_1(1+\lambda_1-\lambda_1\cos 2\alpha-\cos 2\lambda_1\alpha)a_1 \tag{46}$$

and

$$K_{\rm II}^\lambda = \sqrt{2\pi}\xi_1(-1+\xi_1-\xi_1\cos 2\alpha-\cos 2\xi_1\alpha)b_1 \tag{47}$$

2.6 Papkovich-Neuber potentials

In general three-dimensional linear elastic displacements can be described in terms of a stress function F and harmonic functions known as Papkovich-Neuber potentials;[16] thus

$$\begin{aligned}2\mu u_x &= -\frac{\partial F}{\partial x}+4(1-\nu)\Phi_1\\ 2\mu u_y &= -\frac{\partial F}{\partial y}+4(1-\nu)\Phi_2\\ 2\mu u_z &= -\frac{\partial F}{\partial z}+4(1-\nu)\Phi_3\end{aligned} \tag{48}$$

where F is a three-dimensional stress function and Φ_1, Φ_2 and Φ_3 are harmonic functions, i.e.

$$\nabla^2\Phi_1 = \nabla^2\Phi_2 = \nabla^2\Phi_3 = 0 \tag{49}$$

The relationship between the stress function F and the harmonic functions is given[16] by

$$F = \Phi_0 + x\Phi_1 + y\Phi_2 + z\Phi_3 \tag{50}$$

where Φ_0 is also a harmonic function. The relationship between the stress fields and the stress function F can be obtained from the displacements in eqn (48) and Hooke's law to give

$$\sigma_{xx} = \frac{\partial^2 F}{\partial y^2} + \frac{\partial^2 F}{\partial z^2} + 2(1-\nu)\left(\frac{\partial \Phi_1}{\partial x} - \frac{\partial \Phi_2}{\partial y} - \frac{\partial \Phi_3}{\partial z}\right)$$

$$\sigma_{yy} = \frac{\partial^2 F}{\partial x^2} + \frac{\partial^2 F}{\partial z^2} + 2(1-\nu)\left(\frac{\partial \Phi_2}{\partial y} - \frac{\partial \Phi_3}{\partial z} - \frac{\partial \Phi_1}{\partial x}\right)$$

and

$$\sigma_{zz} = \frac{\partial^2 F}{\partial x^2} + \frac{\partial^2 F}{\partial y^2} + 2(1-\nu)\left(\frac{\partial \Phi_3}{\partial z} - \frac{\partial \Phi_1}{\partial x} - \frac{\partial \Phi_2}{\partial y}\right) \tag{51}$$

for the normal stress components, and

$$\sigma_{xy} = -\frac{\partial^2 F}{\partial x \partial y} + 2(1-\nu)\left(\frac{\partial \Phi_1}{\partial y} + \frac{\partial \Phi_2}{\partial x}\right)$$

$$\sigma_{zx} = -\frac{\partial^2 F}{\partial z \partial x} + 2(1-\nu)\left(\frac{\partial \Phi_1}{\partial z} + \frac{\partial \Phi_3}{\partial x}\right)$$

and

$$\sigma_{zy} = -\frac{\partial^2 F}{\partial z \partial y} + 2(1-\nu)\left(\frac{\partial \Phi_2}{\partial z} + \frac{\partial \Phi_3}{\partial y}\right) \tag{52}$$

for the shear stress components. This description is later used in the derivation of some so-called fundamental fields used in the development of weight function concepts.

2.7 The energy principle

The Griffith energy criterion for fracture states[1] that unstable crack growth occurs if the total energy of the body remains constant or decreases as the crack length increases. In an elastic solid, if U is the elastic strain energy contained in the solid and W is the energy required for crack growth, then according to Griffith the necessary condition for crack growth can be expressed as

$$\frac{\mathrm{d}U}{\mathrm{d}a} \geq \frac{\mathrm{d}W}{\mathrm{d}a} \tag{53}$$

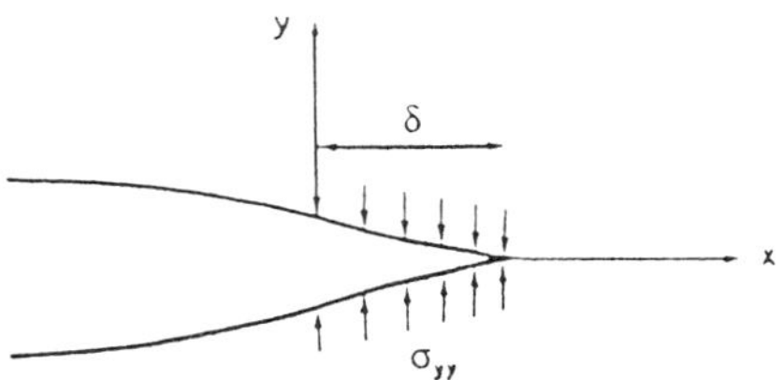

Figure 4: Closing of a crack tip

It is usual to replace dU/da by a so-called 'strain energy release rate' or the 'crack extension force' G and dW/da by R the 'crack resistance'.

It was shown by Irwin[4] that the stress intensity factor K and the strain energy release rate G are related. He considered an infinite sheet containing a crack of length a, subject to forces, which act over an infinitesimal distance δ, and are sufficient to close the crack. In other words, he imagined that the crack had extended by an amount δ as shown in Fig. 4, and calculated the work required to close the tip back to its original position. He equated this amount of work to the product of the energy release rate G and the crack extension increment; thus, for mode I,

$$G_{\mathrm{I}} = 2\lim_{\delta\to 0}\left\{\frac{1}{2\delta}\int_0^\delta \sigma_{yy}u_y \mathrm{d}r\right\} \tag{54}$$

Substituting for σ_{yy} and u_y from eqns (29) and (31) with $\theta = 0$, results in

$$G_{\mathrm{I}} = 2\lim_{\delta\to 0}\left\{\frac{1}{\delta}K_1^2\frac{(1-\nu^2)}{\pi E}\int_0^\delta\left[\frac{r}{\delta - r}\right]^{1/2}\mathrm{d}r\right\}$$

that is

$$G_{\mathrm{I}} = 2\lim_{\delta\to 0}\left\{K_1^2\frac{(1-\nu^2)}{\pi E}\int_0^{\pi/2} 2\sin^2\alpha\mathrm{d}\alpha\right\} \quad (r = \delta\sin^2\alpha)$$

and so

$$G_{\mathrm{I}} = K_1^2(1-\nu^2)/(E) \tag{55}$$

For plane stress this becomes

$$G_{\mathrm{I}} = K_{\mathrm{I}}^2/E \tag{56}$$

Similarly for mode II and mode III

$$G_{\mathrm{II}} = (1-\nu)K_{\mathrm{II}}^2/E \quad \text{and} \quad G_{\mathrm{III}} = (1+\nu)K_{\mathrm{III}}^2/E \tag{57}$$

If the cracking occurs in a combined mode, then the total strain energy release rate is expressed as $G = G_{\mathrm{I}} + G_{\mathrm{II}} + G_{\mathrm{III}}$. The stress intensity factor can also be related to a path independent integral, termed the J integral, described by Rice.[17] This integral is independent of the actual path chosen, provided that the start and end points of the

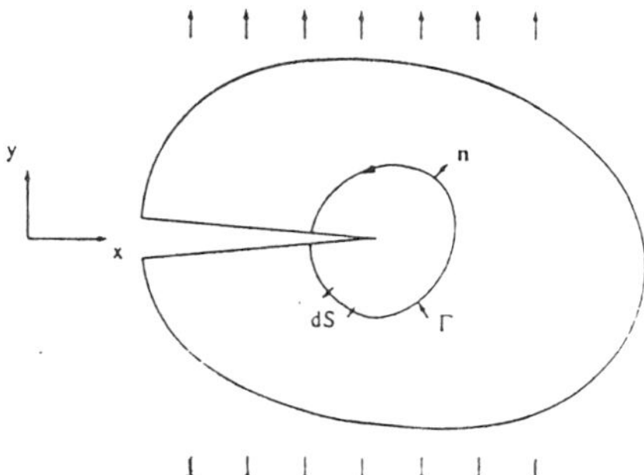

Figure 5: Rice's path independent J-integral

contour Γ (see Fig. 5) are on opposite faces of the crack and that the contour includes the crack tip. If the crack is parallel to the x-axis the J integral can be defined as

$$J = \int_\Gamma (U'\mathrm{d}y - t_i \frac{\partial u_i}{\partial x} \mathrm{d}S) \tag{58}$$

where U' is the strain energy density, t_i are components of the traction vector, u_i are components of the displacement vector (repeated suffix summation is assumed) and $\mathrm{d}S$ is an element of arc along the integration contour Γ. For a linear elastic material it can be shown[17] that $J = G$.

2.8 Elastic-plastic fracture

In practical structural materials, especially metals, non-elastic yielding will occur in the regions around the tip. Providing that this yielded region is sufficiently small relative to the size of the crack and the physical dimensions of the structure, then the growth of the crack can be treated using the principles of linear elastic fracture mechanics. These circumstances usually occur in the fracture of high strength materials or in crack growth under fatigue loading. For the prediction of fracture in medium and low strength materials it is often necessary to consider the effect of yielding at the crack tip using the principles of elastic-plastic fracture mechanics. In what follows only a brief description of some of the models will be given. Particular topics will also be the subject of later sections in this book.

Irwin[18] was the first to develop a procedure for modifying the linear elastic equations to allow for the presence of a plastic zone at the crack tip. He suggested that the occurrence of plasticity makes the crack behave as if it were longer than its physical size and that the displacements are larger and the stiffness lower than in the elastic case. Irwin's model consists of replacing the actual crack length a by a longer 'equivalent' crack length a_e as demonstrated in Fig. 6. Therefore $a_e = a + \Delta a$ where Δa is the addition to the crack length due to the plastic zone. The stress distribution at the tip of the effective crack is limited to the yield stress σ_Y. Consequently Δa must be large enough to preserve the equilibrium conditions; that is the load transmitted across a region ahead of the original crack tip by the stress $\sigma_{yy}(= K_I\sqrt{(2\pi r)})$ must be equal to the load transmitted by the yield stress across the plastic zone of the equivalent crack. This leads to the result[18]

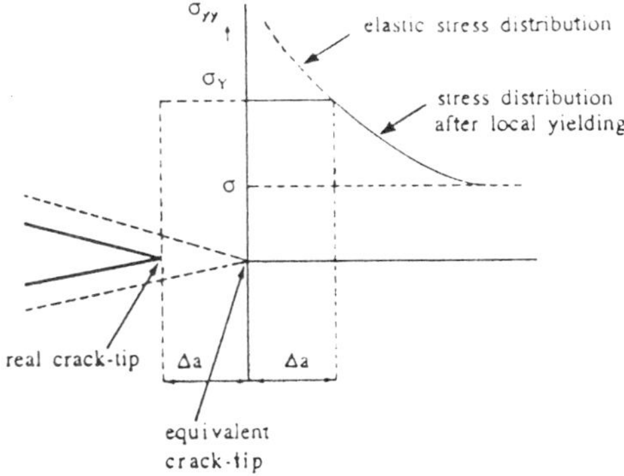

Figure 6: Irwin's equivalent crack and plastic zone site

$$\Delta a = (K_{\rm I}/\sigma_{\rm Y})^2/2\pi \tag{59}$$

The stress ahead of the original crack tip is equal to σ_Y at $r = r_Y$ (say) given by $\sigma_Y = K_{\rm I}\sqrt{(2\pi r_{\rm Y})}$; it therefore follows that Δa is equal to the radius of the plastic zone, that is $\Delta a = r_{\rm Y}$. This approach is suitable only when the plastic zone size at the crack tip is small relative to the elastic region at the crack tip which is dominated by the K-fields given in eqns (32) and (34).

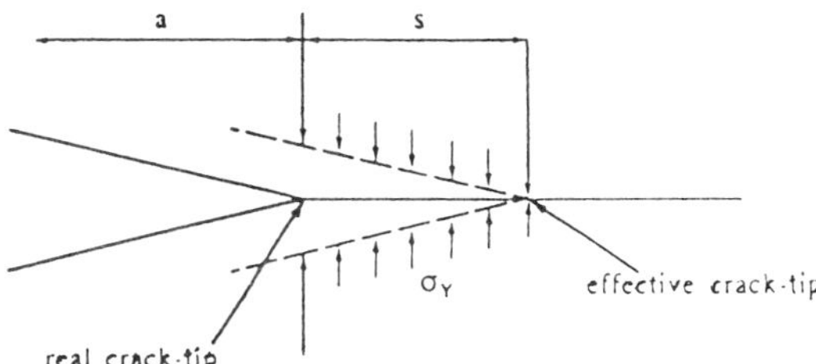

Figure 7: Dugdale's effective crack and plastic zone site

In order to remove this restriction and extend the use of LEFM to more ductile materials, Dugdale[19] proposed a so-called strip yield model as shown in Fig. 7. He also defined an equivalent crack which is longer than the actual crack by extending the crack by a length s, ahead of the original crack tip; the extension is subject to the yield stress $\sigma_{\rm Y}$, which acts to close the crack. The size of s is chosen such that the stress singularity at the tip of the effective crack disappears, i.e. the stress intensity factor is zero. Consider two elastic solutions for a crack of length $a + s$ in an infinite sheet, with the following applied stresses: i) a remote uniform stress σ; or ii) the yield stress $\sigma_{\rm Y}$ distributed on the faces of the crack $a \leq x \leq a + s$. The stress intensity factor for i) is given as

$$K_\sigma = \sigma\sqrt{\pi(a + s)} \tag{60}$$

and for ii) the stress intensity factor is given[20] as

$$K_s = -2\sigma_Y \sqrt{\frac{a+s}{\pi}} \cos^{-1}\left(\frac{a}{a+s}\right) \tag{61}$$

From the condition that $K_\sigma = -K_s$ it follows that

$$\frac{a}{a+s} = \cos\frac{\pi\sigma}{2\sigma_Y} \tag{62}$$

Expanding the cosine in eqn (62) and neglecting the higher order terms in (σ/σ_Y), s is found as

$$s = \frac{\pi^2\sigma^2 a}{8\sigma_Y^2} = \frac{\pi K^2}{8\sigma_Y^2} \tag{63}$$

The crack opening displacement δ at the original crack tip ($x = a$) is for the Dugdale model given as

$$\delta = \frac{8a\sigma_Y}{\pi E}\ln\left[\sec\left(\frac{\pi\sigma}{2\sigma_Y}\right)\right] \tag{64}$$

This model has been used extensively by Bilby, Cotterell & Swinden[20] and others in the calculation of the critical crack opening displacement; in the study of elastic-plastic fracture; and in the failure assessment diagram approach to the prediction of fracture. The model has the advantage that the crack opening displacement can be obtained relatively easy compared to crack models involving incremental elastic-plastic solutions. Several, mostly closed-form, solutions to strip yield zone cracks have been collected together.[21]

2.9 Three-dimensional stress fields

Hartranft & Sih[22] used an asymptotic series expansion in three-dimensions to show that the near-tip behaviour of the three- dimensional field in certain planes is identical to the two-dimensional plane strain field, eqns (29) to (37). The planes are those defined by the normal **n** and the bi-normal **b** at a point on the crack front (see Fig. 8). However the magnitude of the stress intensity factor may now vary with the position s on the crack front.

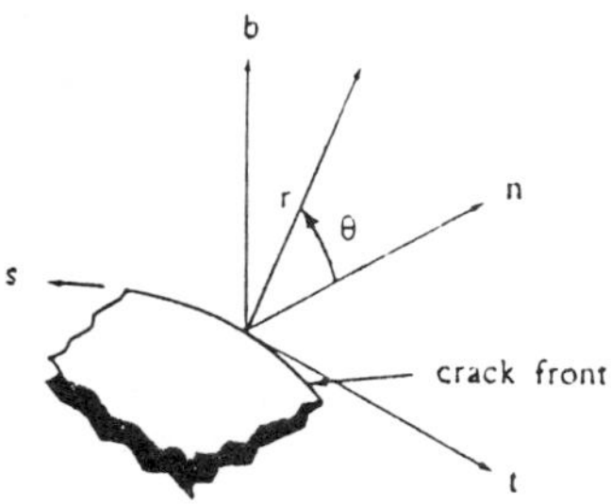

Figure 8: Crack front coordinate system

This type of stress distribution[22] is generally accepted in regions along the crack front far from any intersection with a free surface, where it can be assumed that conditions approximating to plane strain prevail. It has however been shown by Folias[23] that in the vicinity of the intersection of the crack front and a free surface, the stresses are not $0(r^{-1/2})$ but are proportional to $r^{-(1/2+2\nu)}$. Therefore the usual stress intensity factor does not exist.

2.10 Mixed mode fracture

In many practical situations structures are subjected to shear and torsional as well as tensile loadings which will lead to mixed mode cracking. There have been a number of investigations into mixed mode fracture problems, but a generally accepted analysis is yet to be developed. Two criteria for mixed mode loading that allow non-coplanar crack growth have been proposed; one based on the maximum principal stress by Erdogan & Sih[24] and the other on the strain energy density factor proposed by Sih.[25]

The maximum principal stress criterion postulates that a crack will grow in a direction perpendicular to the maximum principal stress. Considering two-dimensional combined mode I and mode II loading the stresses $\sigma_{\theta\theta}$ and $\sigma_{r\theta}$ at the crack tip can be found from expressions in (29) and (35) using the usual transformations; they are given by

$$\sigma_{\theta\theta} = \frac{1}{\sqrt{2\pi r}} \cos\frac{\theta}{2}\left[K_{\mathrm{I}}\cos^2\frac{\theta}{2} - \frac{3}{2}K_{\mathrm{II}}\sin\theta\right]$$

and

$$\sigma_{r\theta} = \frac{1}{2\sqrt{2\pi r}} \cos\frac{\theta}{2}\left[K_{\mathrm{I}}\sin\theta + K_{\mathrm{II}}(3\cos\theta - 1)\right] \tag{65}$$

The stress $\sigma_{\theta\theta}$ will be the maximum principal stress at $\phi = \theta$, where ϕ is defined by $\sigma_{r\theta} = 0$, that is

$$\sigma_{r\theta} = K_{\mathrm{I}}\sin\phi + K_{\mathrm{II}}(3\cos\phi - 1) = 0 \tag{66}$$

Thus the angle ϕ is given by

$$\phi = 2\tan^{-1}\left\{\frac{1}{4}\frac{K_{\mathrm{I}}}{K_{\mathrm{II}}} \pm \frac{1}{4}\sqrt{\left(\frac{K_{\mathrm{I}}}{K_{\mathrm{II}}}\right)^2 + 8}\right\} \tag{67}$$

and the value of the maximum principal stress σ_1 is given by

$$\sigma_1 = \frac{1}{\sqrt{2\pi r}} \cos^2\frac{\phi}{2}\left(K_{\mathrm{I}}\cos\frac{\phi}{2} - 3K_{\mathrm{II}}\sin\frac{\phi}{2}\right) \tag{68}$$

It is now assumed that the crack grows if σ_1 has the same value as σ_c in an equivalent mode I case (see Section 2). The principal stress at fracture for pure mode I is given by

$$\sigma_1 = K_{\mathrm{Ic}}/\sqrt{2\pi r} \tag{69}$$

where K_{Ic} is the critical value of the stress intensity factor at which the crack extends. The fracture condition for mixed mode is obtained by equating eqns (68) and (69); thus

$$K_{\mathrm{I}} \cos^3 \frac{\phi}{2} - 3K_{\mathrm{II}} \cos^2 \frac{\phi}{2} \sin \frac{\phi}{2} = K_{\mathrm{Ic}} \tag{70}$$

The strain energy density criterion[25] states that crack growth takes place in the direction of minimum strain energy density factor S. The strain energy $\mathrm{d}W$ stored in a volume element $\mathrm{d}V$ is given, in the local co-ordinate system (n, b, t), by

$$\frac{\mathrm{d}W}{\mathrm{d}V} = \frac{1}{2E}\left(\sigma_{tt}^2 + \sigma_{nn}^2 + \sigma_{bb}^2\right) - \frac{\nu}{E}\left(\sigma_{tt}\sigma_{nn} + \sigma_{nn}\sigma_{bb} + \sigma_{bb}\sigma_{tt}\right) + \frac{1+\nu}{E}\left(\sigma_{tn}^2 + \sigma_{nb}^2 + \sigma_{bt}^2\right) \tag{71}$$

From the expressions for the stress fields given in previous sections, the strain energy per unit volume near the tip can be written as

$$\frac{\mathrm{d}W}{\mathrm{d}V} = \frac{S(\theta)}{r} + \text{ non singular terms} \tag{72}$$

which becomes singular as $r \to 0$. Therefore the factor S is defined only if $r \neq 0$ and hence the volume element is kept a finite distance from the tip. For plane problems Sih[25] showed that

$$S(\theta) = a_{11}K_{\mathrm{I}}^2 + 2a_{12}K_{\mathrm{I}}K_{\mathrm{II}} + a_{22}K_{\mathrm{II}}^2$$

where

$$a_{11} = \frac{1}{16\mu}[(1 + \cos\theta)(\kappa - \cos\theta)], \quad a_{12} = \frac{1}{16\mu}\sin\theta[2\cos\theta - (\kappa - 1)]$$

and

$$a_{22} = \frac{1}{16\mu}[(1 - \cos\theta)(\kappa + 1) + (1 + \cos\theta)(3\cos\theta - 1) \tag{73}$$

The following hypotheses[25] regarding crack initiation in a two-dimensional stress field have been made:

1) the initial crack growth takes place in the direction of minimum S, so that

$$\frac{\mathrm{d}S}{\mathrm{d}\theta} = 0; \quad \text{at} \quad \theta = \theta_0 \tag{74}$$

where $-\pi < \theta < \pi$;

2) crack initiation occurs when S reaches a critical value S_{cr}, that is

$$a_{11}K_{\mathrm{I}}^2 + 2a_{12}K_{\mathrm{I}}K_{\mathrm{II}} + a_{22}K_{\mathrm{II}}^2 = S_{\mathrm{cr}} \quad \text{for} \quad \theta = \theta_0 \tag{75}$$

where θ_0 is the angle of crack extension. For pure mode I loading S_{cr} is directly related to K_{Ic} as

$$S(\theta = 0) = a_{11}K_{\mathrm{Ic}}^2 \tag{76}$$

therefore

$$S_{cr} = S(\theta = 0) = \frac{(\kappa - 1)}{8\mu} K_{Ic}^2 \tag{77}$$

For mixed modes the fracture criterion becomes

$$\left\{ \frac{8\mu}{(\kappa - 1)} \left(a_{11} K_I^2 + 2a_{12} K_I K_{II} + a_{22} K_{II}^2 \right) |_{\theta=\theta_0} \right\}^{1/2} = K_{Ic} \tag{78}$$

In a similar manner failure criteria for three-dimensional problems have been suggested and implemented.[26]

2.11 Concluding remarks

The basic analytical descriptions of crack problems in linear elasticity have now been formulated. This basis will form the framework within which numerical solution procedures must be developed in order to solve problems with complex geometrical configurations. Analytical solutions of the elasticity equations developed above are usually limited to highly idealised problems which do not adequately model real engineering structures. However, it is often possible to incorporate part of the idealised solutions into real problems. A particular case of this is the use of William's eigenfunction expansion [15] to describe the crack-tip fields in general cases, as will be seen in later sections.

3 Green's functions in fracture mechanics

3.1 Introduction

Many theoretical (analytical and numerical) and experimental methods for evaluating stress intensity factors are now available, they have been described in several reviews.[5,27,28] Techniques involving Green's functions have been used for determining stress intensity factors of cracks under complex loading conditions. These techniques, which offer scope for further development, are discussed in this chapter. A particular application of these techniques is applied to a problem which is of general engineering interest, namely cracks at loaded holes. The Green's functions developed give an insight into the importance of different types of loading and structures and demonstrate the versatility of the technique. Finally two generalisations into what are known as 'weight function' techniques are described.

3.2 Basic principles of Green's functions

The Green's function, first postulated by George Green in 1828, is defined as the response of a system to a standard input. The standard input is usually in the form of an impulse. Stedman[29] has reviewed the use of Green's functions in many fields of mathematical physics. Some examples of classical Green's functions are: the voltage output, as a function of time, of an electronic circuit in response to an input voltage pulse; the dynamic response of a mechanical system set in motion by an impulsive blow; the stress field produced in an elastic body in response to a force acting at a point in the body. If the body in the last example contains a crack, the stress intensity factor at the crack tip which arises in response to the point force may be considered as a special case of a Green's function. The important property of these functions is that, when suitably defined, they contain all the essential information about the system. They can thus be used to obtain the response of the system to any input by considering it as being composed of large numbers of small impulses. The total response is the sum of all the individual responses due to each input impulse acting separately. For the Green's function representation to be valid the system must have the following properties:

i) Causality - if there is no input there is no response

ii) Invariance - the response to a given input is always the same

iii) Linearity - if the response to input I_1 is R_1 and the response to I_2 is R_2 then the response to $I_1 + I_2$ is $R_1 + R_2$

These three conditions lead to the following result for the response $R(\eta)$ to a general input $I(\eta)$:

$$R(\eta) = \int I(\eta')G(\eta - \eta')\mathrm{d}\eta' \qquad (79)$$

where $G(\eta - \eta')$ is defined as the Green's function and is a function of the differences $\eta - \eta'$. The variables η and η' may represent positions and/or time.

3.3 Stress intensity factors as Green's functions

The stress intensity factor is known[30] for the two-dimensional problem of a cracked sheet containing a crack of length $2a$ which has localised forces acting at points on its surfaces (see Fig. 9). If the forces act normal to the crack faces, i.e. a force per unit thickness of P acting on one face and an equal and opposite force acting on the other face, then the opening mode stress intensity factor K_{I} at tip A is given by

$$K_{\mathrm{I}} = \frac{P}{\sqrt{\pi a}}\left[\frac{a+x_0}{a-x_0}\right]^{1/2} \equiv \frac{P}{\sqrt{\pi a}}G(x_0) \tag{80a}$$

where x_0 is the distance of the point of application of the force from the centre of the crack. If forces Q act tangentially to the crack faces then the sliding-mode stress intensity factor K_{II} at tip A is given by

$$K_{\mathrm{II}} = \frac{Q}{\sqrt{\pi a}}\left[\frac{a+x_0}{a-x_0}\right]^{1/2} \equiv \frac{Q}{\sqrt{\pi a}}G(x_0) \tag{80b}$$

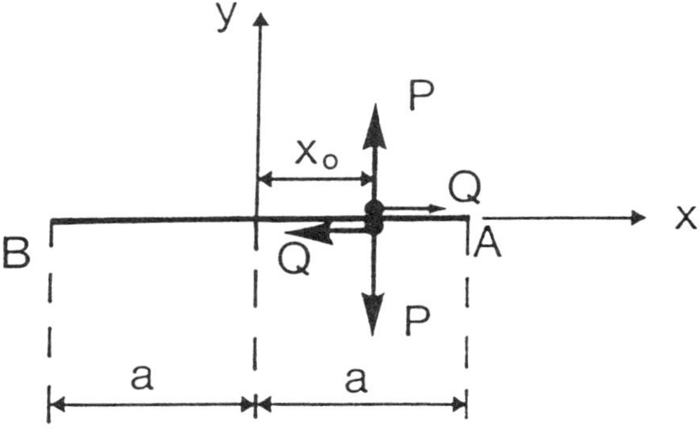

Figure 9: Crack loaded with point force

For tip B at the other end of the crack the stress intensity factors are obtained from eqns (80) and (81) by replacing x_0 with $-x_0$. The function $G(x_0)$ in eqns (80) and (81) can be used as a Green's function to obtain stress intensity factors for cracks subjected to boundary pressures acting on the crack faces. If a pressure $p(x), -a \leq x \leq a$, acts normal to the crack faces then the normal force $\mathrm{d}P$ acting on the face between x and $x + \mathrm{d}x$ is $p(x)\mathrm{d}x$; therefore $\mathrm{d}K_{\mathrm{I}}$, the contribution to the opening-mode stress intensity factor due to $\mathrm{d}P$, is given by eqn (80a) as

$$\mathrm{d}K_{\mathrm{I}} = \frac{\mathrm{d}P}{\sqrt{\pi a}}G(x) \tag{81}$$

Substitution of $\mathrm{d}P$ and subsequent integration of eqn (81) gives

$$K_{\mathrm{I}} = \frac{1}{\sqrt{\pi a}}\int_{-a}^{a} p(x)G(x)\mathrm{d}x \tag{82}$$

For the case of $p(x) = p$ (a constant), eqn (82) gives the well-known result that $K = p\sqrt{(\pi a)}$. A similar result can be obtained for a distribution of shearing forces on the crack. For symmetrical point forces on the crack faces the pressure distribution can be represented by $p(x) = P\delta(x - x_0)$ where $\delta(x - x_0)$ is the Dirac delta function. Substitution of this expression for $p(x)$ reduces eqn (82) to eqn (80a).

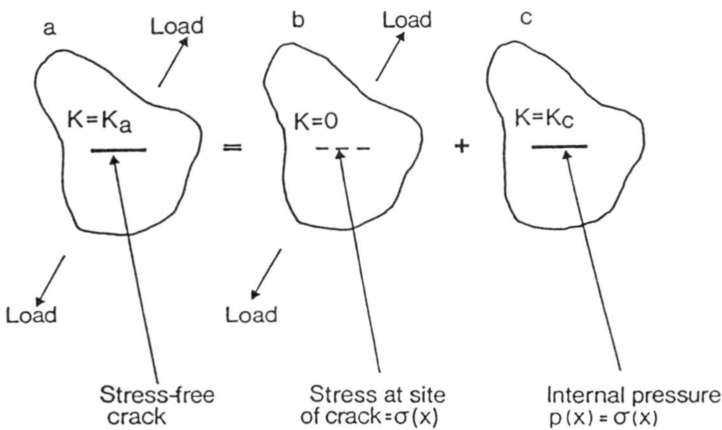

Figure 10: The equivalence of stress intensity factors for external boundary loads and internal pressures; $K_a = K_c$.

Equation (82) can be used together with an important result derived by Bueckner,[31] to obtain opening mode stress intensity factors for cracked bodies subjected to arbitrary forces on their boundaries. Bueckner's result is that the stress intensity factor for a crack in a body subjected to external forces is identical to that for a similar crack, subjected to internal pressure in a similar body which has no external forces acting on it. The internal pressure $p(x)$ acting in the crack is equal to the stress that would exist normal to the crack-line along the crack-site in the uncracked body subjected to the external forces. An analogous procedure exists for shear stresses and the calculation of sliding-mode stress intensity factors. This principle is shown schematically in Fig. 10. In general it is easier to obtain stress fields in an uncracked body, than to calculate the stress intensity factor directly. However for many bodies and external force distributions the stresses along the crack-site may still be difficult to evaluate. Often Green's function techniques can be used in these evaluations and Nisitani[32] has derived stress-field Green's functions for several different bodies, so that the internal stress distributions can be derived for any externally applied forces on these bodies. Many crack Green's functions are available in the standard works cited in the second section.

3.4 Simple methods expressed as Green's functions

Several simple methods[33,34,35] have been proposed for determining stress intensity factors, particularly for the important case of cracks from holes or notches. These methods have been developed by comparing results with known stress intensity factors in particular cases. It will now be shown that some of these simple methods are not arbitrary, but arise from approximations to the Green's function. Results obtained from using them will be compared with those derived by Hsu & Rudd[36] who used a more accurate Green's function. Hsu & Rudd determined stress intensity factors for a symmetrical pair of opposing point forces acting on each of two radial cracks of length ℓ at a hole of radius R (see Fig. 11). They used finite element methods for $x/\ell < 0.9$ and a limiting expression for larger values. The Green's function given by Hsu & Rudd is shown (solid curves) in Fig. 12 for three values of $\ell/R = 0.2$, 1.0 and 3.0. Two features of the Green's function are of special interest; it is a weak function of ℓ/R and it tends to infinity as the point force approaches the crack tip. Green's functions for an edge crack[37,38] depicted in Fig. 13 and an embedded crack[39] depicted in Fig. 14 are also shown in Fig. 12.

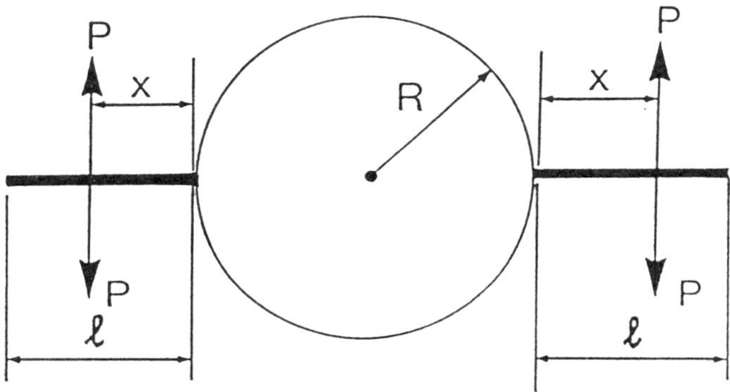

Figure 11: Diametrically opposed radial cracks subjected to symmetrical point forces

The Green's function for the edge crack is given[37] as

$$G_e(X) = \frac{2}{\sqrt{1-X^2}}\{1.2945-0.6857X^2+1.1597X^4+1.7627X^6+1.5036X^8-0.5094X^{10}\} \tag{83}$$

where $X = x/\ell$, and for the embedded crack, from eqn (80), as

$$G_c(X) = \frac{2}{\sqrt{1-X^2}} \tag{84}$$

The approximate methods[33,34,35] can also be expressed as Green's functions thereby establishing their general application to cracks at holes and notches. The Green's function $G_0(x)$ for a point loaded crack at the notch tip in Fig. 15a is shown schematically in Fig. 15b. Three approximations to $G_0(x)$ will be examined, these are also shown in Fig. 15b and are defined as

$$G_1(x) = 1.12\pi\ell\delta(x) \tag{85}$$

$$G_2(x) = 1.12\pi\ell\delta(x-\ell) \tag{86}$$

$$G_3(x) = 1.12\pi[H(x) - H(x-\ell)] \tag{87}$$

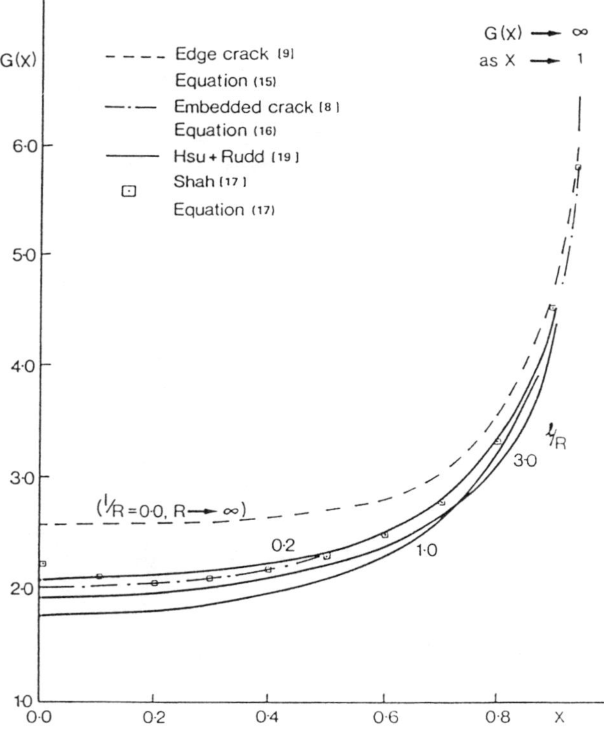

Figure 12: Comparison of Green's functions

where $\delta(x)$ is the Dirac delta function, $H(x)$ is the Heaviside step-function. It will now be shown that the above Green's functions are equivalent to well known results.

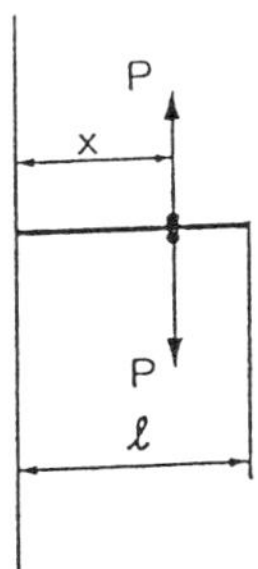

Figure 13: Point loaded crack emanating from a stress free edge

From eqn (82) the stress intensity factors resulting from the Green's functions in eqns (85) to (87), with $p(x)$, set equal to the stress over the crack site $\sigma(x)$, are given by:

$$\text{for } G_1 \quad K_{\mathrm{I}}^{(1)} = \frac{1}{\sqrt{\pi\ell}} \int_0^{\ell} \sigma(x) 1.12\pi\ell\delta(x) \mathrm{d}x = 1.12\sigma(0)\sqrt{\pi\ell} \tag{88}$$

$$\text{for } G_2 \quad K_{\mathrm{I}}^{(2)} = \frac{1}{\sqrt{\pi\ell}} \int_0^{\ell} \sigma(x) 1.12\pi\ell\delta(x-\ell) \mathrm{d}x = 1.12\sigma(\ell)\sqrt{\pi\ell} \tag{89}$$

$$\text{and for } G_3 \quad K_{\mathrm{I}}^{(3)} = \frac{1}{\sqrt{\pi\ell}} \int_0^{\ell} \sigma(x) 1.12\pi[H(x) - H(x-\ell)] \mathrm{d}x$$

$$= 1.12\sqrt{\pi\ell}\left\{\frac{1}{\ell}\int_0^{\ell} \sigma(x)(\mathrm{d})\mathrm{x}\right\} = 1.12\sigma_{\mathbf{mean}}\sqrt{\pi\ell} \tag{90}$$

Equations (88) and (89) are the familiar maximum stress[35] and crack tip stress[33] approximations respectively and eqn (90) is the mean stress method suggested by Williams & Isherwood;[34] several applications of these methods have been considered elsewhere.[5]

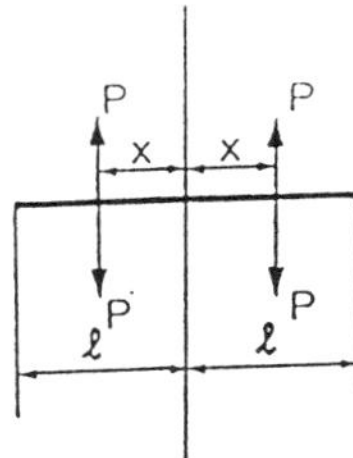

Figure 14: Point loaded crack embedded in an infinite sheet

These approximations are applied to the case of two equal cracks of length ℓ, diametrically opposed at the edge of a hole of radius R in a sheet subjected to a biaxial tensile stress σ, and the results are shown in Table 1; for this case

Table 1: Comparison of stress intensity factors ($K_I/\sigma\sqrt{(\pi\ell)}$)

ℓ/R	Hsu & Rudd Ref. 36	Tip stress eqn (89)	Max stress eqn (88)	Mean stress eqn (90)	Tweed & Rooke Ref. 40
0.00	2.17 (-3)	2.24 (0)	2.24 (0)	2.24 (0)	2.24
0.10	1.99 (-0.5)	2.05 (+2.5)	2.24 (+12)	2.14 (+7)	2.00
0.15	1.90 (-0.5)	1.97 (+3)	2.24 (+17)	2.09 (+9)	1.91
0.20	1.83 (-0.5)	1.90 (+3)	2.24 (+22)	2.05 (+11)	1.84
0.30	1.70 (-1)	1,78 (+3.5)	2.24 (+30)	1.98 (+15)	1.72
0.50	1.54 (-1)	1.62 (+4)	2.24 (+44)	1.87 (+20)	1.56
1.00	1.34 (-2)	1.40 (+2)	2.24 (+64)	1.68 (+23)	1.37

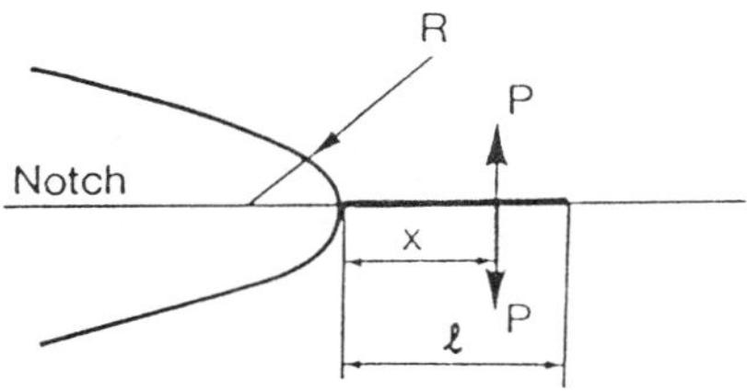

Figure 15a: Point loaded crack at a notch root

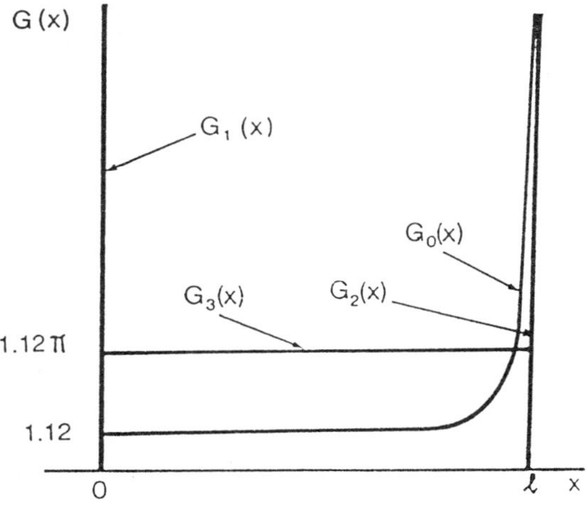

Figure 15b: Schematic Green's functions for a crack at a notch root

$$\sigma(x) = \sigma \left[1 + \frac{R^2}{(R+x)^2} \right] \tag{91}$$

The results derived from Hsu & Rudd[36] are seen to be in close agreement with the accurate solution of Tweed & Rooke;[40] errors for each approximation are shown in parenthesis. The tip stress eqn (89), results in errors of about 3 % showing that the Green's function $G_2(x)$ is a satisfactory approximation to $G_0(x)$. However, despite its limitations the maximum stress method is useful in that it gives an upper limit on the stress intensity factor. Furthermore it is only necessary to know the stress concentration factor for the notch rather than the entire stress field over the crack-site in the uncracked body. The mean stress method also over-estimates the stress intensity factor but in this case it is necessary to know the entire stress field over the crack-site and to determine its mean value at each crack length. Smith[41] has made use of the mean stress and the properties of Green's functions to establish bounds on stress intensity factors of edge cracks when $\sigma(x)$ is a monotonically decreasing function and $\sigma(\ell) > 0$.

3.5 Application of Green's functions

The usefulness of Green's functions in fracture mechanics will now be illustrated by an example. It is chosen to illustrate the sort of practical problem which may be easily and accurately solved by the use of the techniques described in this section.

3.5.1 Effect of pin pressure distribution on a crack at a hole

In this example it is necessary to give a different interpretation to $p(x)$ and $G(x)$ in eqn (82) since in order to solve these problems we need the Green's function which gives the response to a force acting at an arbitrary position in the body (in this case on the hole boundary) rather than that due to a force acting on the surfaces of the crack. For problems of this type the stress intensity factor for an arbitrary stress acting on the body is obtained from the following expression:

$$K_{\text{I}} = \frac{1}{\sqrt{\pi \ell}} \int_{x_1}^{x_2} \sigma(x) G(x) \mathrm{d}x \tag{92}$$

where x_1 and x_2 are positions in the body between which $\sigma(x)$ is the prescribed applied stress and $G(x)$ is the stress intensity factor for a unit force acting at the position denoted by x. As an example of the use of such Green's functions we will consider a radial crack at the edge of a circular hole which is subjected to a point force on its perimeter. This type of problem is frequently encountered in considering cracked-holes in pin-loaded lugs. Often the load transfer between the pin and the hole periphery is not precisely known and it is necessary to investigate various possible load distributions. Green's function techniques are ideal for such investigations since each new distribution just involves changing the stress $\sigma(x)$in eqn (92).

The Green's function required is the stress intensity factor for a radial crack of length ℓ at the edge of a hole of radius R; a radial force per unit thickness of P acts at the edge of the hole in a direction which makes an angle ϑ with the crack (see Fig. 16). This stress intensity factor, obtained by Tweed & Rooke[42] is plotted in Fig. 16

in non-dimensional form as a function of ϑ for various ℓ/R values. The usefulness of Green's functions in describing the response of systems is clearly shown by these results since two important observations can be made from Fig. 16. The first is that the variation of K_I with ϑ increases as the crack length decreases, and the second is that the maximum rate of variation occurs for short cracks at small values of ϑ. These facts lead to important considerations in fracture mechanics applications.

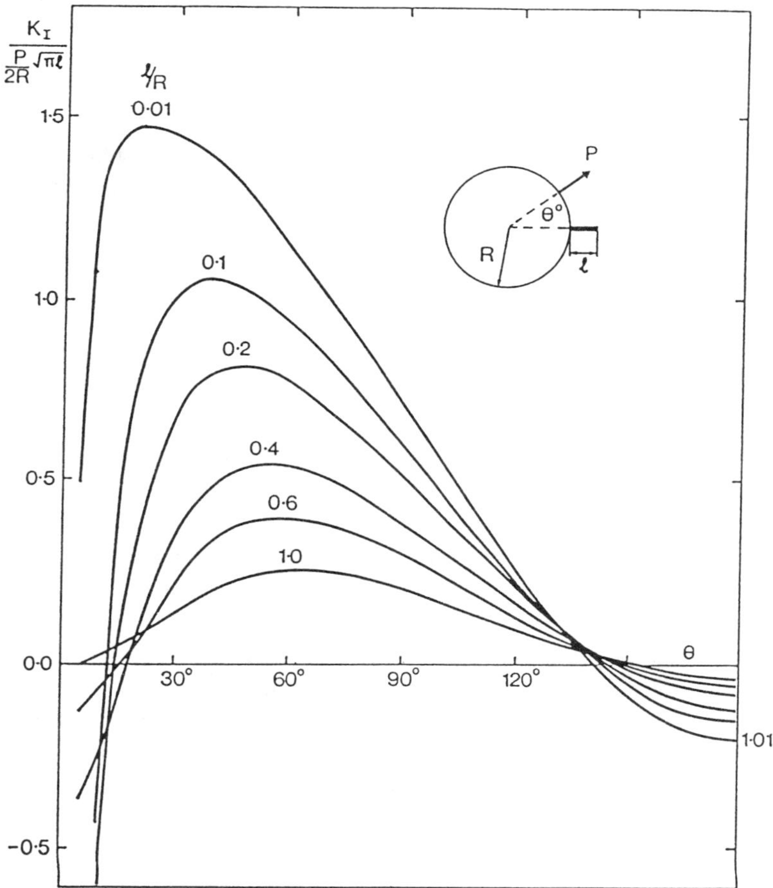

Figure 16: Green's function for a crack at the edge of a loaded circular hole

The first consideration, which is of general importance, is that since most of the lifetime of a fatigue crack is spent while the crack is short, it is necessary to know what influences the stress intensity factor of short cracks. This information is required in order to decide on inspection intervals and maintenance schedules for structures in service. The second consideration is specific to the loaded hole configuration since assumptions have usually to be made about how the load is transferred from the pin to the edge of the hole. It is often assumed that the load is distributed, in some way, between ϑ_1 and ϑ_2 where $\vartheta_1 \sim 0^0$ and $\vartheta_2 \sim 180^0$. The rapid variation in Green's function at small values of ϑ means that an incorrect assumption of the load distribution and the cut-off value ϑ_1 can lead to significant errors in the stress intensity factor. The errors will be largest for the important region of short cracks, so great care must be exercised in simulating the load transfer between the pin and the edge of the hole. The Green's functions for both radial and tangential loads acting on the edge of the hole have been tabulated by Rooke & Hutchins,[43] so that both K_{I} and K_{II} may be calculated for any loading. Some examples of the effect on the stress intensity factor of varying the stress distribution on the edge of the hole have been given by Rooke *et al.*[43,44]

3.6 Generalisation of Green's functions (weight functions)

Rice[45] has derived a more general integral expression for the stress intensity factor. For the special case of a loading $\sigma(x)$ on the crack the expression is a generalisation of eqn (82) and is given by

$$K_{\mathrm{I}} = \int_0^{2a} \sigma(x) H(a,x) \mathrm{d}x \tag{93}$$

The function $H(a,x)$ called a weight-function, is given[45] by

$$H(a,x) = \frac{8\mu}{(1+\kappa)} \cdot \frac{1}{K_{\mathrm{I}}^*} \cdot \frac{\delta u^*}{\delta x}(a,x) \tag{94}$$

where $\kappa = 3-4\nu$ for plane strain and $\kappa = (3-\nu)/(1+\nu)$ for plane stress, with ν and μ the Poisson's ratio and the shear modulus respectively. K_{I}^* is a known stress intensity factor solution for the particular geometric configuration being considered and $u^*(a,x)$ is the displacement on the crack surface due to the loading that corresponded to K_{I}^*. Any simple loading condition may be used to determine K_{I}^* and $u^*(a,x)$. A limitation of this technique is that u^* is not available for many finite configurations. Some approximate techniques for evaluating u^* for a limited range of geometries have been developed by Fett *et al.*[46,47] and Wu.[48] The technique is however versatile and has been extended to more general boundary conditions by Wu & Carlsson[49] and by Parker & Bowie.[50]

An alternative weight function approach has been derived from Betti's reciprocal work theorem by Paris, McMeeking & Tada;[51] they obtained the stress intensity factor for an applied traction $\underline{t}$ as

$$K_N = \frac{E'}{4\sqrt{2\pi}B_N} \int_\Gamma (\underline{t}.\underline{u}^N) \mathrm{d}\Gamma, \quad N = \mathrm{I}, \mathrm{II}, \mathrm{III} \tag{95}$$

where $E' = E$ for plane stress and $E' = E/(1-\nu^2)$ for plane strain, $\underline{t}$ is the applied traction and E is Young's modulus. The displacement $\underline{u}^N$ is the field on the boundary resulting from the presence of a Bueckner singular field, of strength B_N at the crack tip; this field $\underline{u}^N$ is the weight function.[52] The derivation of eqn (95) and the origin of the singular field will be given in detail in a later chapter. Paris *et al.*[51] used a finite element analysis to determine $\underline{u}^I$ and obtained K_{I} from eqn (95). Cartwright & Rooke[53] showed that more accurate values could be obtained more efficiently using boundary element analysis. The method has been extended to mixed mode two-dimensional problems[54] and to three-dimensional problems[55] and recently reviewed in detail.[56]

The relationship between weight functions and Green's functions can be demonstrated by considering the special case of a point load at x_0 normal to the crack surface $y = 0$; that is $\underline{t} = \underline{P}\delta(x - x_0)$.Substitution of this expression into eqn (95) and integrating, using the properties of the delta-function, leads to

$$\frac{K_{\mathrm{I}}}{P_2\sqrt{\pi a}} = \frac{E'\sqrt{a}}{4\sqrt{2}B_{\mathrm{I}}}u_2^{\mathrm{I}}(x_0) \tag{96}$$

where P_2 and u_2^{I} are the force and the displacement respectively in the direction normal to the crack face. From the basic definition of the Green's function in eqn (80) for a point force normal to a crack face, it follows that $u_2(x_0) \propto G(x_0)$, i.e. the weight function for a point force is proportional to the Green's function.

3.7 Conclusions

The application of Green's function techniques to problems in fracture mechanics has been described. It has been shown to be a versatile technique for determining stress intensity factors in a wide variety of problems. Once the Green's function is known it is only necessary to determine the stress distribution along the crack site in the uncracked body. The stress analysis of uncracked bodies is, in general, simpler than the stress analysis of cracked bodies and many experimental and analytical techniques already exist. Several commonly used approximate methods have been given a rational basis; they have been shown to depend on the existence of certain approximate Green's functions. Many important Green's functions have been collected and some of these have been used to solve practical problems in fracture mechanics. The Green's function technique is also applicable to problems in three-dimensions although applications are more limited at present because of the small number of Green's functions available.

4 The compounding method of determining stress intensity factors

4.1 Introduction

The tip of a growing crack is often close to one or more structural boundaries and, as expected, their proximity influences the magnitude of the stress intensity factor. In this section an approximate method is developed which takes into account the effects of near boundaries on the crack. Both the proximity to the crack tip and the shape of the boundary are important. The shape of the boundary is described by the radius of curvature of that part of the boundary which is nearest to the crack.

The development of this method[57] and its application[58,59] to complex geometrical configurations with multiple boundaries enables engineers to extend the use of fracture mechanics to complex engineering structures. Stress intensity factors for many simple configurations are already available[60–63] but these configurations seldom model adequately real engineering structures. The compounding method developed here is a quick and versatile way of extending these solutions to other, more complex, configurations for which the stress intensity factors are not known. An early empirical method which was used by Figge & Newman,[64] Smith[65] and Liu,[66] is a special case of compounding but its generality was not realised or investigated.

In this section only two-dimensional configurations are considered, but solutions to three-dimensional configurations may also be obtained. The compounding method is developed and applied to plane test problems. The compounding method can be applied to stiffened sheets containing cracks if the stiffeners are regarded as boundaries[58,59] in order to calculate their effect on K. The basic theory requires a modification before it can be applied to problems where a boundary (or stiffener) cross a crack. The boundary/crack combination must be replaced by an 'equivalent crack' before interactions with other boundaries can be calculated. This modification is described later and the use of the equivalent crack concept demonstrated for two cracks at the edge of a circular hole. For some hole/crack configurations, boundary-boundary interactions are important and an approximate method is needed to calculate these interactions.

The compounding theory is expressed in terms of non-dimensionalised stress intensity factors since the solutions for the ancillary configurations are usually available in this form (see for example Ref. 60). Configurations with multiple cracks are important because, the presence of other cracks often increases the value of the stress intensity factor and thereby increases the rate at which such cracks will grow under fatigue loading. Thus the fatigue lifetime is shortened which may have important consequences with regard to the safety and economy of operation.

4.2 The compounding method

A configuration containing a crack may have several boundaries, e.g. holes, other cracks or sheet edges; all will influence the stress intensity factor at the tip of the crack under consideration. The principle of the compounding method presented here is to obtain a solution by separating the complex configuration into a number of simpler ancillary configurations which have known solutions. Each ancillary configuration will, usually, contain only one boundary which interacts with the crack. The contributions to the final stress intensity factor are compounded, initially neglecting any effects due to boundary-boundary interactions. The error term, due to neglecting these effects, has been formally derived using the Schwarz alternating technique; it will be demonstrated that the error terms are small for many classes of problem.

4.2.1 Derivation of basic formulae

Consider the configuration shown in Fig. 17(a) containing a crack near to a stress free boundary B_1; the configuration is subjected to an applied stress system S_0 on its boundary B_0 which is remote from the crack. Let the stress intensity factor at one of the crack tips be denoted by K_1. If the stress free boundary B_1 were absent from the configuration, stresses S_1 would occur at the site of B_1; the stress intensity factor, in the absence of internal boundaries, is now given by $\bar{K}$, Fig. 17(b). The original configuration can be obtained by the superposition of the following:

(i) the cracked configuration with applied stress S_0 on B_0 without an internal boundary Fig. 17(b), and

(ii) the cracked configuration with zero stress on B_0 and $-S_1$ on B_1 (see Fig. 17(c)).

Thus the stress intensity factor K_1 is given by

$$K_1 = \bar{K} + K_1^* \tag{97}$$

where K_1^* is the stress intensity factor when the only applied stress is $-S_1$ on B_1. Similarly, for a boundary, B_2 say, the stress intensity factor would be given by

$$K_2 = \bar{K} + K_2^* \tag{98}$$

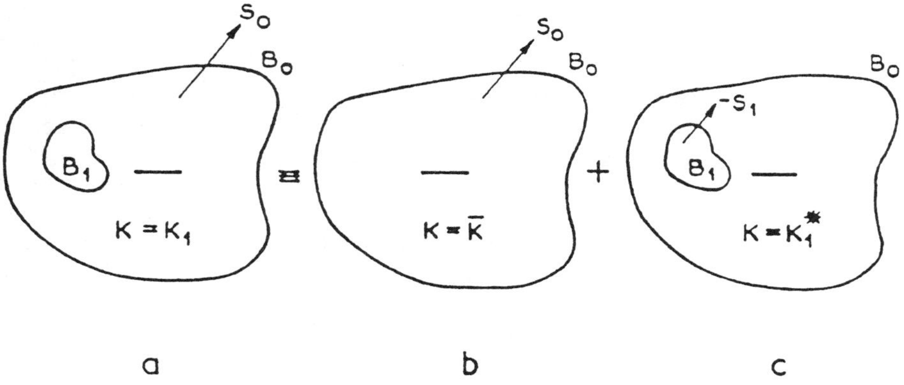

Figure 17: Superposition for a crack near one internal boundary

If the two boundaries B_1 and B_2 are present together (Fig. 18(a)), the resultant stress intensity factor K_r is given by the superposition of Fig. 18(b) and 18(c) as

$$K_r = \bar{K} + K_r^* \tag{99}$$

where K_r^* is the stress intensity factor when the configuration has zero stress on B_0, and stresses $-S_1$ and $-S_2$ on B_1 and B_2 respectively (Fig. 18(c)).

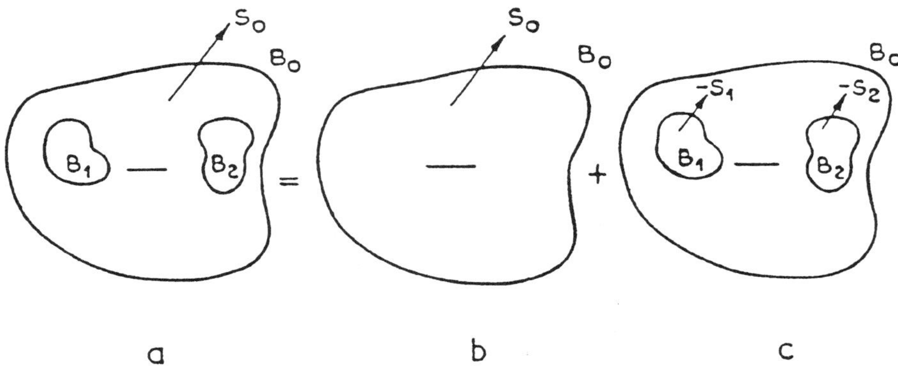

Figure 18: Superposition for a crack near two internal boundaries

If the two boundaries do not interact with each other then, by superposition, K_r^*

$$K_r^* = K_1^* + K_2^* \tag{100}$$

If they do interact there will be an extra term which is denoted by K_e, i.e.

$$K_r^* = K_1^* + K_2^* + K_e \tag{101}$$

Combining eqns (97), (98), (99) and (101) gives

$$K_r = K_1 + K_2 - \bar{K} + K_e \tag{102}$$

Thus the stress intensity factor for a crack in a configuration with multiple internal boundaries can be expressed in terms of stress intensity factors derived from configurations with single internal boundaries, apart from a correction term. In general for N boundaries $B_n(n = 1, 2....N)$, the resultant stress intensity factor is given by

$$K_r = \bar{K} + \sum_{n=1}^{N}(K_n - \bar{K}) + K_e \tag{103}$$

In terms of normalised stress intensity factors Q_r, this becomes

$$Q_r = 1 + \sum_{n=1}^{N}(Q_n - 1) + Q_e \tag{104}$$

where $Q_r = K_r/K, Q_n = K_n/K$ and $Q_e = K_e/K; Q_e$ is the correction term due to the interaction of the N boundaries. If Q_e can be estimated or can be shown to be small

(<< 1), then eqn (104) can be used to build up solutions to complex configurations from known simpler ones.

An empirical method, which has been used[64–66] to obtain approximate stress intensity factors, states that in the case of two boundaries

$$Q_r = Q_1 Q_2 \tag{105}$$

If $Q_1 = 1 + \alpha$ and $Q_2 = 1 + \beta$ where $\alpha, \beta << 1$ then from eqn (105) Q_r is given by

$$Q_r = 1 + \alpha + \beta + \alpha\beta \tag{106}$$

Equation (104) for two boundaries becomes

$$Q_r = 1 + \alpha + \beta + Q_e \tag{107}$$

Thus eqns (106) and (107) are the same but for a small correction term $\alpha\beta$ or Q_e and the empirical expression, eqn (105), is seen to be a special case of eqn (104).

The correction term Q_e can be expressed formally[57] using the Schwarz alternating technique which has been described by Sokolnikoff[67] and used to determine stress intensity factors by Hartranft & Sih[68] and many others. The term Qe can be expected to be small provided that the boundaries are not too close to each other.

4.2.2 Equivalent crack concept

The simple compounding procedure of adding together the effects of the individual boundaries needs to be modified if the crack crosses one of the boundaries, e.g. a crack at the edge of a hole, or a crack beneath a stiffener (which is treated as a boundary). Before the effect of the other boundaries can be considered, the crack plus the boundary it crosses must be replaced[57,58] by an 'equivalent crack' which then interacts with the other boundaries. If the stress intensity factor is K_0 when only the boundary the crack crosses is present, then the equivalent crack is defined in terms of K_0 - two different procedures have been used.[57,58]

In the first procedure[57] the equivalent crack was defined as an isolated crack of length $2a'$ in a sheet with the same remote stress σ as in the original configuration; the length $2a'$ was determined by the condition that the stress intensity factor of the equivalent crack was equal to K_0, that is

$$\sigma\sqrt{\pi a'} = K_0 \tag{108}$$

Since in this case $K = \sigma\sqrt{(\pi a)}$, then a' is given by

$$a' = \left(\frac{K_0}{K}\right)^2 a = Q_0^2 a \tag{109}$$

In the second procedure[59] the equivalent crack was defined as an isolated crack of the same total length $2a$ in a sheet with a remote stress σ', which was determined by the condition that the stress intensity factor was K_0, that is,

$$\sigma'\sqrt{\pi a} = K_0 \tag{110}$$

Therefore

$$\sigma' = Q_0 \sigma \tag{111}$$

The second procedure is essential[69] if there is any loading on the crack faces, and even when the loads are remote from the crack it is as efficient and accurate as the first. Since, in practice, it is easier to use, it is the preferred procedure[59] - the effect of stiffeners should be treated as local loads.

The effects of the other boundaries $B_n (n = 1....N)$ on the original crack plus attached boundary are now considered to be the same as the effects on the equivalent crack in a configuration subjected to an applied stress σ'. The general compounding formula (103) is then modified to

$$K_r = K_0 + \sum_{n=1}^{N} (K'_n - K_0) + K_e \tag{112}$$

where K' is the stress intensity factor for the equivalent crack in the presence of the nth boundary only. Equation (112) can be written in terms of the normalised stress intensity factors, and becomes

$$Q_r = Q_0 \left[1 + \sum_{n=1}^{N} (Q_n - 1) \right] + Q_e \tag{113}$$

where $Q_n = K'_n / K_0$. It is important to note that, since K'_n is proportional to σ' and K_0 is also proportional to σ', see eqn (110), Q_n is independent of σ'.

4.2.3 Boundary-boundary interactions

In some configurations boundary-boundary interactions cannot be neglected, for instance a crack at the edge of a hole that is near another boundary. A measure of the interaction may be obtained from the difference in the *stress concentration factor* K_t at the edge of the hole in the uncracked configuration, with and without other boundaries. If the change in K_t is significant then the boundary-boundary interaction Q_e in the cracked configuration may also be significant particularly for short cracks when the limiting value of the stress intensity factor is proportional to K_t.

In the earlier derivation of the basic compounding method Q_e was shown to arise because the stresses induced on any one boundary site by the presence of the other boundaries, were not allowed for. An approximate technique for allowing for these stresses has been developed and used in the evaluation of the stress intensity factor for a hole with two equal-length radial cracks . The distribution of induced stresses around the hole boundary is represented by two equal and opposite localised forces P_e acting, on the hole perimeter, perpendicular to the crack line. The magnitude of P_e is chosen so that the maximum tensile stress ($\sigma_{\mathbf{max}}$) at the edge of the uncracked hole due to the remote loading and to the forces P_e in the absence of other boundaries, is equal to that in the real configuration. For example for a circular hole of radius R in a finite-width strip with a uniform stress σ, the force P_e would be obtained from

$$\sigma_{\mathbf{max}} 3\sigma + \frac{2P_e}{\pi R} = K_t \sigma \tag{114}$$

where 3σ is the maximum stress at the edge of a circular hole in a uniformly stressed sheet, $2P_e/(\pi R)$ is the stress, at the same position, due to the forces P_e acting on a hole in an infinite sheet[70] and K_t is the stress concentration factor for an uncracked hole in a strip. Stress concentration factors for this configuration, and many others, have been collected by Peterson.[71] The value of Q_e is obtained from the solution of Tweed & Rooke[72] for cracks at the edge of a hole subjected to localised loads and added to the other terms in eqn (113). The resultant stress intensity factor will thus be a good approximation, particularly for short cracks; for a crack of length ℓ, the limiting value is given by

$$\lim_{\ell/R\to 0}\{K_r\} = 1.12\sigma_{\mathbf{max}}\sqrt{\pi\ell} = K_t\sigma\sqrt{\pi\ell} \tag{115}$$

It is important that values of the stress intensity factor are accurately known for short cracks, because the majority of the fatigue life of a structure is spent while the crack is short; this technique for evaluating Q_e ensures accuracy at short cracks. The evaluation of Q_e may need further modification due to the hole/crack interaction with other boundaries (see later).

4.3 Application to plane sheets

In this section approximate compounded solutions are compared with known solutions for three different types of boundary. These boundaries, each characterised by different radii of curvature ρ, are another crack ($\rho = 0$), a circular hole ($\rho =$ hole radius) and a straight edge ($\rho = \infty$). Boundaries with a large radius of curvature will have an effect over a larger distance than boundaries with a small radius.

The approximate formula for N boundaries (omitting Q_e) given by

$$Q_r = 1 + \sum_{n=1}^{N}(Q_n - 1) \tag{116}$$

is used to compound stress intensity factors for configurations with known solutions and, by comparison, it is shown that the omitted term Q_e is small, (i.e. $<< 1$). The magnitude of Q_e will depend on the number, nearness and shape of the boundaries. As errors in Q_r due to using eqn (116) will increase as N increases, multiple boundaries are also considered. Finally the stress intensity factor is obtained for a crack located in a half-plane between two different boundaries, a straight edge and a circular hole.

4.3.1 Test solutions

The configurations shown in Figs 19 to 21 for which solutions are known, are used as test cases to illustrate and assess the method. These represent widely different boundary effects, namely a pair of boundaries of infinite, finite or zero radius of curvature in the path of the crack.

Consider the configuration of a crack in a strip shown in Fig. 19. The required ancillary configurations, studied by Isida,[73] are shown in Fig. 22. The normalised stress intensity factors Q_r for the crack in Fig. 19 are obtained by compounding from eqn (116) with $N = 2$, that is,

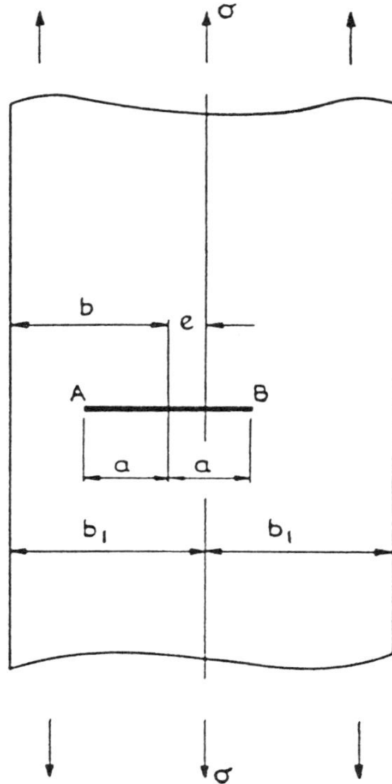

Figure 19: Eccentric crack in a finite width sheet subjected to a uniaxial tensile stress

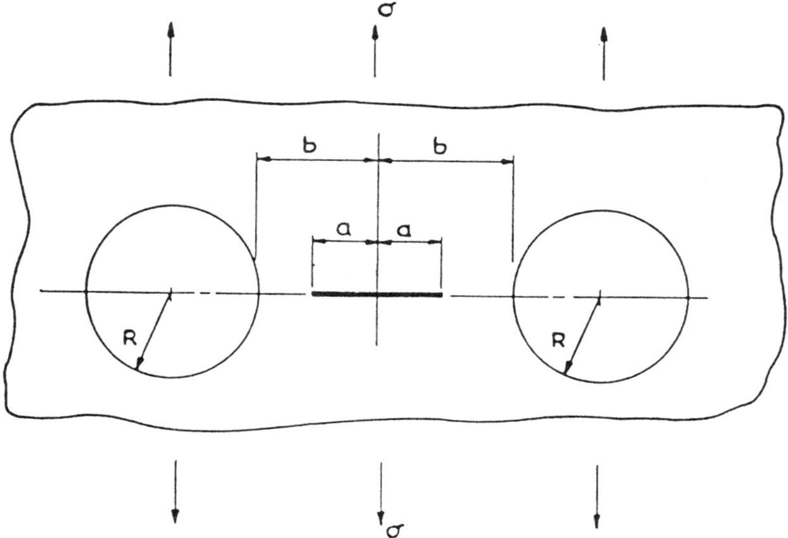

Figure 20: Crack between two holes in an infinite sheet subjected to a uniaxial tensile stress

Table 2: Comparison of values of $K_I/(\sigma\sqrt{\pi a})$ for an eccentric crack in a finite width sheet subjected to a uniaxial tensile stress.

a/b	$e/b_1 = 0.0$		$e/b_1 = 0.0$			
	Tip A and B		Tip A		Tip B	
	Compound	Ref.74	Compound	Ref. 73	Compound	Ref. 73
0.0	1.000	1.000	1.000	1.000	1.000	1.000
0.1	1.005	1.005	1.003	1.003	1.002	1.003
0.2	1.020	1.021	1.011	1.014	1.009	1.012
0.3	1.047	1.05	1.03	1.03	1.02	1.03
0.4	1.09	1.10	1.05	1.07	1.04	1.05
0.5	1.15	1.18	1.09	1.11	1.06	1.08
0.6	1.23	1.29	1.15	1.19	1.08	1.11
0.7	1.34	1.48	1.24	1.30	1.11	1.15

$$Q_e = Q_1 + Q_2 - 1 \tag{117}$$

For tip A in Fig. 20, Q_1 and Q_2 are the normalised stress intensity factors at the left hand tip in Fig. 22a and b respectively and, for tip B, Q_1 and Q_2 are the normalised stress intensity factors for the right hand tip in Fig. 22a and b respectively. Comparison of Figs 19 and 22 shows that $c = b$ and $d = b_1 + e$; in this configuration $\bar{K} = \sigma\sqrt{(\pi a)}$. Values of the opening mode normalised stress intensity factor $Q_r(= K_I/\sigma\sqrt{(\pi a)})$ obtained from eqn (117) are compared in Table 2 with the results of both Benthem & Koiter[74] and Isida[73] for $a/b \leq 0.7$ and $e/b = 0.0$ and 0.8.

In a similar manner solutions are obtained for configurations in Figs 20 and 21 using the work of Isida[73] (a crack near a circular hole) and Tranter[75] (a crack near another crack) to obtain the ancillary solutions. Compounded results for the test case in Fig. 20 ($b/R = 1$) and the test case in Fig. 21 (middle crack of three) are shown in Table 3 and compared with the solutions of Newman[76] and Isida.[73]

The magnitudes of the errors in the stress intensity factors obtained for the configurations in Figs 19 to 21 are summarised in Table 4 and indicate two trends. Firstly errors increase as the crack length increases relative to the distance to a boundary. Secondly at a fixed crack length errors tend to increase with increasing boundary radius. Thus it appears that for a/b up to 0.8 the errors are probably <18 % for straight boundaries, <5 % for circular boundaries (two radii apart) and <1.5 % for other crack boundaries.

As an example of a solution to a problem involving more than two boundaries consider the configuration in Fig. 21 but with the number of cracks increased. Table 5 shows that the compounded normalised stress intensity factor Q_r for the middle crack with $N = 5, 7$ and 11 in eqn (116) obtained from the same ancillary configuration as was used for the three-crack case in Fig. 21. At any fixed crack length, comparison with the results of Isida[73] shows that errors increase with ,he number of cracks; this illustrates the effect of the increasing interaction between boundaries which has not

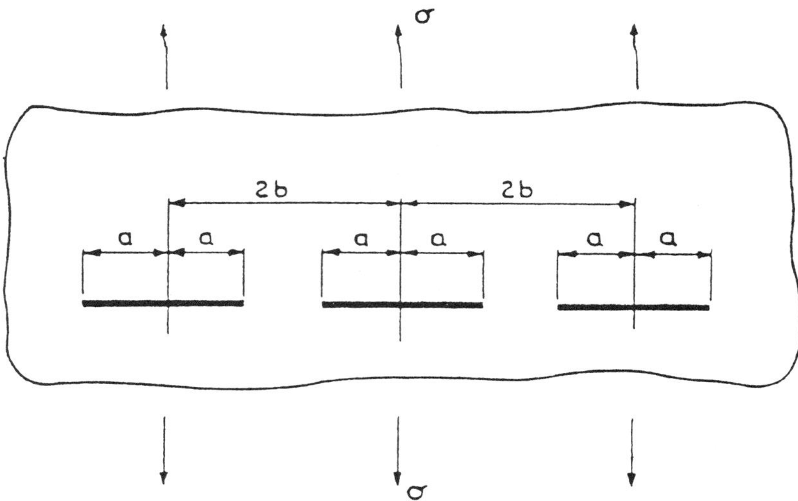

Figure 21: Odd number of collinear cracks in an infinite sheet subjected to a uniaxial tensile stress.

been taken into account in eqn (116).

From the foregoing results, it can therefore be concluded that the errors due to neglecting boundary interactions are usually small (a few per cent), and as such are within the allowable tolerances for many engineering applications.

4.3.2 A solution for two different boundaries

No comparison solution is available for this problem of a crack in the vicinity of a hole in a half-plane subjected to a uniaxial tensile stress (Fig. 23). Ancillary configurations for this problem are obtained from Isida.[73] Compounded stress intensity factors at both crack tips are shown in Fig. 23 for $R/b = 1$. Three ratios of the distance between the crack and the straight boundary c and the distance between the crack and the hole b are considered, namely $c/b = 0.5, 1$ and 2. For $c/b = 0.5$ the tip adjacent to the straight boundary (tip B) has a lower stress intensity factor than tip A for short cracks ($a/b < 0.35$) As a/b increases beyond 0.35 the tip nearer the straight boundary (tip B) has the higher stress intensity factor. This behaviour is due to the tip B approaching close to the boundary: in the limit $K_{\mathrm{B}} \to \infty$ as $a/b \to 0.5$. For the other values of c/b considered the crack tip B adjacent to the hole is more critical for all a/b. If the straight boundary is remote from the hole ($c/b = 2$ say) the stress intensity factor for the tip at B initially reduces slightly as a/b increases since the stress field due to the hole is decreasing. For larger a/b the effect of the straight boundary causes the stress intensity factor to increase. Consideration of the errors estimated previously suggests that this solution is probably accurate to better than 10 %.

Table 3: Comparison of values of $K_I/(\sigma\sqrt{\pi a})$ for the configurations in Figs 20 and 21

a/b	Fig. 20 ($b/R = 1$)		Fig. 21 (middle crack)	
	Compound	Ref.76	Compound	Ref. 73
0.0	1.44	1.47	1.00	1.00
0.1	1.45	1.47	1.00	1.00
0.2	1.46	1.49	1.01	1.01
0.3	1.49	1.52	1.02	1.02
0.4	1.53	1.56	1.05	1.05
0.5	1.59	1.63	1.08	1.08
0.6	1.68	1.72	1.12	1.13
0.7	1.81	1.87	1.19	1.20
0.8	2.02	2.12	1.31	1.33
0.9	2.43	2.64	1.57	1.60

Table 4: Percentage errors for the compounding method

a/b	Percentage error		
	Straight boundaries Fig. 19	Circular boundaries Fig. 20	Crack boundaries Fig. 21
0.0	0.0	2.1	0.0
0.2	0.2	2.1	0.0
0.4	1.6	2.0	0.0
0.6	4.3	2.4	0.5
0.8	17.9	5.0	1.4

Table 5: Comparison of $K/(\sigma\sqrt{(\pi a)})$ for the central crack of an odd number of collinear cracks subjected to a uniform tensile stress.

a/b	5 cracks			7 cracks			11 cracks		
	Present results	Ref. 73	Error %	Present results	Ref. 73	Error %	Present results	Ref. 73	Error %
0.0	1.0	1.00	0.0	1.00	1.0	0.0	1.00	1.00	0.0
0.2	1.01	1.01	0.0	1.01	1.02	0.1	1.02	1.02	0.1
0.4	1.06	1.06	0.3	1.06	1.06	0.4	1.06	1.07	0.7
0.6	1.15	1.16	1.4	1.16	1.17	1.4	1.16	1.19	2.0
0.8	1.36	1.41	4.1	1.37	1.46	5.8	1.38	1.49	7.3

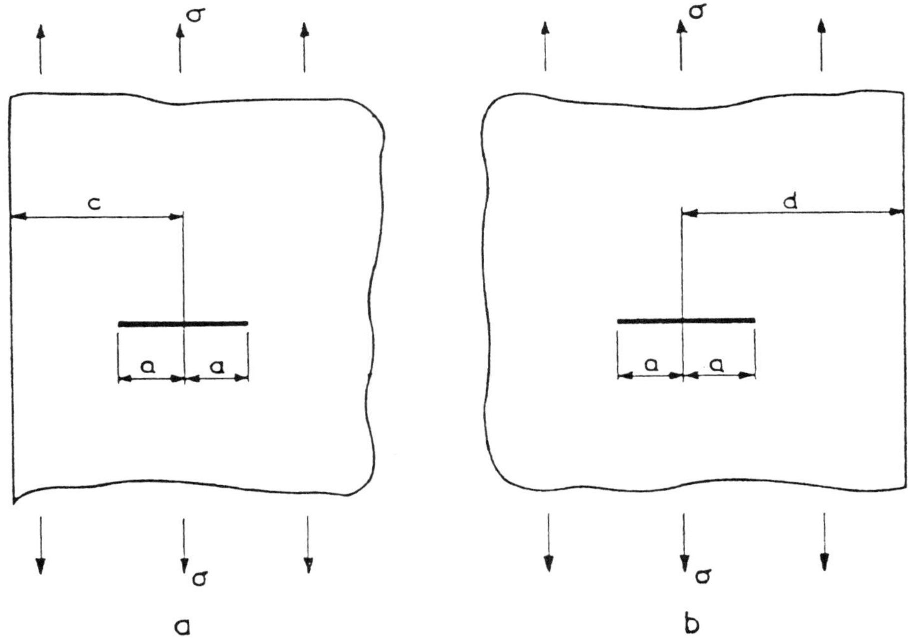

Figure 22: Crack near the edge of a half plane subjected to a uniform tensile stress

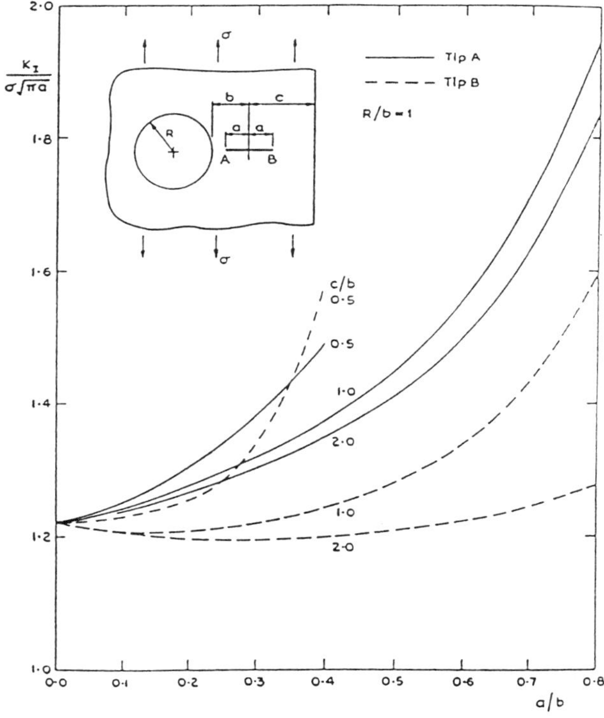

Figure 23: Stress intensity factors for a crack in the vicinity of a hole in a half plane subjected to a uniaxial tensile stress.

4.4 Application to cracks at holes

4.4.1 Two cracks at a circular hole in a strip

The configuration shown in Fig. 24 consists of two cracks of equal length ℓ at opposite ends of a diameter of a circular hole of radius R. The hole is located centrally in a strip of width $2b$ which is subjected to a uniform uniaxial tensile stress σ remote from the hole; the stress acts in a direction parallel to the axis of the strip and perpendicular to the cracks. Also shown in Fig. 24 are the two ancillary configurations required for the determination of the stress-intensity factor.

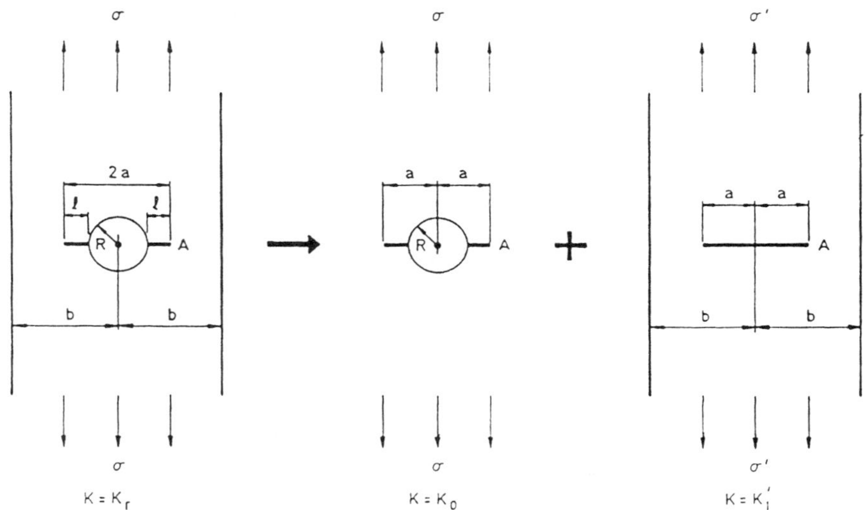

Figure 24: Ancillary configurations for a cracked hole in a strip

They are (i) two cracks at the edge of a circular hole in an infinite sheet which is subjected to a uniform, uniaxial tensile stress σ remote from the hole, and (ii) a crack of length $2a(= 2\ell + 2R)$ centrally located in a strip of width $2b$; the crack is perpendicular to the axis of a strip and to the direction of the uniform, uniaxial tensile stress $\sigma'(= Q_0\sigma)$ which acts remote from the hole.

If the stress-intensity factor for crack tip A in the first ancillary configuration is K_0 and that for crack tip A in the second ancillary configuration is K_1', then the resultant stress-intensity factor K_r is given by eqn (112) as

$$K_r = K_1' + K_e \tag{118}$$

The normalised factor Q_r is given by eqn (113) as

$$Q_r = Q_0 Q_1 + Q_e \tag{119}$$

where Q_0 can be obtained from the work of Tweed & Rooke[72] and Q_1 from Case 1.1.1, Ref. 60. Thus if Q_e can be determined then Q_r can be calculated from eqn (119). The term Q_e is determined, see earlier, by considering the limiting behaviour of eqn (118) as the crack length tends to zero.

Table 6: Parameters for boundary-boundary interactions

R/b	K_t	$Q_1(R/b)$	$\frac{2P_e}{\pi R\sigma}$
0.25	3.24	1.02	0.18
0.50	4.32	1.18	0.78

The limits required are:

$$\lim_{\ell\to 0}\{K_r\} = 1.12 K_t \sigma\sqrt{\pi\ell} \tag{120}$$

where K_t is the stress concentration factor at the crack site in the uncracked configuration;

$$\lim_{\ell\to 0}\{K_1'\} = \lim_{\ell\to 0}\{Q_1 K_0\} = Q_1(R/b)\lim_{\ell\to 0}\{K_0\} \tag{121}$$

where $Q_1(R/b)$ is the normalised stress-intensity factor for a crack of length $2R$ in a strip of width $2b$ subjected to a uniform stress;

$$\lim_{\ell\to 0}\{K_0\} = 1.12\times 3\sigma\sqrt{\pi\ell} \tag{122}$$

since the stress concentration factor for a hole in a uniformly stressed infinite sheet is 3; and

$$\lim_{\ell\to 0}\{K_e\} = 1.12\times\frac{2P_e}{\pi R}\sqrt{\pi\ell} \tag{123}$$

Substitution of eqns (120)-(123) into (118) gives

$$K_t\sigma = 3\sigma Q_1(R/b) + \frac{2P_e}{\pi R} \tag{124}$$

that is

$$\frac{2P_e}{\pi R\sigma} = K_t - 3Q_1(R/b) \tag{125}$$

Thus P_e can now be determined and Q_e obtained from the work of Tweed & Rooke.[72] For the two cases considered here, namely $R/b = 0.25$ and 0.50, the parameters required in eqn (125) are given in Table 6 together with the values of $2P_e/(\pi R\sigma)$.

Values of the normalised stress-intensity factor Q_r have been calculated for several crack lengths for the two test cases. They are compared in Table 7 with the numerical calculations of Newman.[76] The values of Q_r obtained are within 2% of those obtained by Newman[76] throughout the whole range of a/R and R/b. The magnitude of the interaction term Q_e is relatively small, it is 5% and 18% for $R/b = 0.25$ and 0.5 respectively.

Table 7: Values of Q_r for two equal-length cracks at the central hole in a uniformly stressed strip.

a/b	$R/b = 0.25$		$R/b = 0.5$	
	This paper	Newman[76]	This paper	Newman[76]
1.02	-	-	0.65	0.65
1.04	0.67	0.66	0.88	0.88
1.08	0.87	0.85	1.16	1.14
1.20	1.09	1.08	1.52	1.50
1.40	1.18	1.18	1.81	1.82
1.60	1.22	1.22	-	-
2.0	1.27	1.28	-	-

4.5 Conclusions

It has been demonstrated that compounding is a simple and versatile method for evaluating stress intensity factors in complex geometrical components containing cracks. The basic procedure of considering individually the many boundaries that interact with the crack enables the important features to be identified. The results obtained using this method are adequate for many engineering applications.

For many configurations the interaction between boundaries, which would occur even in the absence of a crack, can be neglected. When this interaction is not negligible, e.g. cracks at the edges of holes, a simple approximation has been devised. This approximation ensures accurate results for short cracks, a region of technological importance since the behaviour of short cracks largely determines the fatigue lifetime.

More complex configurations, than the examples considered in this section, can be studied using the compounding technique, for example multiple cracks at multiple stiffeners[58] which may or may not be periodic. The cracks can be asymmetric with respect to the stiffener and of unequal lengths at stiffeners spaced unequal distances apart. A row of holes with unequal length cracks at each hole may also be studied, the holes may be loaded or unloaded. Some further examples have been considered in Ref. 58. The technique can also be used to obtain stress intensity factors for three-dimensional configurations, but is somewhat limited by the small number of ancillary solutions that are available.

5 Residual static strength

5.1 Introduction

This section is concerned with the calculation of the residual strength of a cracked structure. The basic premise in residual strength theory is that a cracked material will fail by rapid unstable crack growth if the applied stress intensity factor is equal to or greater than the fracture toughness of that material. It is a well established fact that the plane strain fracture toughness K_{Ic} applicable to thick sections is a material constant. Examples of its use are given.

For materials in thin sections, the fracture toughness K_{c} is not a constant, but is a function of material thickness, crack length and specimen size. This leads to difficulties in developing a universal procedure for calculating residual strengths. Some of the difficulties are described and also some of the attempts to overcome them. Another important phenomenon which occurs under plane stress conditions is what is called slow stable crack growth, i.e. the crack grows in a stable manner as the load is increased. This implies that the resistance to crack growth (or toughness) is varying. This fact is incorporated into residual strength analysis by using the R-curve approach. This is described and its relationship to the K_{c} parameter explained.

5.2 Plane strain residual strength

The reduction in strength due to the presence of a crack can best be illustrated by an example. Consider a strip of width W and thickness t which has an ultimate tensile strength σ_{u} (see Fig. 25). If the strip is now loaded and the load steadily increased, then failure will occur at an ultimate load L_{u} given by

$$L_{\mathrm{u}} = \sigma_{\mathrm{u}} W t \tag{126}$$

The introduction of cracks will cause failure at much lower loads.

Consider a crack of length a perpendicular to one edge of a strip. If this strip is loaded then failure will occur when the applied stress intensity factor equals the fracture toughness, i.e.

$$\sigma_{\mathrm{c}} \sqrt{\pi a} f\left(\frac{a}{W}\right) = K_{\mathrm{Ic}} \tag{127}$$

where σ_{c} is the critical stress at failure and $f(a/W)$ is a correction for the finite width of the specimen. The failure load L_{c} is given by

$$L_{\mathrm{c}} = \sigma_{\mathrm{c}} W t = \frac{K_{\mathrm{Ic}} W t}{\sqrt{\pi a} f(a/W)} \tag{128}$$

The ratio of the load L_{c} to the ultimate load for the uncracked strip is given by

$$\frac{L_{\mathrm{c}}}{L_{\mathrm{u}}} = \frac{K_{\mathrm{Ic}}}{\sigma_{\mathrm{u}} \sqrt{\pi a} f(a/W)} \tag{129}$$

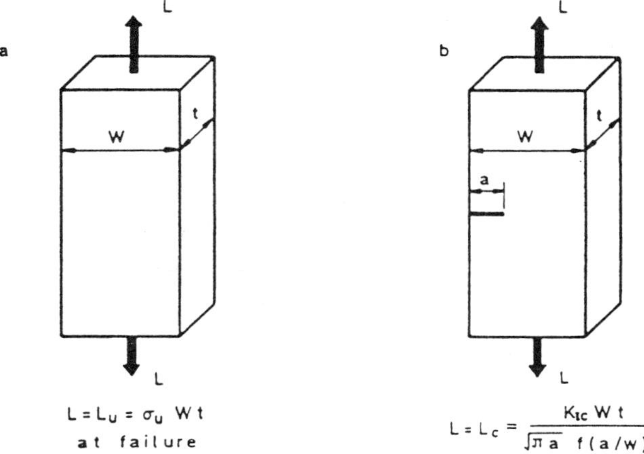

Figure 25: Failure of strip a) uncracked and b) cracked

Table 8: Residual strengths of cracked strip

Material	Tensile strength $\sigma_u(\text{MNm}^{-2})$	Toughness $K_{Ic}(\text{MNm}^{-3/2})$	L_c/L_u
Medium strength steel	850	180	0.64
High strength steel	2000	50	0.08
Aluminium/Copper alloy	450	25	0.16
Aluminium/Zinc alloy	570	29	0.16
Titanium-6Al-4V	900	90	0.30

In order to evaluate this ratio we must consider a specific size of strip and crack and specific materials. Let $W = 250$mm and $a = 25$mm, therefore $a/W = 0.1$ and $f(0.1) = 1.2$. Equation (129) now becomes

$$\frac{L_c}{L_u} = 2.98\frac{K_{Ic}}{\sigma_u} \tag{130}$$

where K_{Ic} is measured in $\text{MNm}^{-3/2}$ and σ_u in MNm^{-2}. The results for this ratio are shown in Table 8 for various materials (typical values are quoted for ultimate strength and toughness).

It can be seen from Table 8 that, in all cases, the strength of the cracked strip is considerably less than that of the uncracked strip. Even in the case of the very tough medium-strength steel there is a reduction of nearly 40% in strength; the high-strength steel with a low toughness has a residual strength of less than one-tenth of the ultimate. It is instructive to compare the effect of introducing a crack whose length is 10% of the width with the effect of reducing the width of an uncracked strip by 10 %. Reducing the width by 10% will, of course, reduce the ultimate load by 10%; Thus the effect of a crack is much more severe than the equivalent loss of area.

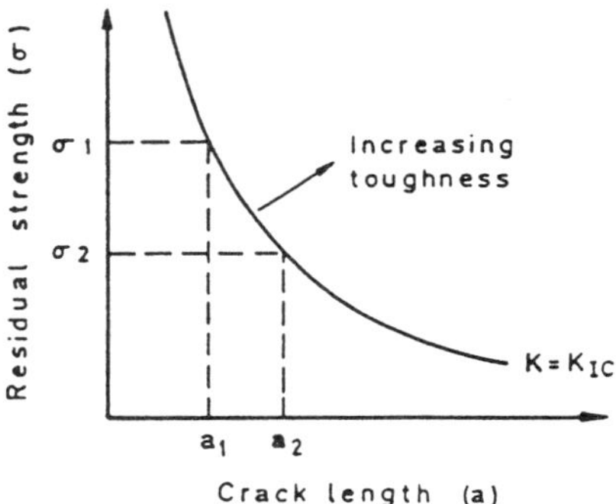

Figure 26: Residual strength of a cracked structure

The general relationship between crack length and residual strength is represented schematically in Fig. 26; this clearly shows that increasing the crack length ($a_1 \rightarrow a_2$) reduces the strength ($\sigma_1 \rightarrow \sigma_2$). It can also be seen that increasing the toughness of a material means that it will have an increased residual strength for the same crack length, or that a longer crack can be tolerated at the same stress level. In practice the residual strength curve in Fig. 26 needs modification if very long or very short cracks are to be considered. For very short cracks eqn (127) predicts stresses in excess of the ultimate strength of the material; in fact $\sigma_c \rightarrow \infty$ as $a \rightarrow 0$ since $f(0)$ is a constant. Other methods of predicting residual strength need to be used in this region as linear elastic fracture mechanics is no longer valid because the radius of the plastic zone is comparable to or larger than the crack length. The maximum crack length is limited by the component dimensions (W say) and therefore if $\sigma_c = 0$ for a finite crack length (W), it follows that $f(1) = \infty$ Thus calculation of residual strength for cracks requires a detailed knowledge of $f(a/W)$ as $a/W \rightarrow 1$. However in this region large stresses are often present on the net section and the size of the plastic zone is such to make linear elastic fracture mechanics inapplicable. A method of calculating residual strength for all values of a/W is described later for plane stress conditions; it is applicable to plane strain conditions also.

5.3 Plane stress residual strength

For structures made from thin sheets, failure due to rapid crack growth occurs under so-called plane stress conditions. The toughness parameter is called K_c; it is not a material constant. It is a function of material thickness, crack length and specimen dimensions. The dependence of K_c on thickness t is shown schematically in Fig. 27. A maximum in K_c occurs at $t = t_0$ when the sheet thickness is approximately twice the plane strain plastic zone correction, i.e.

$$t_0 \approx \frac{K_{\mathrm{Ic}}^2}{3\pi\sigma_{\mathrm{Y}}^2} \tag{131}$$

where σ_{Y} is the yield stress. The minimum value of K_c is K_{Ic} which is obtained for $t > t_1$ where t_1 is given approximately by the empirically determined criterion for the existence of plane strain conditions, namely that the thickness is more than fifty times the plastic zone size; that is,

$$t_1 \approx 2.5 \frac{K_{\mathrm{Ic}}^2}{\sigma_{\mathrm{Y}}^2} \tag{132}$$

The variation of K_c with thickness is important because maximum values of K_c can be several times the value of K_{Ic} and hence greatly affect the residual strength. There are no generally accepted quantitative models for the variation of K_c with thickness; some suggested models have been discussed by Broek.[78]

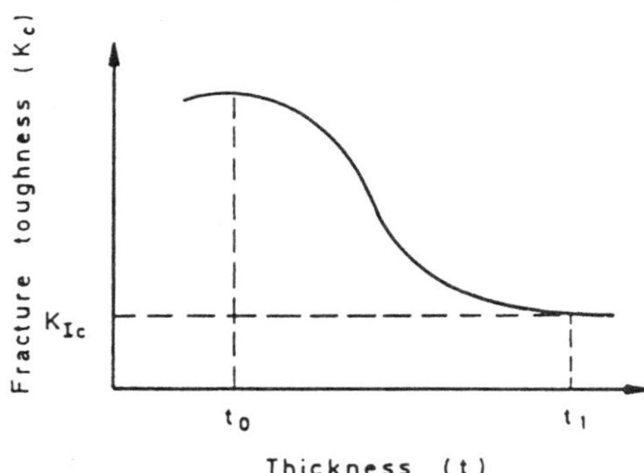

Figure 27: Variation of K_c with thickness

In order to describe the apparent variation of K_c at both short and long crack lengths, Feddersen[79] has suggested a simple model for use in engineering design. His model, which describes failure in sheets with central cracks is supported by much experimental data although the model development is not rigorous. For wide panels (see later), of width W containing a central crack of length $2a$, the relation between residual strength and crack length at constant K_c can be represented by the curve shown in Fig. 28. Also shown is a straight line representing net section yielding; it is assumed that failure would occur by yielding on the net section if the crack did not grow. It can be seen from Fig. 28 that cracks in the shaded regions (short and long cracks) require stresses in excess of the yield stress if failure is controlled by the given K_c. Since stresses above the yield stress will not occur, failure in these regions takes place at stresses lower than those predicted by the constant K_c curve. That is, the apparent toughness of the specimen is less than the given K_c.

Feddersen[79] suggested that two tangents to the K_c curve can be used to obtain a smooth and continuous curve for the residual strength. He showed that such a curve adequately represented the available data. One tangent to the curve is drawn from the point $\sigma = \sigma_Y$ on the strength axis and the other tangent is drawn from the point $2a = W$ on the crack length axis. The slope of a K-curve is given by

$$\frac{\mathrm{d}\sigma}{\mathrm{d}(2a)} = \frac{\mathrm{d}}{\mathrm{d(2a)}} \left(\frac{K_c}{\sqrt{\pi a}} \right) = -\frac{\sigma}{4a} \tag{133}$$

For the tangent at the point $(2a_1, \sigma_1)$ to go through the point $(0, \sigma_y)$ it follows that

$$-\frac{\sigma_1}{4a_1} = -\frac{\sigma_Y - \sigma_1}{2a_1}, \quad \text{i.e. } \sigma_1 = \frac{2}{3}\sigma_Y \tag{134}$$

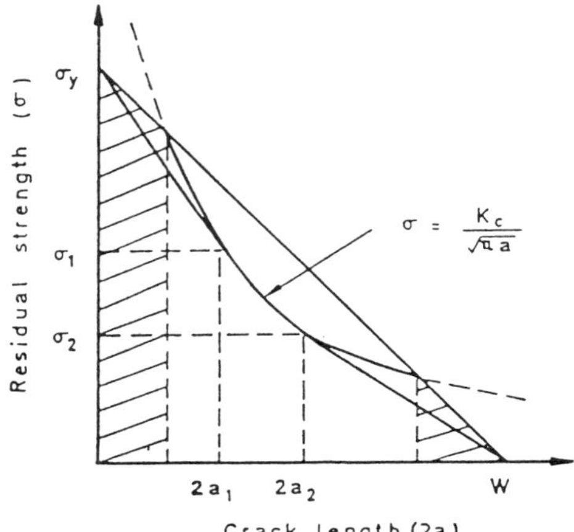

Figure 28: Residual strength of centre-cracked plates

For the tangent at $(2a_2, \sigma_2)$ go through $(W, 0)$ requires

$$-\frac{\sigma_2}{4a_2} = -\frac{\sigma_2}{W - 2a_2} \quad \text{i.e. } 2a_2 = W/3 \tag{135}$$

Thus the two tangent points are constant and independent of K_c, the left-hand point being a function of yield stress and the right-hand a function of specimen width.

If this specimen is used to determine a value of K_c, it follows from the above that K_c can only be determined for test results for which $\sigma < 2\sigma_Y/3$ and $2a < W/3$.Thus K_c will be approximately constant for only a limited range of crack lengths. Since a_1 and a_2 are independent of K_c and of each other, it follows that there is a specimen width (W_{min} say) for which $a_1 = a_2$. From the formula for K it follows that the left-hand tangent point is given by

$$K_c = \frac{2}{3}\sigma_Y\sqrt{\pi a_1} \quad \text{or} \quad 2a_1 = \frac{9K_c^2}{2\pi\sigma_Y^2} \tag{136}$$

The condition for the two points to coincide is $a_1 = a_2$, i.e.

$$\frac{9K_c^2}{2\pi\sigma_Y^2} = \frac{W_{min}}{3} \tag{137}$$

Thus for any W less than W_{min}, i.e.

$$W < \frac{27K_c^2}{2\pi\sigma_Y^2} \tag{138}$$

tangents to the K_c-curve, from the end-points $(0, \sigma_Y)$ and $(W, 0)$ do not exist; and the residual strength is determined by the net section yield criterion and is independent of K_c. Thus residual strength data for which $W < W_{min}$ cannot be used for the prediction of failure in larger specimens. The apparent K_c obtained for small panels will be less than the real K_c and so the strength of larger panels will be underestimated. It follows that K_c can only be measured on specimens for which $W > W_{min}$; Feddersen[79] suggests that $W > 1.5W_{min}$ is necessary. The residual strength for various panel sizes is represented schematically in Fig. 29.

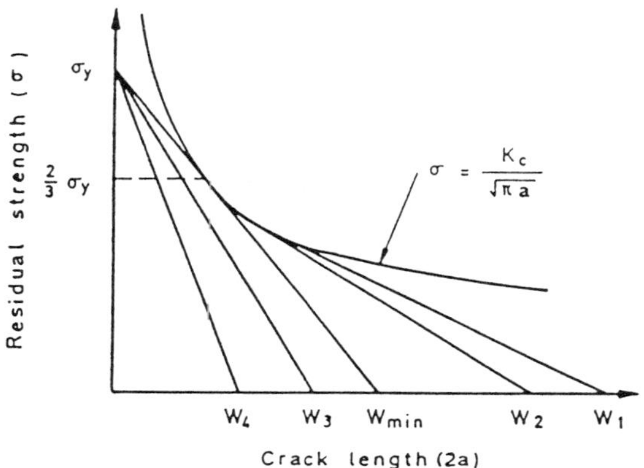

Figure 29: Residual strength for various panel sizes

It must be emphasised that Feddersen's procedure is an engineering approximation and the use of tangents has no sound theoretical basis. Nevertheless the results have been verified by experimental data. This procedure is, however, restricted to plates with central cracks; no generalised procedure applicable to other specimen shapes is yet available.

5.4 The *R*-curve

The above methods for calculating residual strengths do not take any account of stable crack growth which occurs in some materials in thin sections. As the load is increased the crack grows in a stable manner, i.e. if the load is held constant the crack ceases to grow until the load is increased again. A means of describing stable crack growth was first suggested by Krafft, *et al.*[80] Assuming that the resistance to crack extension varies as the crack grows, they explained stable crack growth in the following way. As the crack grows under a rising load the increase in the crack driving force, G, initially balances the increase in the crack growth resistance, R, and growth of the crack is stable. The point of instability is reached when the rate of increase of G equals the rate of increase of R; that is, the curves of G and R as a function of crack length are tangential to each other. The instability condition is $G = R$ and $\mathrm{d}G/\mathrm{d}a = \mathrm{d}R/\mathrm{d}a$ simultaneously. For engineering purposes the R-curve is usually measured in stress intensity factor units, $K_R = \sqrt{(ER)}$; the instability condition becomes $K = K_R$ and $\mathrm{d}K/\mathrm{d}a = \mathrm{d}K_R/\mathrm{da}$. Many workers have derived R-curves based on these concepts; the work on R-curves and their use has been reviewed[81] (see also ASTM standard E561-86).

The relationship between K and K_R is shown schematically as a function of crack length in Fig. 30. If a specimen with an initial crack length a_0 is loaded up to a stress level σ_a a small amount of crack growth will ensue. When the point A is reached no further crack growth will occur at σ_a since K_R becomes greater than K (or in energy terms R becomes greater than G). The stress must be increased. Increasing the stress to σ_b results in further stable growth until the point B is reached. Further increases in stress will result in failure (unstable crack growth) when the stress reaches σ_c. The K and K_R curves are tangential at C and any increase in crack length is unstable since $K > K_R$ beyond C. Different R-curves will, of course, be obtained for different thicknesses of material.

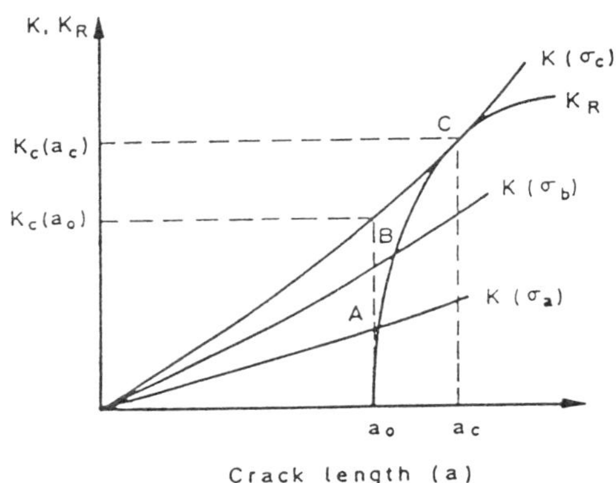

Figure 30: The R-curve

Material toughness parameters (K_c) can be derived from the R-curve. Two possible definitions are shown in Fig. 30. The first, $K_c(a_0)$, is the value of the stress intensity factor obtained from the failure stress and the initial crack length, i.e.

$$K_c(a_0) = \sigma_c\sqrt{\pi a_0} f(a_0/W) \tag{139}$$

where W is a specimen dimension and $f(a_0/W)$ is a correction for the finite size of the specimen. Another definition of toughness is $K_c(a_c)$ which is obtained from the failure stress and the final crack length, i.e.

$$K_c(a_c) = \sigma_c\sqrt{\pi a_c} f(a_c/W) \tag{140}$$

Although $K_c(a_c)$ is a function of the parameters, a_c and σ_c, actually existing at failure, it may be necessary to characterise the material toughness by $K_c(a_0)$ if the crack length at instability cannot be determined.

Krafft *et al.*[80] suggested that the R-curve is independent of the initial crack length a_0; it is a function of the amount of crack growth only. This implies that failure can be predicted for any initial crack length by a simple construction on the R-curve derived for one value of a_0. This is illustrated in Fig. 31 for strips of width W containing cracks of lengths a_1, a_2 and a_3. Because the shape of the K-curve for this configuration varies with a, since $f(a/W)$ is more important for long cracks than for short cracks, it follows that the tangency of K and K_R will not be in the same place for different a_0. This implies that K_c will not be a constant (see Fig. 31). Detailed calculations for this specimen show that K_c has a maximum at $a/W \sim 0.4$ (see Fig. 32).

The variation in the toughness parameter K_c derived from the R-curve is similar to the variation in apparent toughness that was obtained from Feddersen's construc-

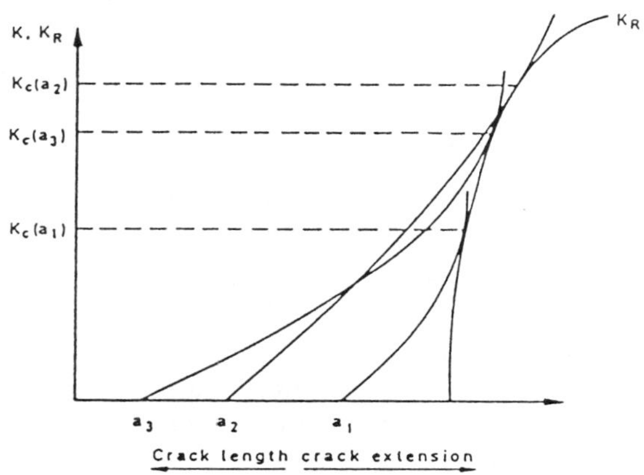

Figure 31: Determination of K_c from the R-curve

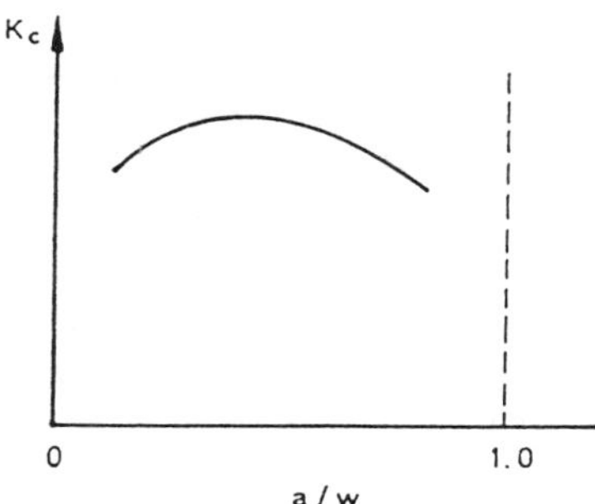

Figure 32: Variation of K_c with crack length

tion above. In that case K_c was a constant (maximum) for a region around $a/W \sim 0.3$ and was reduced for both shorter and longer cracks. Thus both methods reinforce the view that plane stress fracture toughness cannot be characterised by a single materials constant; this is in contrast to plane strain fracture toughness. Despite the shortcomings of the R-curve it has been used, for instance, to calculate residual strengths of cracked panels reinforced with stiffeners.[82,83]

5.5 Elastic-plastic residual strength

Plasticity effects are not important if the plastic zone is small compared to both the crack length and the remaining uncracked section of the structure, but in many practical situations these conditions are not satisfied. If we consider cracks in large sheets, so that the second condition is always fulfilled, then it follows that failure of very short cracks will be dominated by plasticity effects and that failure of long cracks will be described by linear elastic fracture mechanics (LEFM). This is represented schematically in Fig. 33 where failure at short crack lengths will be by plastic collapse at a stress approximately equal to the yield stress, and failure at long crack lengths will be governed by $K_I = K_c$. It is seen that the two curves representing plastic collapse and LEFM form an envelope ABC to the experimental data which was obtained for cracks in large sheets; this envelope can be used as an approximation to the practical failure- stress vs. crack-length curve. This method of calculating residual strength is known as the two criteria approach.[84]

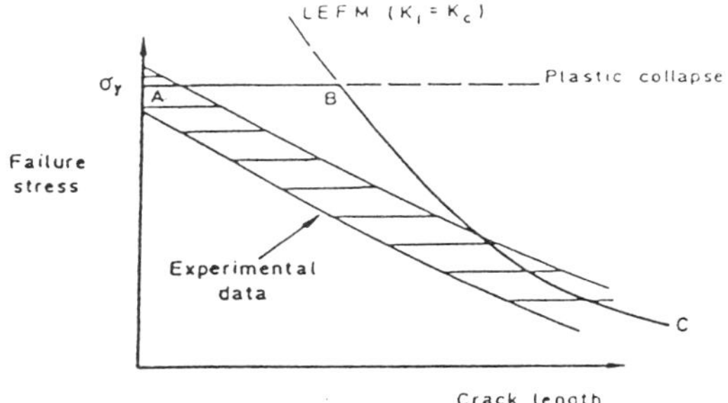

Figure 33: Residual strength of large sheets with short cracks

The major disadvantage of this simple two criteria approach is that large errors can occur in the transition region between plastic collapse and LEFM. Errors can be of order 20-30% in failure stress and more than twice that in critical crack length. The limit of validity of LEFM can be estimated by considering the parameter $\alpha = (K_c/\sigma_Y)^2/\beta$, where β is the crack length or the width of the remaining uncracked section, whichever is smaller. Experimental evidence suggests that good approximations are obtained from LEFM provided that $\alpha < 1$ for plane strain, and that $\alpha < 0.4$ for plane stress.

The simple situation represented in Fig. 33 is more complex if the crack is so long that the plastic zone interacts with structural boundaries, and failure may occur by plastic collapse of the remaining ligament. This is illustrated schematically in Fig. 34, where the failure stress normalised with respect to the yield stress (σ_f/σ_Y is plotted as a function of crack length normalised with respect to the structure width (a/W),

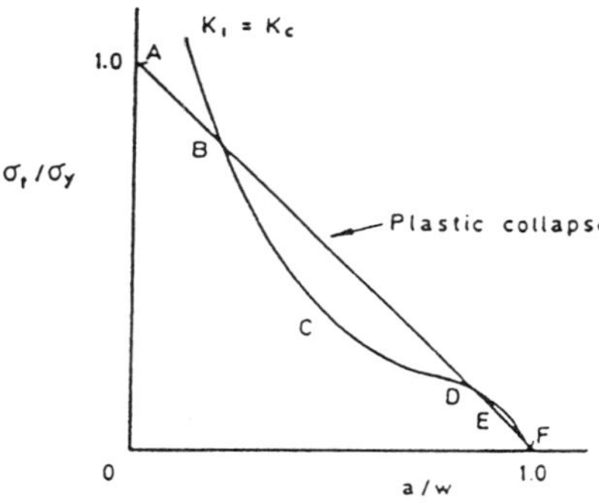

Figure 34: Residual strength of cracked structure

and $K_{\rm I}$ is the opening-mode stress intensity factor for the finite width structure. As can be seen from Fig. 34, at both long and short cracks the stress σ_K will exceed the plastic collapse stress σ_1, where σ_K is given by

$$\sigma_K = \frac{K_{\rm c}}{Y\sqrt{\pi a}} \tag{141}$$

where Y is a geometric parameter, and the collapse stress is given by the straight line AF. Thus in Fig. 34 the failure envelope ABCDEF is determined by

$$\left.\begin{array}{ll} \sigma_{\rm f} = \sigma_K, & \sigma_K < \sigma_1 \\ \sigma_{\rm f} = \sigma_1, & \sigma_K \geq \sigma_1 \end{array}\right\} \tag{142}$$

The failure envelope ADCDEF will be a function of both material properties and structural configuration. To obtain a curve independent of material constants and structural geometry we plot $\sigma_{\rm f}/\sigma_1$ against σ_K/σ_1. This is shown in Fig. 35 where the failure envelope is given by the curve OAD. As can be seen the two extremes, perfectly elastic and perfectly plastic, fit the data well; but there are large errors, particularly in estimates of critical crack lengths, in the transition region.

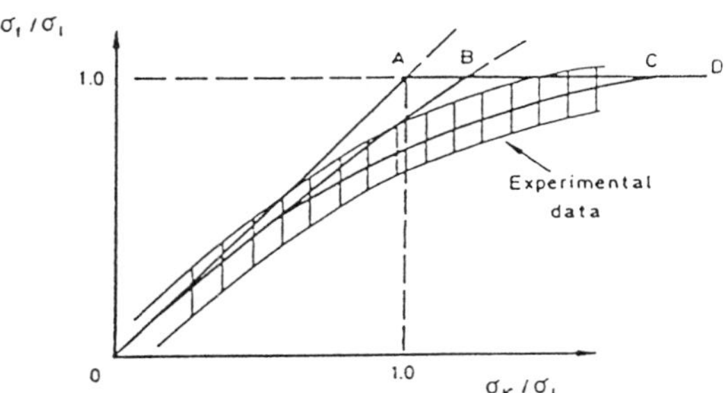

Figure 35: Residual strength models

Predictions of residual strength and critical crack lengths can be improved by incorporating a plastic zone correction into the LEFM curve OA. As shown in section 3, the crack length a can be replaced by an equivalent crack length $a_{\rm e}$ given by

$$a_{\rm e} = a + r_p = a + \frac{1}{2\pi}\left(\frac{K_{\rm I}}{\sigma_{\rm Y}}\right)^2 \tag{143}$$

For a crack in an infinite body the failure criterion

$$K_{\rm I} = \sigma_{\rm f}\sqrt{\pi a_{\rm e}} = K_{\rm c} \tag{144}$$

leads to the following approximation for the toughness:

$$K_{\rm c}^2 = \sigma_{\rm f}^2 \pi a \left[1 + \frac{1}{2}\left(\frac{\sigma_{\rm f}}{\sigma_{\rm Y}}\right)^2\right] \tag{145}$$

This can be combined with eqn (141) to give, for $Y = 1$, the following

$$\left(\frac{\sigma_{\rm f}}{\sigma_{\rm Y}}\right)^2 = \sqrt{1 + 2\sigma_K^2/\sigma_{\rm Y}^2} - 1 \tag{146}$$

This equation is plotted in Fig. 35, since for an infinite sheet $\sigma_1 = \sigma_{\rm Y}$ and it cuts the plastic collapse curve at B. It can be seen that predictions based on the failure curve OBD will be more accurate than those based on the curve OAD.

Further improvements on the prediction of residual strength and critical crack lengths can be achieved by considering a criterion of failure based on the crack tip opening displacement (CTOD), from the Dugdale elastic-plastic model. For a crack in an infinite sheet the CTOD, denoted by δ, is given by

$$\delta = \frac{8\sigma_{\rm Y} a}{\pi E}\ell n\left[\sec\left(\frac{\pi\sigma}{2\sigma_{\rm Y}}\right)\right] \tag{147}$$

So that the above is compatible with results for small scale yielding ($\sigma << \sigma_{\rm Y}$), it follows that

$$\delta = K_{\rm I}^2/(\sigma_{\rm Y} E) \tag{148}$$

The failure criterion

$$\delta = \delta_{\rm c} = K_{\rm c}^2/(\sigma_{\rm Y} E) \tag{149}$$

combined with eqn (147) gives

$$\sigma_{\rm f} = \frac{2\sigma_{\rm Y}}{\pi}\cos^{-1}\left[\exp\left(-\frac{\pi K_{\rm c}^2}{8a\sigma_{\rm Y}}\right)\right] \tag{150}$$

The failure stress can be expressed as a function of σ_K by using eqn (141) in eqn (150) to give

$$\frac{\sigma_{\rm f}}{\sigma_{\rm Y}} = \frac{2}{\pi}\cos^{-1}\left[\exp\left(-\frac{\pi^2\sigma_K^2}{2\sigma_{\rm Y}^2}\right)\right] \tag{151}$$

Since for an infinite sheet $\sigma_{\rm Y} = \sigma_1$, eqn (151) can be plotted on Fig. 35 to give the curve OCD. It can be seen that this curve fits the experimental data very well over the whole range from elastic to plastic conditions. The critical crack length at failure is given by

$$a_c = \frac{\pi K_c^2}{8\sigma_Y^2 \ell n \left[\sec\left(\frac{\pi\sigma_f}{2\sigma_Y^2}\right)\right]} \tag{152}$$

For an infinite sheet the value of the linear-elastic stress intensity factor K at failure is given by

$$K_I^f = \sigma_f\sqrt{\pi a_c} \tag{153}$$

Combining eqns (152) and (153) gives

$$K_r = \left\{\frac{8}{\pi^2 S_r^2}\ell n \left[\sec\left(\frac{\pi S_r}{2}\right)\right]\right\}^{-1/2} \tag{154}$$

where $K_r = K_I^f/K_c$ and $S_r = \sigma_f/\sigma_Y$. This equation forms the basis of the failure assessment diagram developed at the CEGB.[84,85]

Since the analysis leading to eqn (154) is based on the theory of cracks in infinite sheets, it is necessary to make some empirical modifications in the interpretation of the ratios K_r and S_r, before the analysis can be applied to the failure of real structures. The factor K_I^f in K_r is to be re-interpreted as the linear elastic stress intensity factor for the actual structure being analysed, and the yield stress σ_Y in S_r, is to be replaced by the plastic collapse stress σ_1 for the cracked structure. These modifications ensure that for small scale plasticity the LEFM limit is approached, and for large scale plasticity failure occurs at the plastic collapse load. It must be remembered that K_c appearing in the ratio K_r is not, in general, the linear elastic fracture toughness, because it is defined by eqn (149). Since the crack tip opening displacement is related to the J-integral (see section 3) it follows that

$$K_c^2 = E\sigma_Y\delta_c = EJ_c \tag{155}$$

Thus K_c may be obtained from measurements of either the critical value of the CTOD or the critical value of the J-integral. In the limit of small-scale yielding K_c will tend to the elastic fracture toughness.

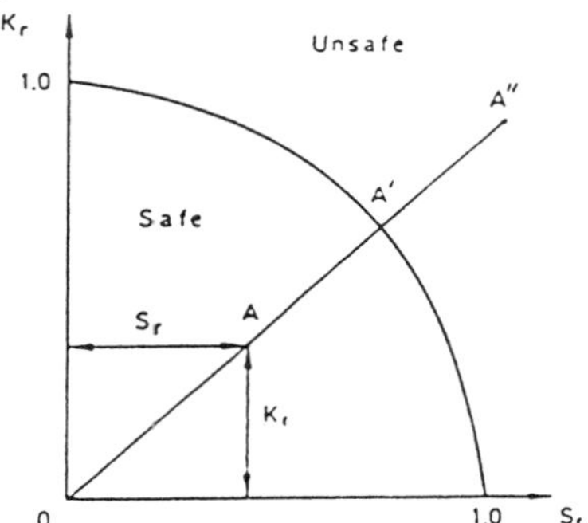

Figure 36: Failure assessment diagram

A schematic representation of the failure assessment diagram is shown in Fig. 36; the bounding curve of the 'SAFE' region is the locus of the failure points given by eqn

(154). To assess a given structure, the parameters K_r and S_r are evaluated for the loads and crack lengths likely to the encountered in practice. If the point (S_r, K_r) lies within the boundary, at A say, then the structure is considered safe; if the point is on or outside the boundary, A′ or A″, then the structure is considered unsafe. Since both K_r and S_r are proportional to the applied stress σ, it follows that OA is also proportional to the applied stress. The safety factor F on the applied stress is therefore given by OA′/OA and the failure stress σ_f is $F\sigma$.The failure assessment diagram permits the investigation of such parameters as material, loading, defect geometry, design details, etc. Its use has been experimentally validated and it is discussed more fully in Ref. 85.

5.6 Conclusions

Residual strength calculations for plane strain conditions are straightforward and are based on the fact that the fracture toughness K_{Ic} is a materials constant. Care must be exercised when applying the procedures to both very short and very long cracks; the existence of large plastic zones may invalidate linear elastic fracture mechanics. No well-established universal procedure exists for the calculation of residual strengths in plane stress conditions. Semi-empirical methods do exist and are extensively used, but more work is needed in this area of fracture mechanics.

6 Fracture control

6.1 Introduction

The ultimate aim of fracture mechanics is fracture control.[86,87] Whatever the conditions, elastic or plastic, static or dynamic and whatever the materials, the aim is to develop procedures for controlling fracture in practical engineering components. The potential cost of uncontrolled fracture is enormous: in 1983 the US National Bureau of Standards estimated[88] that the cost of uncontrolled fracture was approximately 4% of the Gross National Product. Furthermore they estimated that this cost could be halved by the proper use of existing technology and the pursuit of fracture-related research. Apart from the economic cost it is also important to realise that in some situations uncontrolled fracture can put human life at risk.

Many fracture control plans already exist, indeed in some industries there are statutory requirements for such plans. The development of a particular plan will depend on the type of structure being considered, and the likely consequences of fracture to the operators, the owners, the public, the environment, etc. In all cases the following requirements are fundamental for the development of an effective fracture control plan:

1) to identify the factors that may contribute to failure by fracture under given service conditions;
2) to establish the relative importance of these factors;
3) to investigate the possibilities of minimising the potential for fracture, by choices of material, structural and service parameters at the design stage; and
4) to recommend specific rules for materials selection, structural detail, maximum design stresses, fabrication procedures and inspection intervals in service.

The development of fracture control plans requires a knowledge of two basics, the residual strength of the cracked structure and the rate of growth of a crack in the structure. The first is required in order to determine whether the structure is safe or not, and the second to determine how long it remains safe (see section 2). The second provides a time frame within which the various options for fracture control must operate. Frequently fracture control is exercised by means of regular inspections and safety is guaranteed by detecting the crack, before it becomes dangerous, and repairing or replacing the component. So-called damage tolerance analysis is often specified as the means of establishing the inspection procedures in a rational way.

The successful application of damage tolerance analysis relies on the detection and measurement of cracks. There are many physical methods of detecting cracks, but none of them are 100% reliable - there is always some chance that an existing crack will be missed during an inspection. The statistical nature of crack detection must, therefore, be taken into account when devising fracture control plans. The probability of crack detection will depend on many variables such as the crack length, the position in the structure, the type (through-the-thickness, surface, interior), the detection method, the experience of the inspector, etc. The importance of detecting a crack of a given length will vary from structure to structure because the crack growth characteristics are a function of material and structural properties. These features will be discussed in some detail in this section.

6.2 Basic information

The residual strength curve for the structure under consideration will be of the form shown in Fig. 37. The maximum service load σ_{max} is set equal to σ_d/f where σ_d is the uncracked design strength, and f is a safety factor ($f > 1$). From the residual strength curve, the critical crack length σ_c at which failure will occur, can be determined for the maximum service load. It is usually a requirement that the lowest acceptable strength σ_p in the presence of cracks should lie between σ_{max} and σ_d. Thus the maximum acceptable crack size $a_p$is less than a_c, thus ensuring a margin of safety at the maximum service load.

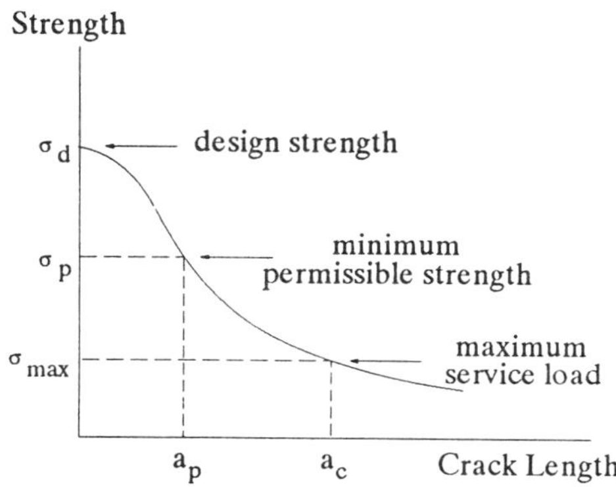

Figure 37: Residual strength diagram: permissible crack size a_p

Even if a structure is safe at any given time it may not remain so, because of crack growth which lowers the residual strength. The effect of crack growth must be estimated from a crack-growth curve such as that shown in Fig. 38. The initial crack size a_i at time t_i is probably unknown because it is usually less than the minimum detectable crack a_0 which is reached at a time t_0. Therefore any life available while the crack is growing from a_i to a_0, that is $t_0 - t_i$, cannot be utilised in design, because it cannot be reliably calculated. The lifetime that is reliably available is the time T for the crack to grow from a length a_0 to a_p, that is $t_p - t_0$. This time is the basis for the fracture control strategy which determines the necessary repair or replacement procedures needed to ensure safety.

Fracture control can be exercised in many ways, the following are some of those ways:

(1) periodic inspection - repair upon crack detection;
(2) fail safe design - repair upon partial failure;
(3) safe life design - replacement after time $\leq T$;
(4) periodic proof tests - repair after failure in proof test.

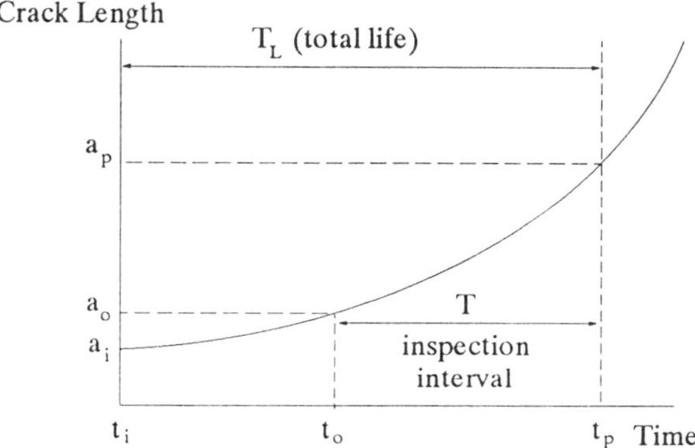

Figure 38: Crack growth diagram: inspection interval(T)

In many cases the type of structure or its use coupled with any existing statutory requirements will determine the fracture control program. For example, in the aircraft industry types (1) and (2) are common for civil aeroplanes and types (1) and (3) for military aeroplanes.

6.2.1 Periodic inspection

This is a very common procedure in fracture control and safety is ensured by detecting and then eliminating cracks before they reach the maximum permissible size a_p. There are many different inspection procedures in use, and some of the more common ones are summarised in Table 9.

Whatever inspection technique is used it is essential that the crack is detected in the time T it takes to grow from a_0 to a_p. The maximum permissible crack size a_p is determined from the residual strength curve, but a_0 the minimum reliably detectable crack will depend on where it is and what detection method is being used. The smaller a_0 the more time is available for detection. Crack detection is not an exact science, and there is always a finite probability that any crack may not be found in a given inspection. The statistics of crack detection will be discussed later in this section. Because of the possibility of missing a crack, even if it is longer than a_0, it is usual to ensure that more than one inspection takes place in the period T. For example if inspections take place every $T/2$, then there will always be two while the crack lies between a_0 and a_p. In general if inspections take place every T/n, there will always be n inspections during this growth period. This is illustrated (for $n = 2$) in Fig 39. If the inspection interval is too short for practical convenience, it can be lengthened by using a better inspection technique which will detect smaller cracks (a_0). Better techniques are usually more costly, but frequent inspections are also costly.

6.2.2 Fail-safe design

In this design method the structure is designed for tolerance to larger scale damage and individual small cracks do not necessarily have to be found (or inspected for). Fail safety can be achieved in many ways. For instance, crack arresters which will stop large cracks can be incorporated into the design. Multiple load paths can be provided, so that should one member fail, the remaining members can sustain the load. In some

Table 9: Available inspection techniques for crack detection

Inspection method	Physical principles	Remarks
Visual	Naked eye with hand held aids	Cracks must be in easily accessible region
Dye penetrant	Dye penetrates into crack to produce coloured line	
Magnetic particle	Liquid of magnetic particles applied. Cracks indicated by magnetic field lines	Magnetic materials only. Not suitable for in situ inspection
X-ray	Transmission of X-rays - cracks recorded on film	Access to both sides. Difficult to detect small surface flaws
Ultrasonic	High frequency sound waves reflected by cracks	Many variants, widely used where cracks expected
Eddy current	Eddy currents induced in specimen are altered by presence of crack	Relatively cheap and easy method for conducting materials
Acoustic emission	Stress waves emitted from crack tips due to plastic deformation under load	Inspection under load, but difficult to interpret

structures special features can be incorporated such as leak-before-break. Such a feature is often designed into pressure vessels so that cracks grow through the thickness and a detectable leak occurs before structural failure. An important requirement is that the large-scale failure is immediately detectable and is then repaired. Although such a failure is not immediately catastrophic, the structure will be severely weakened.

6.2.3 Safe life design

For some structures it is not convenient or not possible for inspections to be carried out during service. For such structures the interval T must be estimated, often empirically, and the component replaced before T is reached in service. It is common for a safety factor of at least two to be introduced. That is, the component is replaced at $T/2$ or less. To make this procedure viable, T must be a long period and so over-conservative designs can result in order to ensure this. It may also be wasteful in the sense that components may be thrown away at $T/2$ which do not, in fact, contain cracks, and could therefore still be used safely. A major problem with this approach is that it depends heavily on the assumed initial crack size (needed to determine T), with no opportunity to update or modify in the light of inspections, because none are carried out.

6.2.4 Proof testing

In the case of a critical crack length a_p which is less than the detectable crack size a_0, inspections would not be of any help. If the structure is very large and critical crack sizes also large, then regular inspection is impractical. In both these cases proof testing may be suitable. A proof test consists of applying a proof stress σ_{pr} (say) which is larger than the minimum permissible strength. If the component withstands the proof stress, that implies that no cracks whose length is $\geq a_{pr}$ exist; where a_{pr} is the size of the crack that would cause failure at σ_{pr}. Since $\sigma_{pr} > \sigma_p$ it follows that $a_{pr} < a_p$. The period available for safe operation is at least the time T for a crack to grow from a_{pr} to a_p. Successive proof tests at intervals less than T will ensure safety in service; if a crack grows to a length greater than a_{pr}, the structure will fail during the proof test.

Pressure vessels, pipelines or components that can easily be removed and appropriately loaded are suitable structures for proof testing. A pressure vessel or a pipeline which in normal use may carry dangerous gases or liquids would usually be proof tested with water, so that failure results in nothing more dangerous that a water leak. In some cases it may be convenient to conduct the proof test at an artificially low temperature. At lower temperatures the toughness is lower and hence smaller proof stresses will be required to guarantee that no cracks greater than a_{pr} exist.

6.3 Crack detection

The lifetime T and the inspection intervals can readily be determined once a_0, the minimum detectable crack size is known. However obtaining this knowledge is not straightforward, but will depend on many factors, such as material properties, type of structure, position in structure, accessibility and the inspection technique used. Observation of Fig. 38, shows that T is largely determined by a_0, and small changes

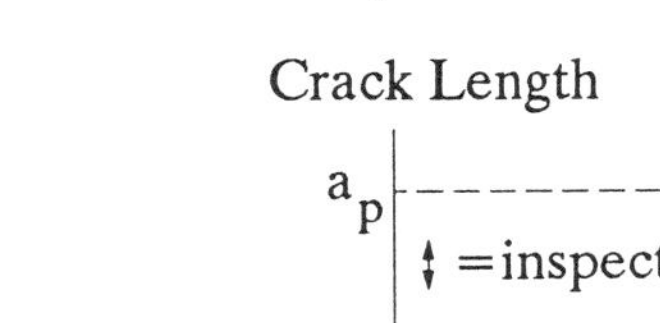

Figure 39: Two inspections between a_0 and a_p for any crack start

in a_0 can result is large changes in T, and hence the time available for safe operation. A major difficulty in determining a_0 is that crack detection of any crack by any method cannot be guaranteed; there is always a finite possibility a crack which should be detected by the technique in use, will not, in fact, be detected. The reasons for this are various, but the consequence is that crack detection must be considered in a statistical sense. The probability of detection will, in general, increase with size. The conditions of the inspection (see the factors listed above) will also affect the probability of detection. The type of results expected are shown schematically in Fig. 40. Good conditions such as a sophisticated technique on a simple structure in a laboratory test will produce high probabilities of detection for quite small cracks. The probabilities will be less under poor conditions such as a field inspection of relatively inaccessible parts of a large structure with an insensitive technique.

The actual shape of the crack growth curve also has a big influence on the detectability of cracks during regular inspections. Consider two crack growth curves A and B, shown in Fig. 41 which have the same lifetime T to grow from a_0 to a_p. The actual shapes of A and B will be determined by both the material and the structural configuration. It can be seen from Fig. 41 that at both inspections the crack will be larger in A than in B, which means that the probability of crack detection is higher for curve A than curve B. Thus the structure with curve A is safer (less chance of a crack being undetected at a regular inspection) than one with curve B. To ensure the same degree of safety, the cumulative probability of crack detection must be the same in each case: this would require shorter inspection intervals in case B. Similar considerations of other factors that affect crack detection suggest that a rational procedure for the determination of inspection intervals should be based on the concept that equal cumulative probabilities implies equal levels of safety (see for instance Broek).[89]

If the probability of detecting a crack at the first inspection is p_1, then the probability that the crack is undetected is $q_1 = 1 - p_1$. Similarly at the second inspection

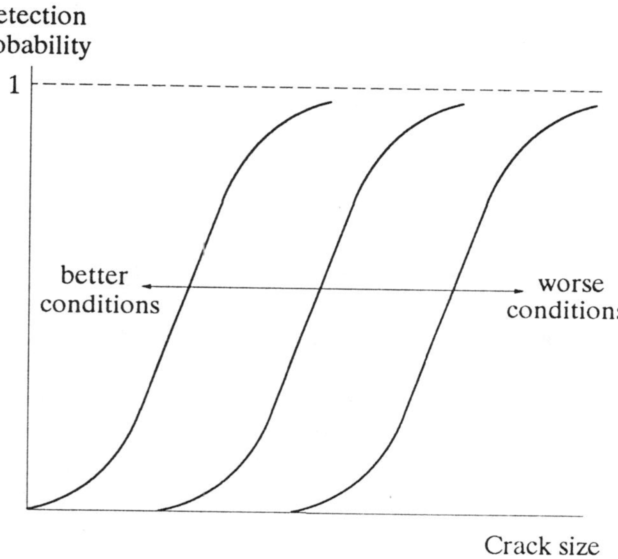

Figure 40: Probability of crack detection in an inspection

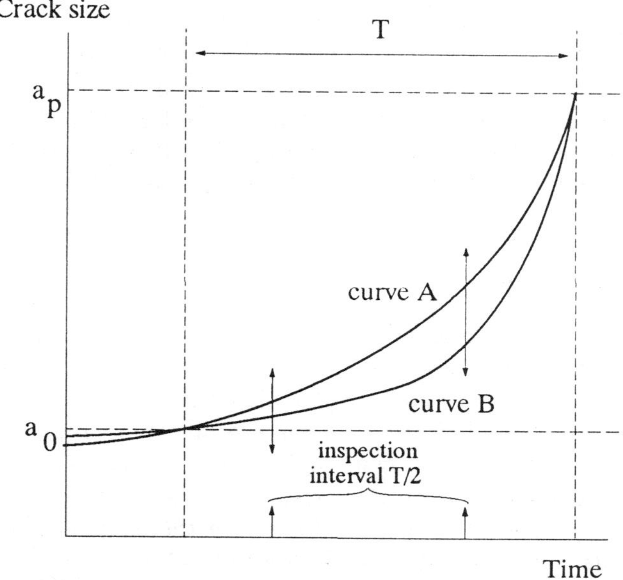

Figure 41: Effect of different crack growth curves on crack detection

the probability of detection is p_2 and the probability of missing the crack is $q_2 = 1 - p_2$. The probability Q that the crack will be missed in n successive inspections is given by the product of all the q's, that is

$$Q_n = \prod_{i=1}^{n} q_i = \prod_{i=1}^{n} (1 - p_i) \tag{156}$$

The cumulative probability of detection after n inspections P_n is therefore given by

$$P_n = 1 - Q_n = 1 - \prod_{i=1}^{n} (1 - p_i) \tag{157}$$

It is clearly desirable that P_n should be as large as possible. In any given study this can usually be done in several ways. For instance, the data given in Fig. 41 imply that the same inspection procedures (interval and technique) would give lower crack detection rates for curve B than for curve A; that is $p_i^{\mathrm{B}} < p_i^{\mathrm{A}}$ since crack sizes for B are always smaller than A during the period of inspection T. This means that the chance of missing a crack, Q_n, is greater for curve B than A, that is

$$Q_n^{\mathrm{B}} > Q_n^{\mathrm{A}} \quad \text{or} \quad P_n^{\mathrm{B}} < P_n^{\mathrm{A}} \tag{158}$$

Two possibilities can be considered to make P_n^{B} equal to P_n^{A}, and therefore have the same degree of safety in the two cases. The individual detection probability p_i^{B} could be increased by using an improved detection system (albeit usually more costly); or the number of inspections could be increased such that $n^{\mathrm{B}} > n^{\mathrm{A}}$. Either of these strategies can be used to make $P_n^{\mathrm{B}} \approx P_n^{\mathrm{A}}$.

Similar arguments would need to be considered for all other factors that may affect crack detection probabilities. It is thus obvious that in order to 'guarantee' a level of safety in any given structure requires a considerable amount of damage-tolerance analysis and a significant database of experimental test results as well as a sound engineering appreciation of the structural complexities. However, avoidance of fracture events is most desirable, because the cost of fracture, in all its aspects, can be very high indeed.

6.4 Repair or replacement

Once a crack is detected, then ideally some remedial action should be taken immediately. It may be possible to allow the structure to continue in service until the next major structural overhaul takes place, but increased surveillance of the known crack would be advised. The remedial action, ideally, would be to replace the cracked part by an uncracked part. If this is not possible a repair may be considered. Replacement should mean that no modification to the fracture control plan is required as the structure has been returned to the state for which the original plan was deduced, a plan which succeeded in detecting a crack in a safe manner.

The situation if the structure is repaired may be quite different. Ideally a new fracture control plan is required for the new repaired cracked structure, a structure which would probably not have been considered in the original plan. A common repair scheme is to place a patch over the region containing the crack, so that some of the load is carried by the patch away from the crack region; thus the crack growth curve will be different The presence of the patch is likely to reduce accessibility, making crack monitoring more difficult. In addition extra cracking may result as a direct consequence of the presence of the repair patches. This is particularly likely if the patches are attached to the original structure by bolts or rivets, a common practice in the repair of metallic structures.

A patch attached, by local fasteners, over a crack in a sheet is illustrated in Fig. 42. The holes through which the bolts or rivets pass will act as stress concentrations which may act as crack initiators. Such cracks, not present in the original structure, may grow fast because of the high stress near the holes. Such an eventuality must be taken into account in the new fracture control plan. Another possible site of new cracking is the edge of the patch; the edges of the patches are often chamfered to obviate this effect.

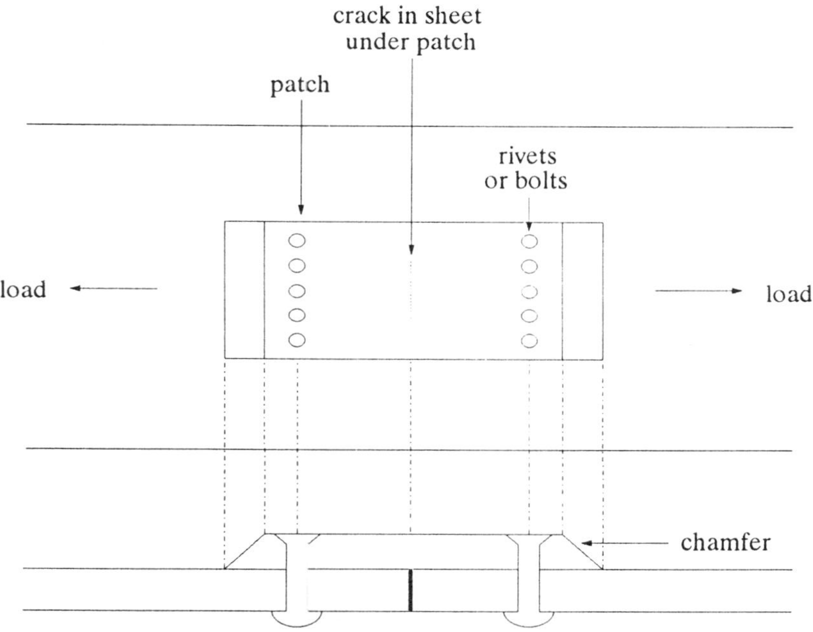

Figure 42: Locally attached repair patch

In recent years the use of adhesively bonded repair patches is being increasingly studied and practised. They are usually made from several layers of fibre reinforced laminates. Such patches have several advantages:

1) no holes are required for fixing;
2) the individual laminates are thin enough to follow any curvature of the underlying structure;
3) the elastic properties can be tailored to suit the particular application; and
4) they are easily fabricated.

An example is shown in Fig. 43. Such patches will again require a new fracture control plan. A new feature in the plan must be a consideration of the fracture properties

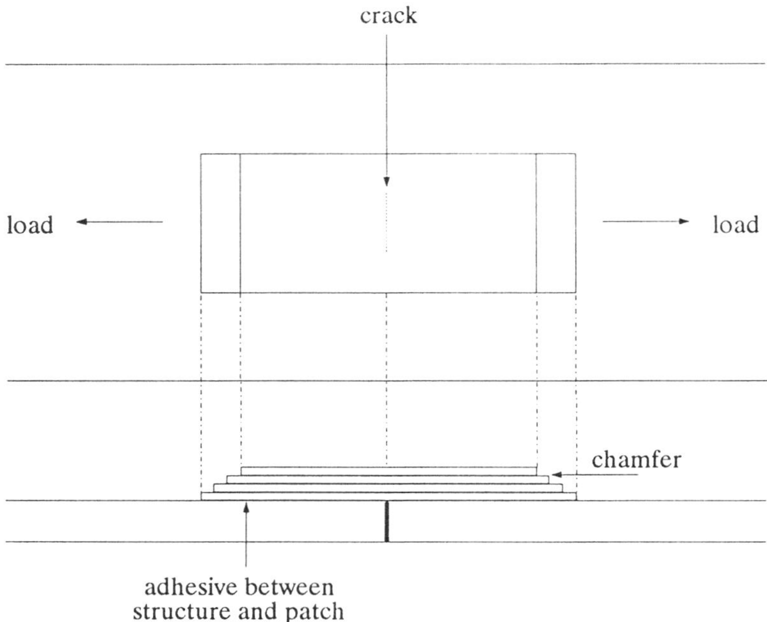

Figure 43: Adhesively bonded repair patch

of the adhesive bond layer. Boundary element analysis has been developed[90] to facilitate the damage tolerance study of bonded patches, as have finite element analysis.[91] Several practical examples of adhesively bonded structures are given in Ref. 91.

6.5 Conclusions

In this section the concept of fracture control has been explained. In particular it has been emphasised that the real purpose of fracture mechanics calculation, especially damage tolerance analysis, is to ensure that structures do not fail by fracture, and can be operated safely within the limitations of the specified fracture control plan. Most plans are based on a regular (periodic) inspection procedure and are usually specific. That is the plan applies only to a particular structure (or part of a structure) under specific loading conditions subject to a well defined inspection procedure with specified techniques.

All control plans depend on the detectability of the crack. There is ample experimental evidence to show that crack detection must be considered in a statistical sense - there is always a finite probability that a crack, normally detectable, may be missed in any given inspection. It is shown that the statistical concept of cumulative probability of detection forms a sound basis for the development of inspection based fracture control plans. Finally the consequences are discussed of adopting a replacement or repair philosophy after crack detection and two types of repair patching techniques are outlined.

References

1. A.A. Griffith, The phenomena of rupture and flow in solids, *Phil. Trans. A.*, 1920, **221**, 163-198.

2. Irwin, G.R. Fracturing of Metals, ASM Symposium, Chicago, Cleveland, ASM, 1948.

3. Orowan, E. Fundamentals of brittle behaviour in metals, *Fatigue and Fracture of Metals*, ed. W.M. Murray, Wiley, New York, 1952, pp. 139.

4. Irwin, G.R. Analysis of stresses and strains near the end of a crack traversing a plate, Trans. ASME, *J. Appl. Mech.*, 1957, **24**, 361-364.

5. Rooke, D.P., Baratta, F.I. & Cartwright, D.J. Simple methods of determining stress intensity factors, *Engng Fracture Mech.*, 1981, **14**, 397-426.

6. Paris, P.C. The fracture mechanics approach to fatigue, *Fatigue: An Interdisciplinary Approach*, ed. J.J. Burke, N.L. Reed & V. Weiss, Syracuse University Press, 1964, pp. 107-132.

7. Brown, B.F. Fundamentals, Stress-Corrosion Cracking in High-Strength Steels and in Titanium and Aluminium Alloys, ed. B.F. Brown, Naval Research Lab., Washington D.C., 1972.

8. Timoshenko, S.P. *Theory of Elasticity* (3rd Edition), McGraw Hill, 1970.

9. Mushhelishvili, N.I. *Some Basic Problems of the Mathematical Theory of Elasticity*, Noordhoof, Leyden, 1953.

10. Aliabadi, M.H. & Rooke, D.P. *Numerical Fracture Mechanics*, Computational Mechanics Publications, Southampton, 1991.

11. Westergaard, H.M. Bearing pressures and cracks, *J. Appl. Mech.*, 1939, **6**, A49-53.

12. Eftis, J., Subramonian, N. & Liebowitz, H. Crack border stress and displacement equations revisited, *Engng Frac. Mech.*, 1977, **9**, 189-210.

13. Sih, G.C. On the Westergaard method of crack analysis, *Int. J. Fract. Mech.*, 1966, **2**, 628-631.

14. Eftis, J. & Liebowitz, H. On the modified Westergaard equations for certain plane crack problems, *Int. J. Fract. Mech.*, 1972, **8**, 383-392.

15. Williams, M.L. Stress singularities resulting from various boundary conditions in angular corners of plates in extension, *J. Appl. Mech.*, 1952, **19**, 526-528.

16. Neuber, H. *Theory of Notch Stresses*, Trans. J. W. Edwards, Ann Arbor, Michigan, 1946.

17. Rice, J.R. A path independent integral and the approximate analysis of strain concentration by notches and cracks, Trans. ASME, *J. Appl. Mech.*,1968, **35**, 379-386.

18. Irwin, G.R. Fracture, *Handbuch der Physik.*, Vol. 1, Springer-Verlag, Berlin, 1958, pp. 551-590.

19. Dugdale, D.S. Yielding of steel sheets containing slits, *J. Mech. Phys. Solids*, 1960, **8**, 100-104.

20. Bilby, B.A., Cottrell, A.H. & Swinden, K.H. The spread of plastic yield from a notch, *Proc. Roy. Soc.*, 1963, **A272**, 304-314.

21. Tada, H., Paris, P.C. & Irwin, G.R. *The Stress Analysis of Cracks Handbook*, (2nd edition), Paris Production Inc., St. Louis, Missouri, 1985.

22. Hartranft, R.J. & Sih, G.C. The use of eigenfunction expansions in the general solution of three-dimensional crack problems, *J. of Mathematics and Mech.*, 1969, **19**, 123-138.

23. Folias, E.S. On the three-dimensional theory of cracked plates, *J. Appl. Mech.*, 1975, **42**, 663-674.

24. Erdogan, F. & Sih, G.C. On the crack extension in plates under plane loading and transverse shear, *J. Basic. Engng*, 1963, **85**, 519-527.

25. Sih, G.C. Strain energy density factor applied to mixed mode crack problems, *Int. J. Fract.*,1974, **10**, 305-321.

26. Kassir, M.K. & Sih, G.C. Three dimensional crack problems, *Mechanics of Fracture 2*, Noordhoff, Leyden, 1975.

27. Sih, G.C. (Ed) Methods of analysis and solutions of crack problems, *Mechanics of Fracture*, Vol. 1, Noordhoff, Leyden, 1973.

28. Cartwright, D.J. &Rooke, D.P. Evaluation of stress intensity factors, *A General Introduction to Fracture Mechanics*, Mech. Eng. Publications, London, 1978, pp. 54-73.

29. Stedman, G.E. Green's functions, *Contemp. Phys.*, 1968, **9**, 49-69.

30. Paris, P.C. Stress analysis of cracks, *Fracture Toughness and its Applications*, STP 381 ASTM, 1965, pp. 30-83.

31. Bueckner, H.F. The propagation of cracks and the energy of elastic deformation, *Trans ASME*, 1958, **80E**, 1225-1230.

32. Nisitani, H. Solutions of notch problems by body force method, stress analysis of notch problems, *Mechanics of Fracture*, Vol. 5, ed. G.C. Sih, Noordhoff Alpen Aan Den Riijn, 1978, pp. 1-68.

33. Cartwright, D.J. Stress intensity factors and residual static strength in certain structural elements, Ph.D. Thesis, Mech. Engng. Dept., University of Southampton, 1971.

34. Williams, J.G. & Isherwood, D.P. Calculation of the strain energy release rates of cracked plates by an approximate method, *J. Strain Anal.*, 1968, **3**, 17-22.

35. Rooke, D.P. Asymptotic stress intensity factors for fatigue crack-growth calculations, *Int. J. Fatigue*, 1980, **2**, 69-75.

36. Hsu, T.M. & Rudd, J.L. Green's function for thru-crack emanating from fastener holes, *Proceedings of the Fourth International Conference on Fracture*, Vol. **3**, University of Waterloo Press, Canada, 1977, pp. 139-148.

37. Hartranft, R.J. & Sih, G.C. Alternating method applied to edge and surface crack problems, Chapter 4, pp. 179-238 in Reference 1.

38. Tada, H., Paris, P. & Irwin, G. *The Stress Analysis of Cracks Handbook*, Del Research Corp, Hellertown, Pa., 1973.

39. Erdogan, F. On the stress distribution in plates with collinear cuts under arbitrary loads, *Proc. 4th U.S. Nat. Congress of Applied Mechanics*, Vol. 1, 1962, pp. 547-553.

40. Tweed, J. & Rooke, D.P. The elastic problem for an infinite solid containing a circular hole with a pair of radial edge cracks of different lengths, *Int. J. Engng Sci.*, 1976, **14**, 925-933.

41. Smith, E. Simple approximate methods for determining the stress intensification at the tip of a crack, *Int. J. Fracture*, 1977, **13**, 515-518.

42. Tweed, J. & Rooke, D.P. The stress intensity factor for a crack at the edge of a loaded hole, *Int. J. Solids Struct.*, 1979, **15**, 899-906.

43. Rooke, D.P. & Hutchins, S.M. Stress intensity factors for cracks at loaded holes - effect of load distribution, *J. Strain Anal.*, 1984, **19**, 81-96.

44. Rooke, D.P. & Tweed, J. Stress intensity factors for a crack at the edge of a pressurized hole, *Int. J. Engng Sci.*, 1980, **18**, 109-121.

45. Rice, J.R. Some remarks on elastic crack-tip stress fields, *Int. J. Solids Structures*, 1972, **8**, 751-758.

46. Fett, T., Matthek, C. & Munz, D. On the calculation of crack opening displacement from the stress intensity factor, *Engng Fract. Mech.*, 1987, **27**, 697-715.

47. Fett, T., Stamm, H. & Walz, G. Weight-function for finite strip with double edge notches, *Theor. Appl. Fract. Mech.*, 1988, **10**, 227-230.

48. Wu, X.R. Approximate weight functions for centre and edge cracks in finite bodies, *Engng Fract. Mech.*, 1984, **20**, 35-49.

49. Wu, X.R. & Carlsson, J. Generalized weight function method for crack problems with mixed boundary conditions, *J. Mech. Phys. Solids*, 1983, **31**, 485-497.

50. Parker, A.P. & Bowie, O.L. The weight function for various boundary condition problems, *Engng Fract. Mech.*, 1983, **18**, 473-477.

51. Paris, P.C., McMeeking, R.M. & Tada, H. The weight function method for determining stress intensity factors, *Cracks and Fracture*,1976, **STP 601**, ASTM, Philadelphia, pp. 471-489.

52. Bueckner, H.F. Field singularities and related integral representations, *Methods of Analysis and Solutions of Crack Problems, Mechanics of Fracture I*, ed. G.C. Sih, Noordhoff, 1973 pp. 239-314.

53. Cartwright, D.J. & Rooke, D.P. An efficient boundary element model for calculating Green's functions in fracture mechanics, *Int. J. Fracture*, 1985, **27**, R43-R50.

54. Aliabadi, M.H., Rooke, D.P. & Cartwright, D.J. Mixed-mode Bueckner weight functions using boundary element analysis, *Int. J. Fracture*, 1987, **34**, 131-147.

55. Rooke, D.P., Cartwright, D.J. & Aliabadi, M.H. Boundary elements combined with singular fields for three- dimensional cracked solids, *Proc. 4th Int. Conf. on Numerical Methods in Fracture Mechanics*, ed. A.R. Luxmore *et al.*, Pineridge Press, Swansea, 1987, pp. 15-26.

56. Rooke, D.P. & Aliabadi, M.H. Weight functions for crack problems using boundary element analysis, *Engng Analysis*, 1989, **6**, 19-29.

57. Cartwright, D.J. & Rooke, D.P. Approximate stress intensity factors compounded from known solutions, *Engng Fracture Mech.*, 1974, **6**, 563-571.

58. Rooke, D.P. Compounding stress intensity factors: applications to engineering structures, *Research Reports in Materials Science*, Parthenon Press, 1986.

59. Rooke, D.P. An improved compounding method for calculating stress-intensity factors, *Engng Fracture Mech.*, 1986, **23**, 783-792.

60. Rooke, D.P. & Cartwright, D.J. *Compendium of Stress Intensity Factors*, HMSO, London, 1976.

61. Tada, H., Paris, P.C. & Irwin, G. *The Stress Analysis of Cracks Handbook*, Del Research Corp., Hellertown, 1973.

62. Sih, G.C. *Handbook of Stress Intensity Factors*, Lehigh University, 1973.

63. Murakami, Y. *Stress Intensity Factors Handbook* (2 Vols.), Pergamon, 1987.

64. Figge, I.E. & Newman, J.C. Jr. Fatigue crack propagation in structures with simulated rivet forces, *ASTM STP*, 1967, **415**, 71-93.

65. Smith, F.W. Stress intensity factors for a semi-elliptical surface flaw, Boeing Company, Structural Development Research Memo 17, 1966.

66. Liu, A.F. Stress intensity factors for a corner flaw, *Engng Fracture Mech.*, 1972, **4**, 175-179.

67. Sokolnikoff, I.S. *Mathematical Theory of Elasticity*, McGraw-Hill, 1956, pp. 318-327.

68. Hartranft, R.J. &Sih, G.C. Alternating method applied to edge and surface cracks problems, methods of analysis and solutions of crack problems, *Mechanics of Fracture*, Vol. 1, ed. G.C. Sih, Noordhoff, Leyden, 1973.

69. Rooke, D.P. Compounded stress intensity factors for cracks at fastener holes, *Engng Fracture Mech.*, 1984, **19**, 359-374.

70. Timoshenko, S. & Goodier, J.N. *Theory of Elasticity*, 2nd edition, McGraw-Hill, New York, 1951.

71. Peterson, R.E. *Stress Concentration Factors*, 2nd edition, Wiley, New York, 1974.

72. Tweed, J. & Rooke, D.P. The elastic problem for an infinite solid containing a circular hole with a pair of radial edge cracks of different lengths, *Int. J. Engng. Sci.*, 1976, **14**, 925-933.

73. Isida, M. Method of Laurent series expansion for internal crack problems, Methods of analysis and solutions of crack problems, *Mechanics of Fracture*, Vol.1, ed. G.C. Sih, Noordhoff, Leyden, 1973, Chapter 2.

74. Benthem, J.P. & Koiter, W.T. Asymptotic approximations to crack problems, Methods of analysis and solutions of crack problems, *Mechanics of Fracture*, Vol. 1, ed. G.C. Sih, Noordhoff, Leyden, 1973, Chapter 3.

75. Tranter, C.J. The opening of a pair of coplanar Griffith cracks under internal pressure, *Quart. J. Mech. Appl. Math.*, 1961, **24**, 283-292.

76. Newman, J.C. Jr., An improved method of collocation for the stress analysis of cracked plates with various shaped boundaries, NASA TN-D-6376, 1971.

77. Rooke, D.P. & Cartwright, D.J. The compounding method applied to cracks in stiffened sheets, *Engng. Fracture Mech.*, 1976, **8**, 567-573.

78. Broek, D. *Elementary Engineering Fracture Mechanics*, 4th edition, Leyden, Noordhoff, 1974.

79. Feddersen, C.E. Evaluation and prediction of the residual strength of center cracked tension panels, *Damage Tolerance in Aircraft Structures*, ed. M.S. Rosenfield, ASTM STP 486, 1971.

80. Krafft, J.M., Sullivan, A.M. & Boyle, R.W. Effect of dimensions on fast fracture instability of notched sheets, *Proceedings of the Crack Propagation Symposium*, Cranfield, 1961.

81. Schwalbe, K.H. & Setz, W. R-curve and fracture toughness of thin sheet materials, *J. Test and Eval.*, 1981, **9**, 182-194.

82. Broek, D. *et al.* Fail-safe design procedures, *Fracture Mechanics of Aircraft Structures*, ed. H. Liebowitz. AGARDograph, 1974, No. 173, pp. 120-369.

83. Vlieger, H. *et al.* Built-up structures, *Practical Applications of Fracture Mechanics*, ed. H. Liebowitz. AGARDograph, 1980, No. 257, pp. 3-1 - 3-113.

84. Chell, G.G. Elastic-plastic fracture mechanics, *Developments in Fracture Mechanics-I*, ed. G.G. Chell, Applied Science Pub., London, 1979, Chapter 3, pp. 67-105.

85. Milne, I. Failure assessment, *Developments in Fracture Mechanics-I*, ed.G.G. Chell, Applied Science Pub., London, 1979, Chapter 8, pp. 259-301.

86. Broek, D. *The Practical Use of Fracture Mechanics*, Kluwer Academic Publishers, Dordrecht, 1988.

87. Rolfe, S.T. & Barsom, J.M. *Fracture and Fatigue Control in Structures*, Prentice Hall, New Jersey, 1977.

88. US National Bureau of Standards, Economic effect of fracture in the United States, Special Publication 647-1, NBS, Washington D.C., 1983.

89. Broek, D. Fracture control by periodic inspection with fixed cumulative probability of crack detection, *Structural Failure, Product Liability and Technical Insurance*, ed. Rossmanith, Interscience Enterprises Ltd., USA, 1987, pp. 238-258.

90. Young, A., Rooke, D.P. & Cartwright, D.J. Analysis of patched and stiffened cracked panels using the boundary element method, *Int. J. Solids Structures*, 1992, **29**, 2201-2216.

91. Baker, A.A. & Jones, R. (Eds). *Bonded Repair of Aircraft Structures*, Martinus Nijhoff, Dordrecht, 1988.

Chapter 2

The boundary element method

C.A. Brebbia

Wessex Institute of Technology, Ashurst Lodge, Ashurst, Southampton SO4 2AA, UK

Introduction

The best numerical method for fracture mechanics is the boundary element method (BEM). The technique not only requires a simple discretization of the surface rather than the volume of the component but also offers the accuracy that is required for the computation of quantities such as stress intensity factors. In addition, BEM is ideally suited for crack propagation studies, for which the extension of the fracture can be modelled simply by adding a few more elements. Because of this it is not surprising that BEM is becoming the most widely used numerical method in damage tolerance studies, displacing to a great extent the classical finite element method (FEM).

Boundary elements have emerged as a powerful alternative to finite elements not only in cases where better accuracy is required but also where the domain extends to infinity. The BEM mesh needed in these cases only models the surface of the crack and elements are usually not required at infinity. FEM by contrast requires the discretization of the whole volume and the use of an artificial boundary at a finite distance from the crack where some boundary conditions are imposed. The need to have this artificial boundary can lead to serious errors in the numerical results, particularly in elastodynamics where waves can be reflected by that boundary.

The future of boundary elements hinges on its acceptance by practicing engineers as a design tool. In this regard, the ease with which boundary elements can be integrated in the Computer Aided Engineering process explains the growing awareness of design engineers to the advantages of the new technique. The most important advantage of boundary elements over finite elements is that it requires only the discretization of the surface rather than the volume. Hence boundary element codes are easier to use with existing solid modellers and mesh generators. Meshes can be easily generated by the designer and changes do not require a complete remeshing. These advantages are even more marked when using discontinuous elements. These elements (Fig. 1) permit sudden changes in mesh grading and are unique to boundary elements. Discontinuous elements offer many advantages in terms of alternatives of meshes and general versatility and are an essential part of any well written boundary element code.

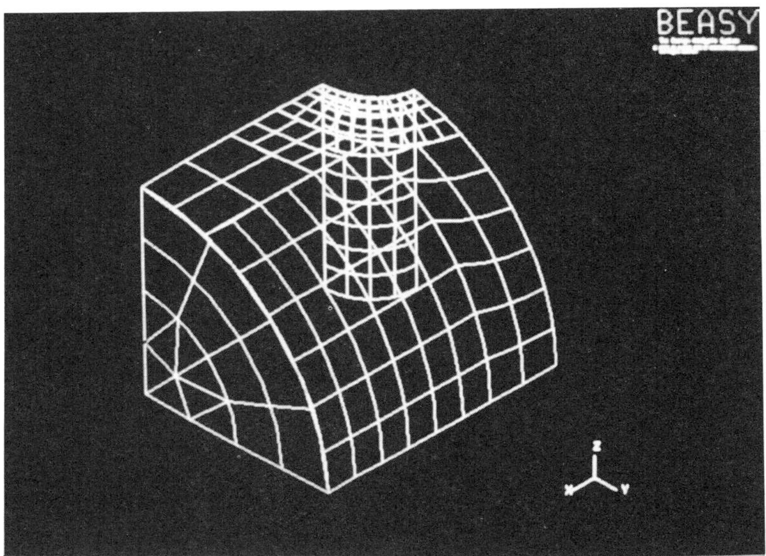

Figure 1: Discontinuous and continuous boundary element mesh

It is no exaggeration to say that BEM is the most exciting new development in numerical methods for fracture mechanics. BEM not only offers a high degree of versatility but produces very accurate results and allows to model problems extending to infinity without incurring errors associated with the fictitious boundaries required in FEM. It also provides the only feasible numerical method to analyse three dimensional problems. More recently the development of hypersingular elements has allowed us to study problems such as crack propagation which can now be solved using discontinuous boundary elements.

1 Basic equations of linear elastostatics

In what follows the basic equations for linear elasticity problems will be summarized. These equations assume that the material behaves linearly and that the changes of orientation of the body in the deformed state are negligible. The latter assumption leads to linear strain displacement relations and allows the equilibrium equations to be referred to the undeformed geometry.

The indicial notation will be used throughout the chapter for simplicity in addition to the matrix notation where appropriate.

In solid mechanics one needs to consider the forces or state of stress in the body and the deformation or state of strain. Both states are interrelated by applying the material behaviour or constitutive equations, i.e. those that relate stresses and strains.

1.1 State of stress

The state of stress at a point is defined in terms of stress components, in principle 9 of them for three dimensional elastostatics (Fig. 2), i.e.

$$\sigma_{ij} \quad \text{where} \begin{cases} i = 1,2,3 \\ j = 1,2,3 \end{cases} \tag{1}$$

The components are not independent but related through the equilibrium equations which are of two types, i.e. i) Moment equations and ii) Direct components equations.

The momentum equilibrium equations are written by taking moments of the stress components with respect to a point of the differential element. In the limit they produce the complementary shear relationships, i.e.

$$\sigma_{21} = \sigma_{12}; \quad \sigma_{31} = \sigma_{13}; \quad \sigma_{32} = \sigma_{23} \tag{2}$$

which reduce the 9 stress components to 6 independent ones. Equilibrium of the forces in any of the x directions produces the well known force equilibrium equations to be satisfied throughout the interior of the body or domain Ω, i.e.

$$\sigma_{ij,j} + b_i = 0; \quad \begin{cases} i = 1,2,3 \text{ in } \Omega \\ j = 1,2,3 \end{cases} \tag{3}$$

where the comma indicates derivative with respect to that x-component.

When the stress components are projected into a differential on the boundary $d\Gamma$ they produce the surface forces or tractions which are denoted by t_i such that

$$t_i = \sigma_{ij}\, n_j, \quad \begin{cases} i = 1,2,3 \\ j = 1,2,3 \end{cases} \quad \text{on } \Gamma \tag{4}$$

n_j are the direction cosines of the outward normal n with respect to the x axis, i.e.

$$n_j = \cos(n, x_j) \quad j = 1,2,3 \quad \text{on } \Gamma \tag{5}$$

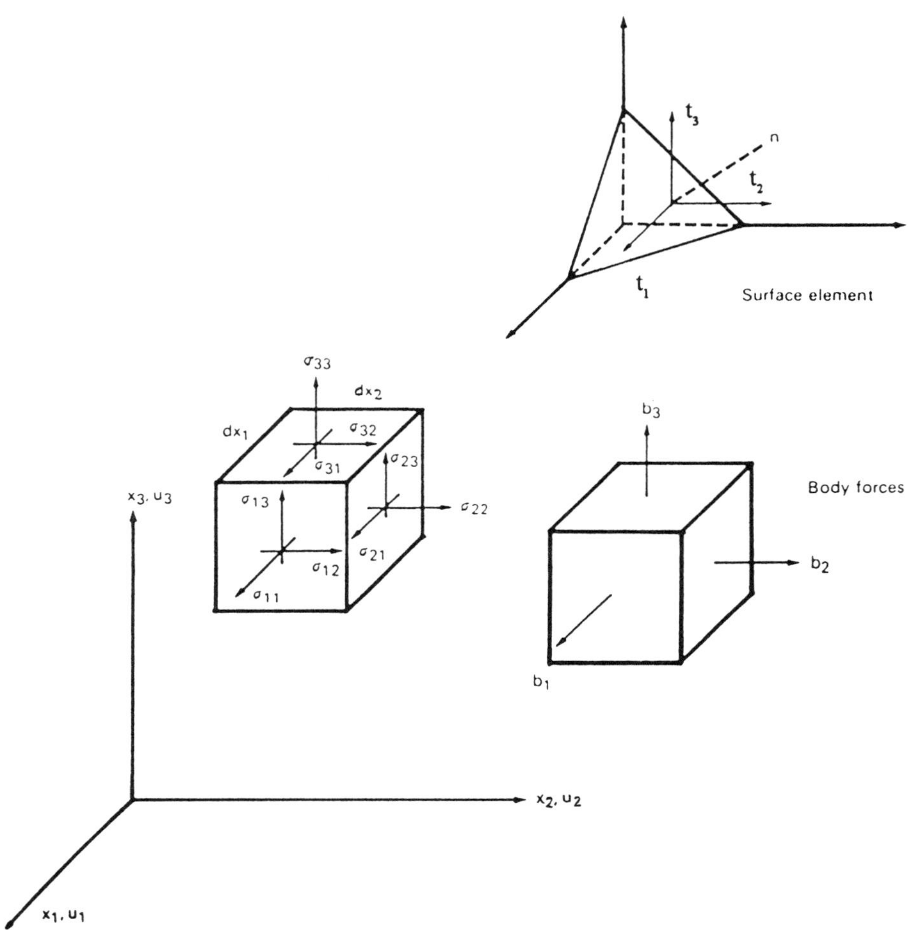

Figure 2: Notation for tractions, stresses and displacements

The tractions are assumed to be known on Γ_2 part of the boundary ($\Gamma = \Gamma_1 + \Gamma_2$) and they are the forces or 'natural' boundary conditions for the problem. In this case

$$t_i = \sigma_{ij}\, n_j = \overline{t}_i \tag{6}$$

where the dash indicates a known quantity.

1.2 State of strain

The deformations of the boundary are function of the displacements which have the components u_i ($i = 1,2,3$) at any point.

They produce strains which for the linear case can be written as

$$\varepsilon_{ij} = \frac{1}{2}(u_{i,j} + u_{j,i}) \tag{7}$$

Notice that by definition $\varepsilon_{21} = \varepsilon_{12}$; $\varepsilon_{31} = \varepsilon_{13}$; $\varepsilon_{32} = \varepsilon_{23}$.

It is easy to apply the boundary conditions in terms of displacements rather than strains. Hence on the Γ_1 part of the boundary the following displacements or 'essential' boundary conditions can be defined.

$$u_j = \overline{u}_j, \quad j = 1,2,3 \quad \text{on } \Gamma_1 \tag{8}$$

where $\overline{u}_j$ are the prescribed values.

1.3 Material properties

The states of stress and strain in a body are related throughout the strain-stress relations or constitutive equations for the material. For a linear elastic isotropic material one can define two constants only, called Lame's constants μ and λ which are associated with the volumetric and shear components. The strain-stress relationship can be written as

$$\sigma_{ij} = \lambda\, \delta_{ij}\, \varepsilon_{kk} + 2\mu\, \varepsilon_{ij} \tag{9}$$

where δ_{ij} is the Kronecker delta ($\equiv 1$ for $i = j$ and $\equiv 0$ for $i \neq j$).

Notice that ε_{kk} has only internal indexes and hence it implies the sum of the three direct strain components, and because of this is called the volumetric strain, i.e.

$$\varepsilon_{kk} = \varepsilon_{11} + \varepsilon_{22} + \varepsilon_{33} \tag{10}$$

The inverse of eqn (9) can be written as

$$\varepsilon_{ij} = -\frac{\lambda\, \delta_{ij}}{2\mu(3\lambda + 2\mu)}\, \sigma_{kk} + \frac{1}{2\mu}\, \sigma_{ij} \tag{11}$$

The Lame's constants are sometimes expressed in terms of the more familiar shear modulus G, Modulus of Elasticity, E and Poisson's ratio ν as follows

$$\mu = G = \frac{E}{2(1+\nu)}; \quad \lambda = \frac{\nu E}{(1+\nu)(1-2\nu)} \tag{12}$$

The strain and stress components in terms of E and ν can be written as

$$\varepsilon_{ij} = -\frac{\nu}{E}\,\sigma_{kk}\,\delta_{ij} + \frac{1+\nu}{E}\,\sigma_{ij} \tag{13}$$

$$\sigma_{ij} = \frac{E}{(1+\nu)}\left[\frac{\nu}{(1-2\nu)}\,\delta_{ij}\,\varepsilon_{kk} + \varepsilon_{ij}\right] \tag{14}$$

The material properties for general isotropic materials in three dimensions involve 21 different constants and the relationship is usually expressed by

$$\sigma_{ij} = d_{ijkl}\,\varepsilon_{kl} \tag{15}$$

The equations of equilibrium (3), strain displacement relations (6) and constitutive equations (6) give a complete system of equations from which the components of stress (6), the displacements (3) and the strains (6) can be determined.

2 Fundamental solution

The formulation of the boundary integral equations for elastostatics to be described in section 3 requires the knowledge of the solution of the elastic problem with the same material properties as the body under consideration but corresponding to an infinite domain with a concentrated point load at a particular location. This solution is called fundamental or Kelvin's solution in elastostatics.

Consider the equilibrium equations corresponding to this domain extending to infinity where the quantities will be denoted by an asterisk to differentiate from the actual solution, i.e.

$$\sigma^*_{ij,j} + b^*_j = 0 \tag{16}$$

These equations are sometimes expressed in terms of displacement components and refer to in that case as Navier's equations, i.e.

$$\frac{1}{1-2\nu}\, u^*_{j,jl} + u^*_{l,jj} + \frac{1}{\mu}\, b^*_l = 0 \tag{17}$$

Kelvin's solution is obtained by solving eqn (16) or (17) when a concentrated unit load is applied at point 'i' in the direction of the unit vector e_l (Fig. 3), i.e.

$$b^*_l = \Delta^i e_l \tag{18}$$

where Δ^i is the Dirac delta.

The solution to these problems is well known[1] and in terms of displacements is given by

$$u^*_{lk} = \frac{1}{16\pi\mu(1-\nu)r}\,[(3-4\nu)\delta_{lk} + r_{,l}\, r_{,k}] \tag{19}$$

for three dimensional problems. The variable r represents the modulus of the vector r or distance between the point i of application of the force and any other field point. Its derivatives are

$$r_{,l} = \frac{\partial r}{\partial x_l}; \quad r_{,k} = \frac{\partial r}{\partial x_k} \tag{20}$$

which are the vectors between the projection of r in the x_1, x_2 and x_3 directions which can be called r_l ($l = 1,2,3$) and the length of r (Fig. 4), i.e.

$$r_{,l} = \frac{r_l}{r} \tag{21}$$

For the two dimensional plane strain problem, the fundamental solution in terms of displacements is given by

$$u^*_{lk} = \frac{1}{8\pi\mu(1-\nu)}\left[(3-4\nu)\ \ln\left(\frac{1}{r}\right)\delta_{lk} + r_{,l} r_{,k}\right] \tag{22}$$

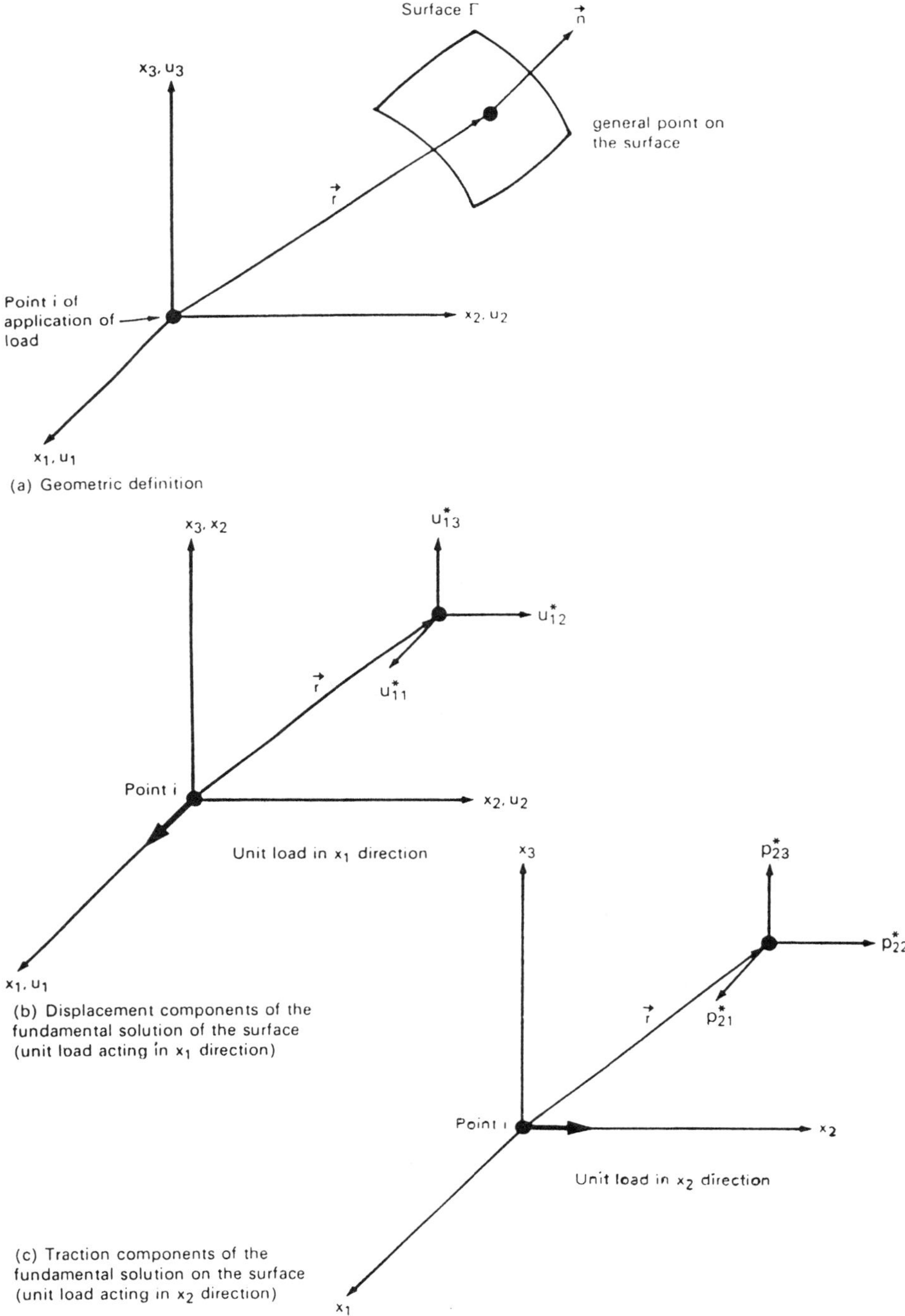

Figure 3: Geometrical interpolation of the components of the fundamental solution

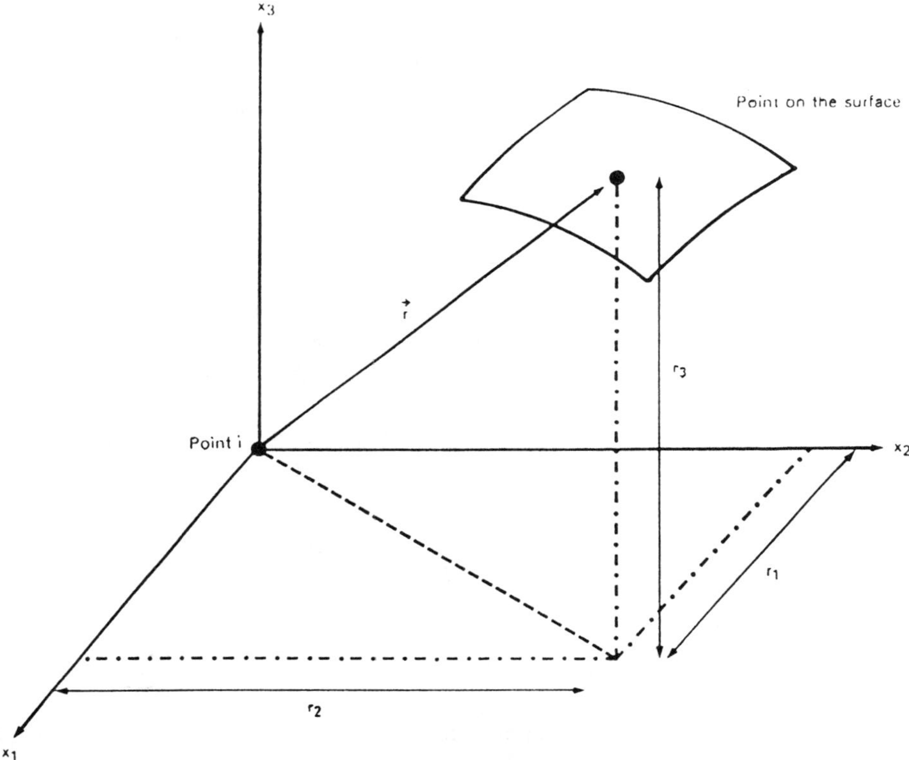

Figure 4: Interpolation of the components of the distance vector r

Stresses at internal points can be written using the strain-displacement relations (7) and the strain-stress eqns (9). They can be expressed as

$$\sigma^*_{kj} = S^*_{lkj}\, e_l \tag{23}$$

where the kernal S^*_{lkj} has been obtained from operating on u^*_{lk} is given in full in Refs 1 and 2.

The tractions or surface forces on the boundary containing a normal n can be written through (4) and (23) as

$$t^*_l = t^*_{lk}\, n_l \tag{24}$$

where the traction components for the three dimensional case are

$$t^*_{lk} = -\frac{1}{8\pi(1-\nu)r^2}\left[\frac{\partial r}{\partial n}\left[(1-2\nu)\delta_{lk} + 3r_{,l}r_{,k}\right] + (1-2\nu)\left(n_l r_{,k} - n_k r_{,l}\right)\right] \tag{25}$$

n_l and n_k are the direction cosines of the normal with respect to x and x_k. $\partial r/\partial n$ is the derivative of the distance vector r with respect to the normal.

For two dimensional plane strain problems one has

$$t^*_{lk} = -\frac{1}{4\pi(1-\nu)r}\left[\frac{\partial r}{\partial n}\left[(1-2\nu)\delta_{lk} + 2r_{,l}r_{,k}\right] + (1-2\nu)\left(n_l r_{,k} - n_k r_{,l}\right)\right] \tag{26}$$

Notice that in two dimensional boundary elements one works in principle in plane strain as the fundamental solution is known for this type of problem. Plate stretching problems can be solved by replacing the E and ν material constants by two new values, i.e.

$$E' = (1-\nu^2)E; \quad \nu' = \frac{\nu}{1+\nu} \tag{27}$$

This is the opposite of what is usually done in finite elements where the programs are written for plate stretching and used in plane strain by the application of the inverse of the above relationships.

3 Boundary integral formulation

The governing integral equations for elastostatics will be deduced using conditions of error minimization in a weighted residual way. The explanation given in this section is a shorter version of the one shown in Refs 1 or 2 which the reader may be interested to consult.

Consider first that one desires to minimize the errors involved in the numerical approximation of the equilibrium equations, i.e.

$$\sigma_{kj,j} + b_k = 0 \quad \text{in } \Omega \tag{28}$$

which have to satisfy the following boundary conditions.

i) Essential or displacement conditions

$$u_k = \overline{u}_k \ \text{ on } \Gamma_1 \tag{29}$$

ii) Natural or traction conditions

$$t_k = \overline{t}_k \ \text{ on } \Gamma_2 \tag{30}$$

Consider first that we are only interested in minimizing (28). To this end we can 'weight' each of these equations by a displacement type function u_k^* and orthogonalize the product, i.e.

$$\int_\Omega (\sigma_{kj,j} + b_k) u_k^* \, \mathrm{d}\Omega = 0 \tag{31}$$

If we carry out the integration by parts of the first term of the equation and group the corresponding terms together, one finds the following expression,

$$-\int_\Omega \sigma_{kj}\varepsilon_{kj}^* \, \mathrm{d}\Omega + \int_\Omega b_k u_k^* \, \mathrm{d}\Omega = -\int_\Gamma t_k u_k^* \, \mathrm{d}\Gamma \tag{32}$$

Noticing that

$$\int_\Omega \sigma_{kj}\varepsilon_{kj}^* \, \mathrm{d}\Omega = \int_\Omega \sigma_{kj}^*\varepsilon_{kj} \, \mathrm{d}\Omega \tag{33}$$

and integrating by parts again one finds

$$\int_\Omega \sigma_{kj,j}^* u_k \, \mathrm{d}\Omega + \int_\Omega b_k u_k^* \, \mathrm{d}\Omega = -\int_\Gamma t_k u_k^* \, \mathrm{d}\Gamma + \int_\Gamma t_k^* u_k \, \mathrm{d}\Gamma \tag{34}$$

Notice that the solution with the asterisk can be assumed to be the fundamental solution, in which case

$$\sigma_{lj,j}^* + \Delta^i \, e_l = 0 \tag{35}$$

This allows to write the first term of (34) as follows

$$\int_\Omega \sigma^*_{kj,j} u_k \, d\Omega = \int_\Omega \sigma^*_{lj,j} u_l \, d\Omega = -\int_\Omega \Delta^i e_l u_l \, d\Omega = -u_l^i \, e_l \tag{36}$$

where u_l^i represents the l component of the displacement at the point i of application of the unit load.

Equation (34) can now be written to represent the three separate components of the displacement at i by taking the three directions of the point load at 'i' independently, i.e.

$$u_l^i + \int_\Gamma t^*_{lk} u_k \, d\Gamma = \int_\Gamma u^*_{lk} t_k \, d\Gamma + \int_\Omega u^*_{lk} b_k \, d\Omega \tag{37}$$

Notice that one needs still to aply the boundary conditions on Γ_1 and Γ_2 in accordance with formula (29) and (30). This can be done however at a later stage, i.e. when working with the resulting matrices, to simplify the problem.

Equation (37) is sometimes called Somigliani's identity and gives the value of the displacement at any internal points in terms of the boundary values. u_k and t_k; the forces throughout the domain and the known fundamental solution. Equation (37) is valid for any particular 'i' point in the domain where the forces are applied.

3.1 Boundary points

Somigliani's identity gives the displacements at any internal point once u_k and t_k are known at all boundary point and consequently only after the boundary value problem has been solved can the values at the internal points be calculated. However, a boundary integral expression of (37) can be obtained by taking the point 'i' to the boundary to produce a system of equations which once solved give the boundary values as will be shown shortly.

When 'i' is taken to the boundary however, the integrals present a different type of singularity and one needs to analyse this behaviour. This has been done in detail in Ref. 1 where it is shown that eqn (37) becomes,

$$c^i_{lk} u^i_k + \int_\Gamma t^*_{lk} u_k \, d\Gamma = \int_\Gamma u^*_{lk} t_k \, d\Gamma + \int_\Omega u^*_{lk} b_k \, d\Omega \tag{38}$$

where the integrals are now in the sense of Cauchy principal value. If Γ is smooth at 'i' the values of $c_{lk} = 1/2 \; \delta_{lk}$. When '$i$' is at a point where the boundary is not smooth however, the value of the c coefficients are different and it is usually difficult to obtain a general expression for elastostatics. Fortunately, explicit calculations of this value are not usually necessary as it can be obtained using rigid boundary considerations as will be shown in section 4.

Boundary eqn (38) permits to solve the general boundary value problem of elastostatics. If the displacements are known over the whole boundary, eqn (38) produces an integral equation of the first kind; if the tractions are known over all the boundary an integral equation of the second kind is obtained and finally a combination of both types of boundary conditions results in a mixed integral equation.

4 Boundary element formulation

In order to solve the integral equations numerically, the boundary will be discretized into a series of elements over which displacements and tractions are written in terms of their values at a series of nodal points. Writing the discretized form of (38) for every nodal point, a system of linear algebraic equations is obtained. Once the boundary conditions are applied the system can be solved to obtain all the unknown values and consequently an approximate solution to the boundary value problem is obtained.

It is more convenient from now on to work with matrices rather than carry on with the indicial notation. To this effect one can start by defining the u and t functions which apply over each element 'j' (see Fig. 5), i.e.

$$\mathbf{u} = \boldsymbol{\phi}\, \mathbf{u}^j; \qquad \mathbf{t} = \boldsymbol{\phi}\, \mathbf{t}^j \tag{39}$$

where $\mathbf{u}^j$ and $\mathbf{t}^j$ are the element nodal displacements and traction vectors of dimensions $2 \times q$ for two dimensions and $3 \times q$ for three dimensions, q being the number of nodes of the elements. $\mathbf{u}$ and $\mathbf{t}$ are the displacements and tractions at any point on the boundary, i.e.

$$\mathbf{u} = \begin{Bmatrix} u_1 \\ u_2 \\ u_3 \end{Bmatrix}; \quad \mathbf{t} = \begin{Bmatrix} t_1 \\ t_2 \\ t_3 \end{Bmatrix} \tag{40}$$

Notice that the BEM formulation is in terms of displacements and tractions, i.e. is a 'mixed' formulation and consequently interelement continuity of these variables is not required. This property allows for the use of a wide variety of discontinuous, semidiscontinuous and continuous elements (Fig. 6).

The interpolation matrix $\boldsymbol{\phi}$ is a $3 \times 3q$ (or a $2 \times 2q$ in two dimensions) array of shape functions, i.e.

$$\boldsymbol{\phi} = [\boldsymbol{\phi}_1 \;\; \boldsymbol{\phi}_2 \;\cdots\; \boldsymbol{\phi}_q] = \begin{bmatrix} \phi_1 & 0 & 0 & \phi_2 & 0 & 0 & \cdot & \cdot & \cdot & \phi_q & 0 & 0 \\ 0 & \phi_1 & 0 & 0 & \phi_2 & 0 & \cdot & \cdot & \cdot & 0 & \phi_q & 0 \\ 0 & 0 & \phi_1 & 0 & 0 & \phi_2 & \cdot & \cdot & \cdot & 0 & 0 & \phi_q \end{bmatrix} \tag{41}$$

These functions are similar to the standard finite element type function but can also allow for interelement discontinuities in tractions and displacements.

Notice that the body forces at any point on the Ω domain can also be expressed in vector form in function of the three components, i.e.

$$b = \begin{Bmatrix} b_1 \\ b_2 \\ b_3 \end{Bmatrix} \tag{42}$$

The fundamental solution coefficients can be expressed as (Fig. 3)

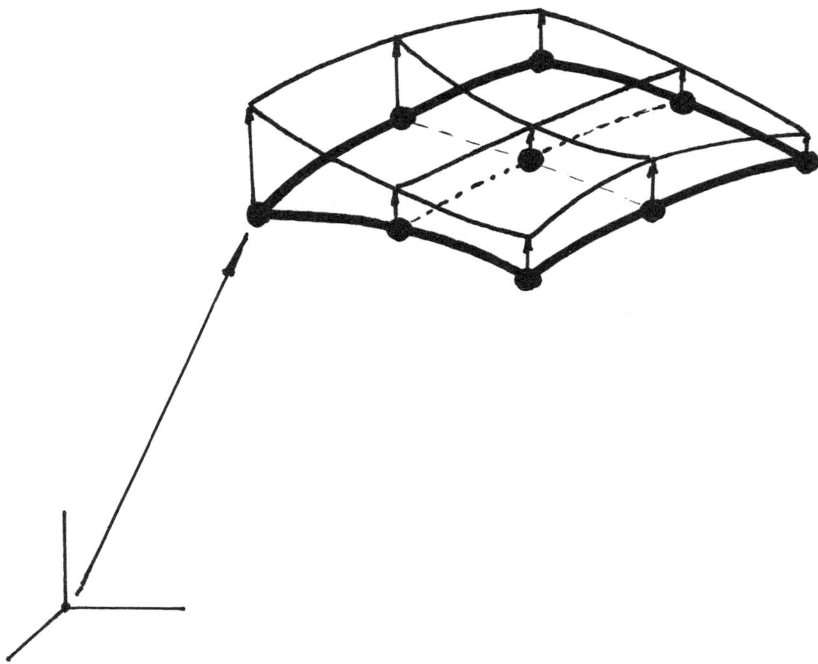

Figure 5: Interpolation functions over a quadratic element

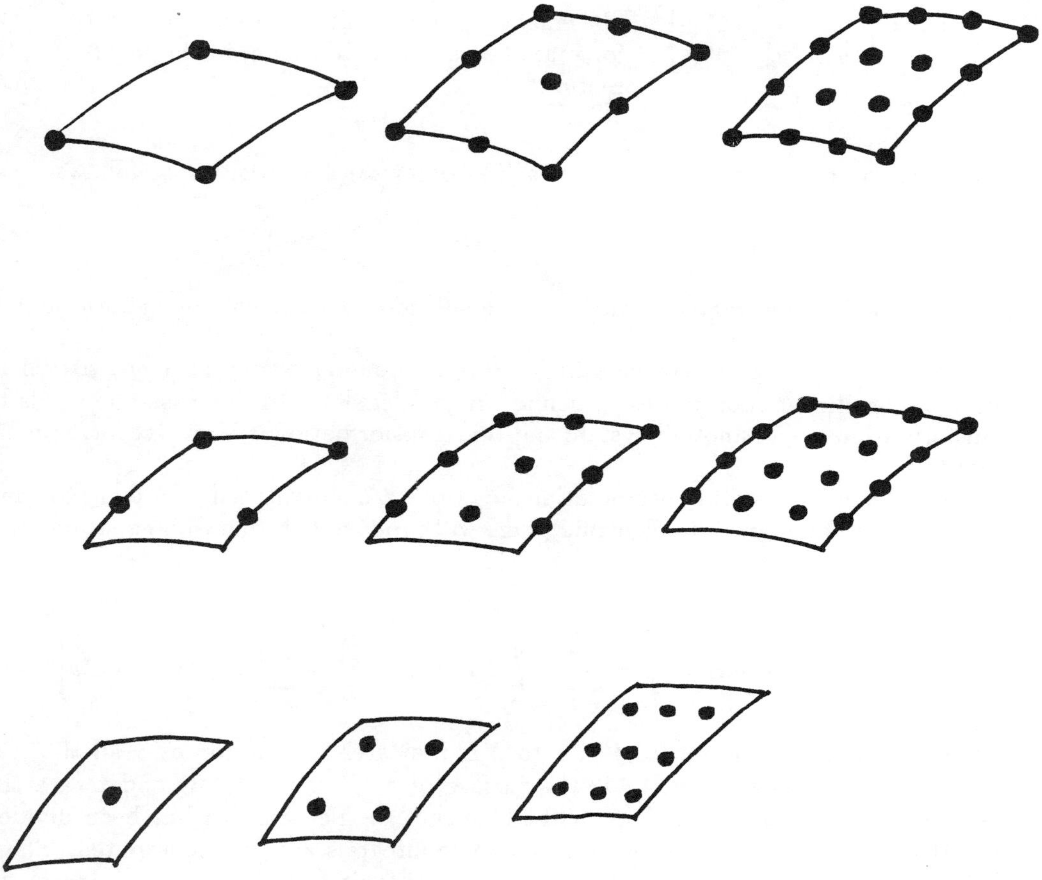

Figure 6: Some different types of continuous and discontinuous elements

$$\mathbf{t}^* = \begin{bmatrix} t^*_{11} & t^*_{12} & t^*_{13} \\ t^*_{21} & t^*_{22} & t^*_{23} \\ t^*_{31} & t^*_{32} & t^*_{33} \end{bmatrix}$$ Matrix whose coefficients t^*_{lk} are the tractions in k direction due to a unit force applied at point 'i' acting in the 'l' direction (43)

$$\mathbf{u}^* = \begin{bmatrix} u^*_{11} & u^*_{12} & u^*_{13} \\ u^*_{21} & u^*_{22} & u^*_{23} \\ u^*_{31} & u^*_{32} & u^*_{33} \end{bmatrix}$$ Matrix whose coefficients u^*_{lk} are the displacements in the 'k' direction due to a unit force applied at point 'i' acting in the 'l' direction (44)

With this notation, eqn (38) valid for each 'i' point can be rewritten as follows

$$\mathbf{c}^i\mathbf{u}^i + \int_\Gamma \mathbf{t}^*\mathbf{u}\, d\Gamma = \int_\Gamma \mathbf{u}^*\mathbf{t}\, d\Gamma + \int_\Omega \mathbf{u}^*\mathbf{b}\, d\Omega \tag{45}$$

where $\mathbf{c}^i$ is a diagonal matrix with 1/2 on the diagonal for the case of smooth boundaries.

Notice that the cartesian coordinates of the boundary may also be written in terms of curvilinear coordinates to define curved elements. In this case one needs to transform from one to another system and this transformation will involve introducing a Jacobian.

Consider now that the interpolation functions are substituted into eqn (45) and discretize the boundary into elements. This will produce the following equation for a nodal point 'i'.

$$\mathbf{c}^i\mathbf{u}^i + \sum_{j=1}^{NE}\left\{\int_{\Gamma_j}(\mathbf{t}^*\boldsymbol{\phi})\, d\Gamma\, \mathbf{u}^j\right\} = \sum_{j=1}^{NE}\left\{\int_{\Gamma_j}(\mathbf{u}^*\boldsymbol{\phi})\, d\Gamma\, \mathbf{t}^j\right\} + \sum_{s=1}^{M}\left\{\int_{\Omega_s}(\mathbf{u}^*\mathbf{b})\, d\Omega\right\} \tag{46}$$

Notice that the summation for $j = 1$ to NE indicates summation over all the NE elements on the surface and Γ_j is the surface of a 'j' element. $\mathbf{u}^j$ and $\mathbf{t}^j$ are the nodal displacement and tractions in the element 'j'. The domain has been divided into M internal cells over which the body force integrals are to be computed. These domain integrals can in most cases be avoided by taking the body force integrals to the boundary using particular solutions or more general techniques such as the Dual Reciprocity Method.[3,4,5]

The integrals in (46) are usually solved numerically, particularly if the elements are curved, as it is difficult to integrate them analytically in this case. The interpolation functions $\boldsymbol{\phi}$ tend to be expressed in a homogeneous or curvilinear system of coordinates.

Equation (46) corresponds to a particular node 'i' and once integrated can be written as

$$\mathbf{c}^i\mathbf{u}^i + \sum_{k=1}^{N}\hat{\mathbf{H}}^{ik}\mathbf{u}^k = \sum_{k=1}^{N}\mathbf{G}^{ik}\mathbf{t}^k + \sum_{s=1}^{M}\mathbf{B}^{is} \tag{47}$$

where N is the number of nodes, $\mathbf{u}^k$ and $\mathbf{t}^k$ are the displacements and tractions at node 'k'.

Calling

$$\begin{aligned} \mathbf{H}^{ij} &= \hat{\mathbf{H}}^{ij} && \text{if} \quad i \neq j \\ \mathbf{H}^{ij} &= \hat{\mathbf{H}}^{ij} + \mathbf{c}^i && \text{if} \quad i = j \end{aligned} \tag{48}$$

eqn (47) for node 'i' becomes

$$\sum_{j=1}^{n} \mathbf{H}^{ij}\mathbf{u}^j = \sum_{j=1}^{N} \mathbf{G}^{ij}\mathbf{t}^j + \sum_{s=1}^{M} \mathbf{B}^{is} \tag{49}$$

The contribution for all 'i' nodes can be written together in matrix form to give the global system equations, i.e.

$$\mathbf{HU} = \mathbf{GT} + \mathbf{B} \tag{50}$$

Notice that the elements of $\mathbf{c}^i$ will be a series of 3×3 submatrices on the diagonal of $\mathbf{H}$ (or 2×2 in two dimensional cases). The elements of these submatrices can be very cumbersome to calculate. Fortunately this is not required as they can be found by consideration of rigid body movements as shown below.

The vectors $\mathbf{U}$ and $\mathbf{T}$ represent all the boundary values of displacements and tractions before applying the boundary conditions. These conditions can be introduced by reorganizing the columns of $\mathbf{H}$ and $\mathbf{G}$, passing all unknowns to a vector $\mathbf{X}$ on the left hand side and all known terms, including those on $\mathbf{B}$ to a vector $\mathbf{F}$ on the right hand side. This gives the final system of equations, i.e.

$$\mathbf{AX} = \mathbf{F} \tag{51}$$

Solving the above system all boundary values are fully determined.

4.1 Rigid body considerations

As it was pointed out before, the diagonal submatrices $\mathbf{H}^{ii}$ in $\mathbf{H}$ are given by adding in $\hat{\mathbf{H}}^{ii}$ and $\mathbf{c}^i$. Difficulties appear when trying to compute explicitly those terms particularly at corners due to the singularity of the fundamental solution. Assuming a rigid body displacement in the direction of one of the cartesian coordinates the traction and body forces vector in a closed system must be zero and hence from (50)

$$\mathbf{H}\,\mathbf{I}^q = 0 \tag{52}$$

where $\mathbf{I}^q$ is a vector that for all nodes has a unit displacement along the 'q' direction (q = 1,2 or 3) and zero displacements in any other direction. Since (52) has to be satisfied for any rigid body displacement one can write

$$\mathbf{H}^{ii} = -\sum_{j=1}^{N} \mathbf{H}^{ij} \quad (\text{if } i = j) \tag{53}$$

which gives the diagonal submatrices in terms of the rest of the terms of the **H** matrix.

The above considerations are strictly valid for closed domains. When dealing with infinite regions, eqn (53) must be modified. If the rigid body displacement is prescribed for a boundless domain the integral

$$\int_{\Gamma_\infty} \mathbf{t}^* \mathbf{I}^q \, d\Gamma = \left\{ \int_{\Gamma_\infty} \mathbf{t}^* \, d\Gamma \right\} \mathbf{I}^q \tag{54}$$

over the external boundary Γ_∞ will not be zero at infinity. Since the tractions $\mathbf{t}^*$ are due to a unit point load, this integral must be

$$\int_{\Gamma_\infty} \mathbf{t}^* \, d\Gamma = -\mathbf{I} \tag{55}$$

where **I** is a 3×3 (or 2×2 in two dimensions) identity matrix. The diagonal submatrices for this case are

$$\mathbf{H}^{ii} = \mathbf{I} - \sum_{j=1}^{N} \mathbf{H}^{ij} \quad (\text{for } i = j) \tag{56}$$

4.2 Internal points

Somigliani's identity (37) gives the displacement at any internal point in terms of the boundary displacements and tractions. Considering its boundary element representation one has

$$\mathbf{u}^i = \sum_{j=1}^{NE} \left\{ \int_{\Gamma_j} \mathbf{u}^* \phi \, d\Gamma \right\} \mathbf{t}^j - \sum_{j=1}^{NE} \left\{ \int_{\Gamma_j} \mathbf{t}^* \phi \, d\Gamma \right\} \mathbf{u}^j + \sum_{s=1}^{M} \left\{ \int_{\Omega_s} \mathbf{u}^* \mathbf{b} \, d\Omega \right\} \tag{57}$$

where Γ_j is the surface corresponding to element j and 'i' is now an internal point. The internal point displacements in terms of the displacements and tractions on the boundary can be written in the following form, i.e.

$$\mathbf{u}^i = \sum_{j=1}^{N} \mathbf{G}^{ij} \mathbf{t}^j - \sum_{j=1}^{N} \mathbf{H}^{ij} \mathbf{u}^j + \sum_{s=1}^{M} \mathbf{B}^{is} \tag{58}$$

The terms $\mathbf{G}^{ij}$ and $\mathbf{H}^{ij}$ consist of integrals over the elements to which node j belongs. Those integrals do not contain any singularity and can be easily computed using numerical integration. Terms like $\mathbf{B}^{is}$ however will contain a singularity, (notice that they are domain terms and the point i is now in the domain) and special care should be taken when computing them numerically. Being domain integrals their order of singularity is one less than the integrals on the boundary and consequently can be more accurately computed using numerical integration.

For an isotropic medium the stress can be computed by differentiating the displacements at internal points and introducing the corresponding strains into the stress-strain relationships, i.e.

$$\sigma_{ij} = \frac{2\mu\nu}{1-2\nu}\delta_{ij}u_{k,k} + \mu(u_{i,j} + u_{j,i}) \tag{59}$$

After carrying out the derivatives inside the integral equations and computing the derivatives of the fundamental solution one obtains,

$$\sigma_{ij} = \int_\Gamma D_{kij}t_k \,\mathrm{d}\Gamma - \int_\Gamma S_{kij}u_k \,\mathrm{d}\Gamma + \int_\Omega D_{kij}b_k \,\mathrm{d}\Omega \tag{60}$$

where the third order tensor components D_{kij} and S_{kij} are

$$D_{kij} = \frac{1}{r^\alpha}\left\{(1-2\nu)\{\delta_{ki}\,r_{,j} + \delta_{kj}\,r_{,i} - \delta_{ij}\,r_{,k}\} + \beta\,r_{,i}\,r_{,s}\,r_{,j}\right\}\frac{1}{4\alpha\pi(1-\nu)} \tag{61}$$

$$\begin{aligned} S_{k,j} \;=\; & \frac{2\mu}{r^\beta}\left\{\beta\frac{\partial r}{\partial n}\left[(1-2\nu)\delta_{ij}\,r_{,k} + \nu\left(\delta_{ik}\,r_{,j} + \delta_{jk}\,r_{,i}\right)\right.\right. \\ & -\gamma\,r_{,i}\,r_{,j}\,r_{,k}\big] + \beta\nu\left(n_i\,r_{,j}\,r_{,k} + n_j\,r_{,i}\,r_{,k}\right) \\ & +(1-2\nu)\left(\beta\,n_k\,r_{,i}\,r_{,j} + n_j\,\delta_{ik} + n_i\,\delta_{jk}\right. \\ & \left.\left. -(1-4\nu)n_k\,\delta_{ij}\,\frac{1}{4\alpha\pi(1-\nu)}\right\} \end{aligned} \tag{62}$$

The above formulae are applicable for 2 or 3 dimensions. For the former case $\alpha = 1$; $\beta = 2$ and $\gamma = 4$ and for the latter $\alpha = 2$, $\beta = 3$ and $\gamma = 5$.

All the derivatives indicated by commas are taken at the boundary point x_i^B of Fig. 7, i.e.

$$r_{,i} = \frac{\partial r}{\partial x_i} = \frac{r_i^B}{r^B} \tag{63}$$

These derivatives are equal and opposite in sign to those taken at an internal point.

Equation (60) can be discretized by dividing the Γ boundary into a summation over all the boundary elements and assuming the corresponding interpolation function for $\mathbf{u}_k$ and $\mathbf{t}_k$ as explained earlier.

The values obtained for the internal stresses using the above formulae are in general more accurate than those computed using other numerical methods. The same can be said of the internal displacements computed through (58). However, special care needs to be taken when the internal point is very close to the boundary (say less than 1/4 of the length of the nearest element). This is because of the peak in the fundamental solution, and in this case special numerical integration technique schemes have to be used to obtain accurate stresses and displacements.

Notice that although all values of displacements at the boundary are known from the integral equation solution the same cannot be said of the surface stresses since the stress vector has more terms than the known surface traction vector. The problem is studied in what follows.

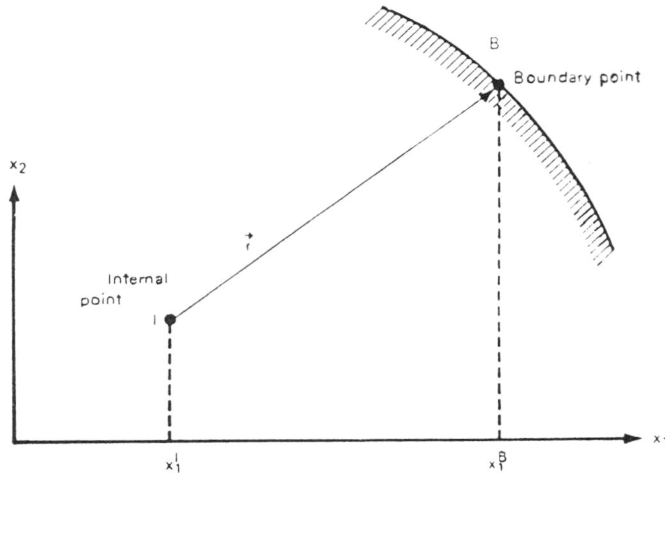

$$r_{,j} = \frac{\partial r}{\partial x_j} = \frac{x_j^B - x_j^I}{| r |}$$

Figure 7: Definition of derivatives required in the internal stress formulae (two dimensional case).

4.3 Stresses on the boundary

Although the solution of the system of equations in BEM produces all the surface tractions, one needs to know in many cases the boundary stresses. In order to compute them one can take eqn (60) to the boundary but this produces higher order singularity which is difficult to compute. For three dimensional problems the D^{kij} and S^{kij} terms contain singularities of order $1/r^2$ and $1/r^3$ respectively and taking eqn (60) to the boundary requires special consideration of how to compute the principal values.

The simplest way of determining the stress and tensor at the boundary points is to compute its component from the known boundary tractions and displacements. This is the way most computer codes used in industry operate. Assume for the three dimensional case a local cartesian coordinate system at the boundary points whose stresses are to be computed (Fig. 8). It is easy to see that

$$\begin{aligned} \sigma'_{13} &= \sigma'_{31} = t'_1 \\ \sigma'_{23} &= \sigma'_{32} = t'_2 \\ \sigma'_{33} &= t'_3 \end{aligned} \qquad (64)$$

where the prime indicates local coordinates. In addition to those tracts a discrete expression for the boundary displacements over the element can also be proposed as given by (39) and a transformation matrix **R**, i.e.

$$u' = \mathbf{R}^T \phi \ \mathbf{u}^j \tag{65}$$

where **R** transforms the displacements from the global to the local system. Four components of the strain tensor can be computed by differentiating u' as follows

$$\varepsilon'_{ij} = 1/2(u'_{j,i} + u'_{i,j}) \quad i, j = 1, 2 \tag{66}$$

where $\varepsilon'_{12} = \varepsilon'_{21}$ gives only three independent components.
Notice that the ε'_{ij} will depend on the derivatives of the shape functions and the nodal displacements. If constant elements are used, the displacement derivatives can be computed using a finite different approximation between adjacent nodes.

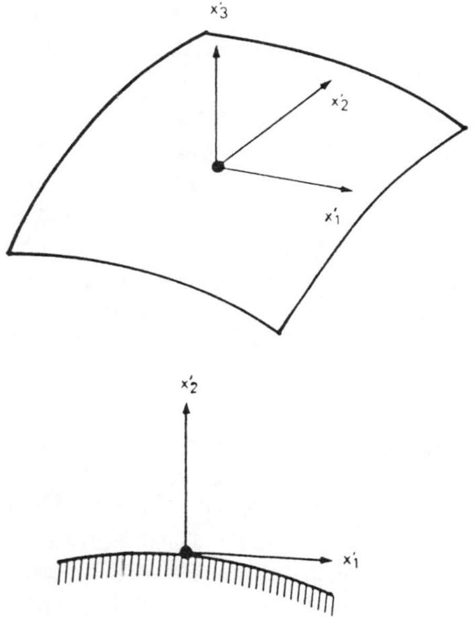

Figure 8: Local system of coordinates over the element

The rest of the terms of the stress tensor can then be computed from the constitutive equations as follows

$$\begin{aligned}
\sigma'_{12} &= \sigma'_{21} = 2\mu \, \varepsilon'_{12} \\
\sigma'_{11} &= \frac{1}{1-\nu} \left[\nu\sigma'_{33} + 2\mu(\varepsilon'_{11} + \nu\varepsilon'_{22}) \right] \\
\sigma'_{22} &= \frac{1}{1-\nu} \left[\nu\sigma'_{33} + 2\mu(\varepsilon'_{22} + \nu\sigma'_{11}) \right]
\end{aligned} \tag{67}$$

The procedure is analogous for 2 dimensional problems. The independent components of the stress tensor are obtained from the tractions, i.e.

$$\begin{aligned}
\sigma'_{12} &= \sigma'_{21} = t'_1 \\
\sigma'_{22} &= t'_2
\end{aligned} \tag{68}$$

and the other component needed is computed from the surface displacements as follows

$$\varepsilon'_{11} = u'_{1,1} \tag{69}$$

The stress component for plane strain is then given by

$$\sigma'_{11} = \frac{1}{1-\nu}\left[\nu\sigma'_{22} + 2\mu\varepsilon'_{11}\right] \tag{70}$$

4.4 Treatment of corners

The case of non-smoooth boundaries produces corners in two dimensions and edges or solid angles in three dimensional problems. At these points or lines discontinuity of tractions will occur. This implies that the number of possible unknown tractions at a node on these places will be larger than the number of equations.

To understand better what occurs, consider the case of a two dimensional corner for simplicity (Fig. 9). When the tractions are known at both sides of the corner node, only the two components of the nodal displacements are unknown and no special treatment of the corner node is required. It may also happen that the displacement and one of the tractions either 'before' or 'after' the node are known, then the other traction 'after' or 'before' the node is unknown and the problem can be solved without difficulty. The problem occurs when the values of both tractions (i.e. 'before' and 'after' the node) are unknown and only the displacement is known.

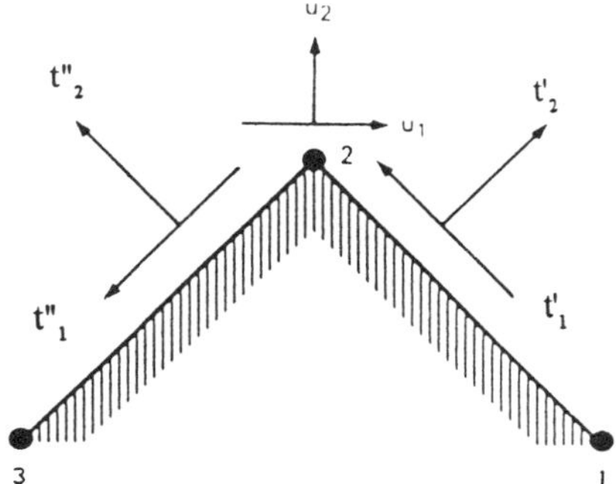

Figure 9: Continuous corner node

The simplest way of solving the problem is then by doubling the corner node. One way of doing this is by slightly modifying the geometry of the corners (Fig. 10). The problem can be solved by the standard procedure. The distance between the two corners must be very small in this case and is limited by the numerical problems which may originate by the existence of two sets of equations whose coefficients are very close to each other. In practice excellent results are obtained if the distances are not too small. When the corner is doubled a small gap may be left between the two nodes, as seen in Fig. 10a or a small element may be assumed between the two nodes, as illustrated in Fig. 10b. In the latter case, the tractions over the small element are assumed to be equal to t'_1 t'_2 and t''_1 t''_2 for nodes $2'$ and $2''$ respectively.

Another interesting version of the double node approach is the use of discontinuous elements. This consists of displacing inside the elements the nodes that meet or that would meet at corners or edges (Fig. 10c). Elements like this require using different interpolation functions than the continuous elements but the approach is very simple and effective and has the added advantage that it can model better corners with high stress concentrations. When used to model singularities discontinuous elements have converged well to the correct solution, while the previous approaches cannot be used in those cases. Discontinuous elements are also important when changing mesh density and to allow for arbitrary mesh densities to meet at edges in three dimensional problems. They make mesh generation with BEM a very simple task. Another advantage of discontinuous elements is that they can be used to increase the mesh densities in a particular region without having to redesign the whole mesh. Their importance in boundary elements seems to have been underestimated and it is still surprising to see that only some commercial codes have a library of discontinuous and semicontinuous as well as continuous elements. They are essential in fracture mechanics to be able to model crack propagation as will be seen in subsequent chapters.

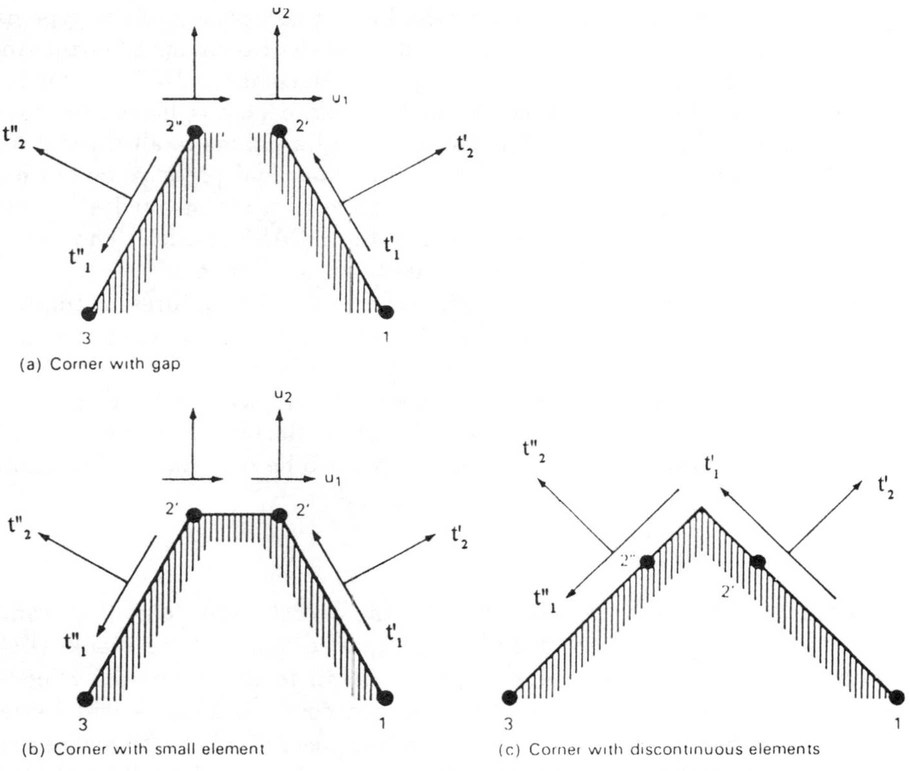

Figure 10: Corner tractions and displacements modelling

5 Applications

The following examples demonstrate the versatility of the boundary element method. The applications are for a series of aerospace and automotive engineering problems which are difficult to solve using other techniques. They are

i) Double lug component analysis

ii) Full crankshaft study

iii) Crankshaft throw

iv) Double lug component with elliptical crack

Example 1

The example shown in Fig. 11 represents a double lug component of the type used in aerospace engineering. The example was solved using the Boundary Element Analysis System called BEASY developed by Computational Mechanics, UK.[6] The lug is made of two separate plates having a common base. Each plate has two holes through which a loading is applied. The base is restrained against all motions in all three directions across the back plane. The specified loading is a sinusoidal pressure across half the inner area of the hole. The pressure can be applied as a set of quadratic functions over the various elements defining that area in the BEASY model. The pressure is considered to be acting in the direction normal to these elements.

The model was divided into five distinct zones or substructures to improve the analysis performance. This substructuring destroys the full character of the boundary element matrices, producing zero submatrices off the diagonal.

The model was discretized using 718 elements which gave 12,018 degrees of freedom. Notice the presence of discontinuous elements. Results in terms of Von Mises surface stresses are shown in Fig. 12 and they appear to be reasonable when compared to other calculations.

Example 2

The model in Fig. 13 has been used to study the overall behaviour of a crankshaft under different loadings. Once this study is completed, parts of the crankshaft, such as an individual throw can be studied in more detail to determine the effect of fillets, oil holes, etc. on the stress distribution and concentration as will be seen in Example 4. The crankshaft shown in the figure has been divided into 8 zones or substructures to improve the computational efficiency of the boundary element solution. The discretization involves 1908 linear elements with 9,813 degrees of freedom. Notice the presence of discontinuous elements and the concentration of them in regions where high stresses are expected. The discontinuity property is an essential feature of boundary elements which adds to the versatility and ease of use of the method. Figure 14 shows the Von Mises surface stresses over part of the crankshaft with some colour areas indicating the high stress concentration existing around the fillets.

Example 3

Figure 15 shows part of a crankshaft or throw being subjected to a shear loading. The interesting thing about the boundary element model shown in the figure is that it allows for the inclusion of an oil duct represented by 32 elements. The view shown in the figure is without some of the foreground surface elements in order to show the cylindrical oil duct. Modelling of this problem using finite elements would be much more difficult as internal meshes will be required. The finite element mesh would need to satisfy continuity throughout and this implies problems with matching different element size grids. Notice that the boundary element mesh shown in the figure presents discontinuity in many regions.

The figure also shows the Von Mises surface stress distribution for the particular load case superimposed on the surface mesh, although this may be difficult to appreciate in a black and white photograph.

The model consisted of 564 linear elements with 3,234 degrees of freedom. Only one zone needs to be considered in this case. The analysis was carried out using the BEASY code and the results were displayed through the IDEAS system.

Example 4

The same double lug component as studied in Example 1 was analysed assuming the existence of an elliptical crack. Figure 16 shows the mesh used for this model. Notice the new mesh at the top right hand side part of the figure. The more detailed mesh near the crack is shown in Fig. 17 and the Von Mises stress distribution near the crack can be seen in Fig. 18. The number of elements in this case is 875 most of them linear with quadratic elements on the crack surface and near the crack tip and the increase over the previous example is due to the need of modelling the crack more accurately.

The boundary element method, because of its high accuracy, is an ideal tool for computing stress intensity factors and because of this has gained rapid acceptability in fracture mechanics. The values of the stress intensity factors in this example were calculated using Irwin's displacement formulae to compute their variation along the crack front.

Results compare well with other solutions obtained using J integral.[7]

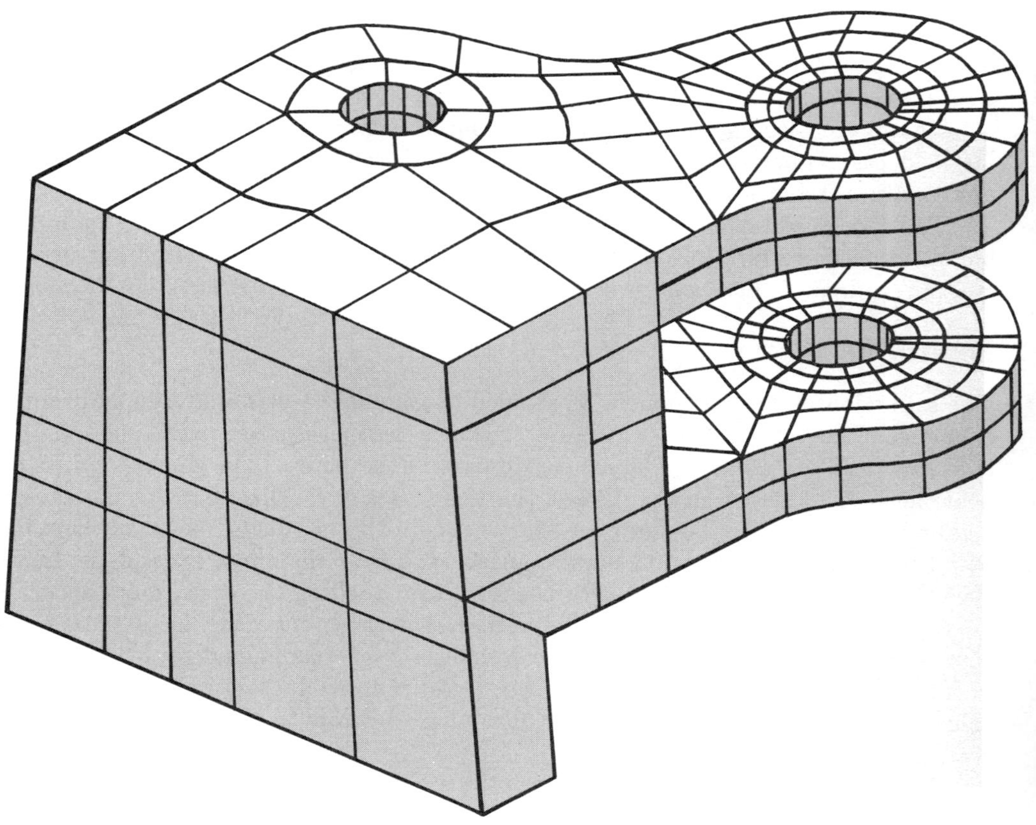

Figure 11: Boundary element model for a double lug component

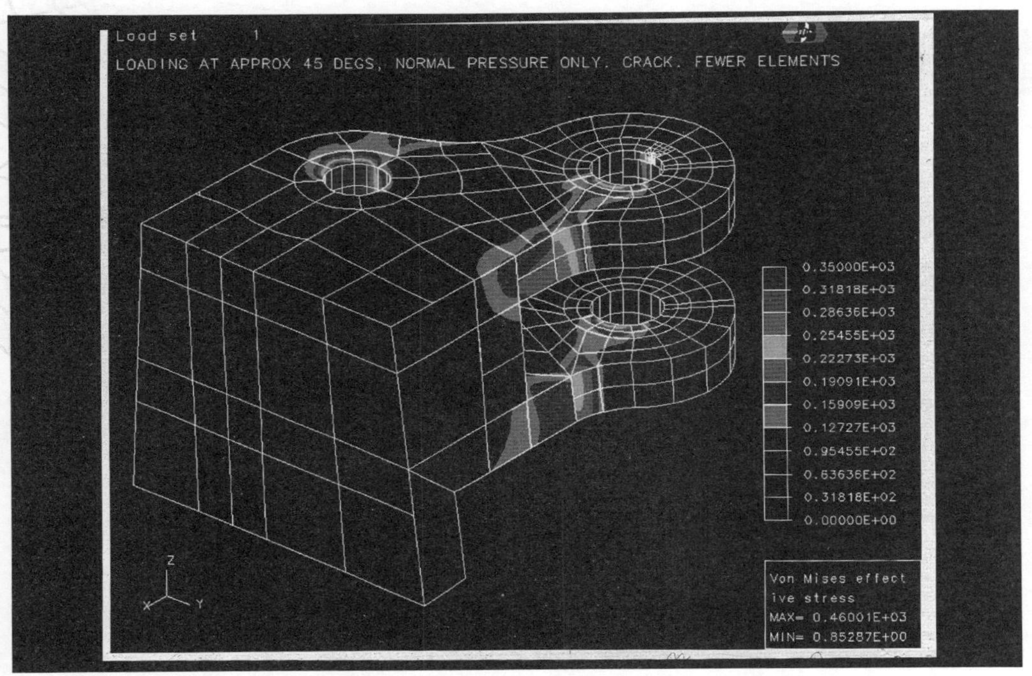

Figure 12: Display of Von Mises stress contours on the double lug component

Figure 13: Crankshaft discretization using boundary elements

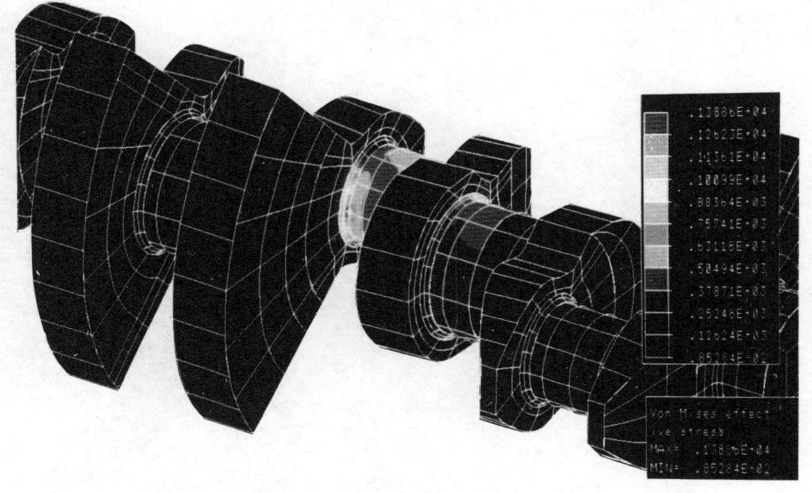

Figure 14: Von Mises surface stresses for part of the crankshaft

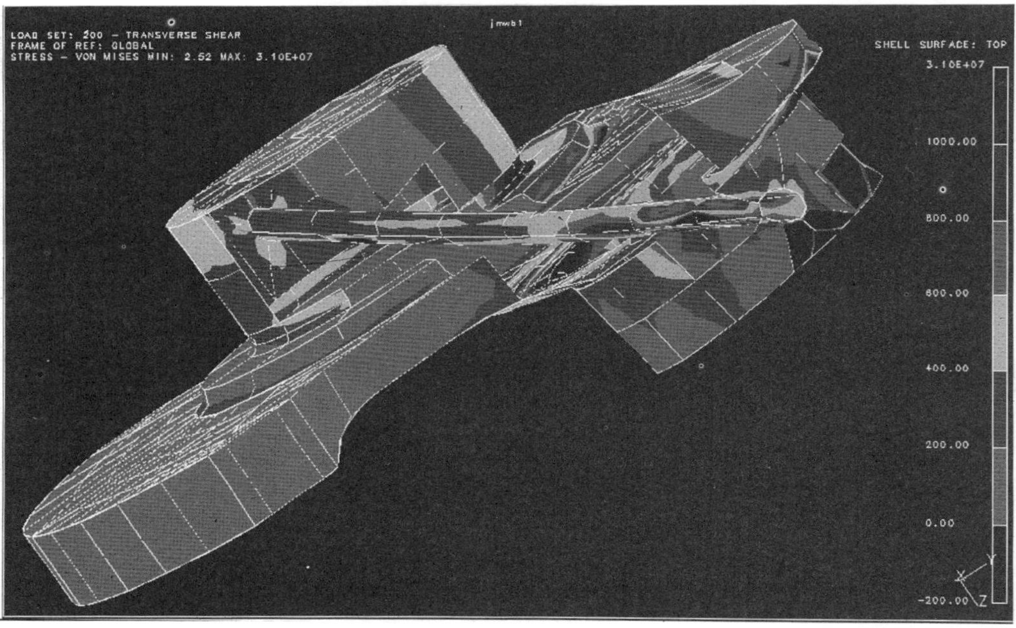

Figure 15: Crankshaft throw with oil duct. Results displayed are Von Mises surface stresses

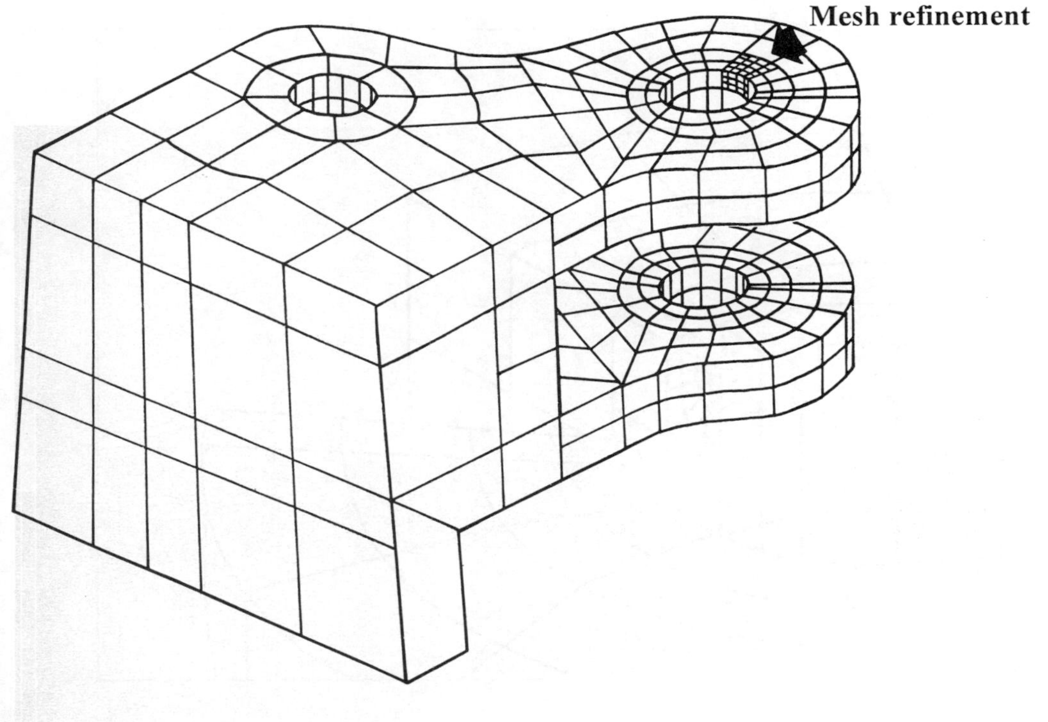

Figure 16: BEASY model for the double lug component showing mesh refinement near the hole with crack.

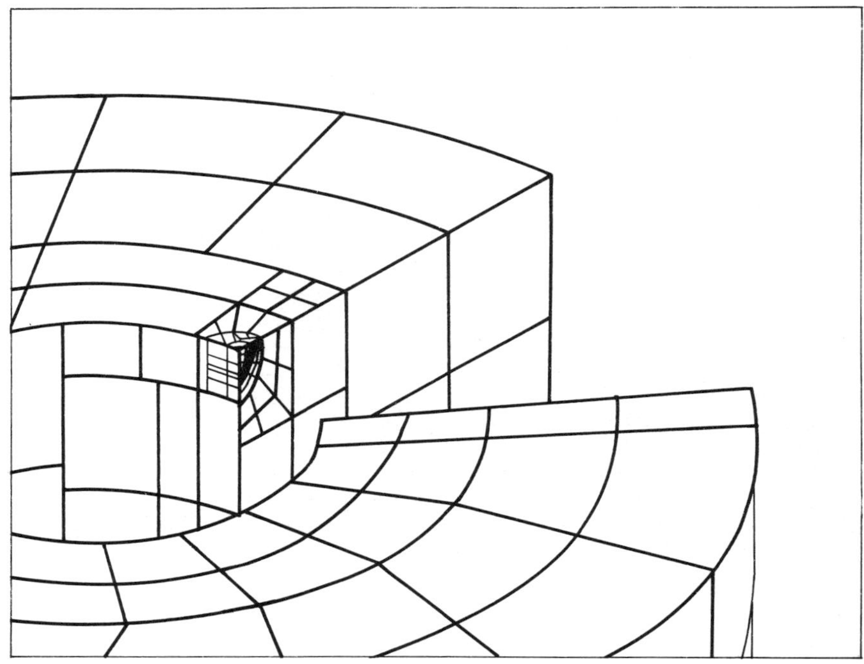

Figure 17: Detailed mesh near the crack

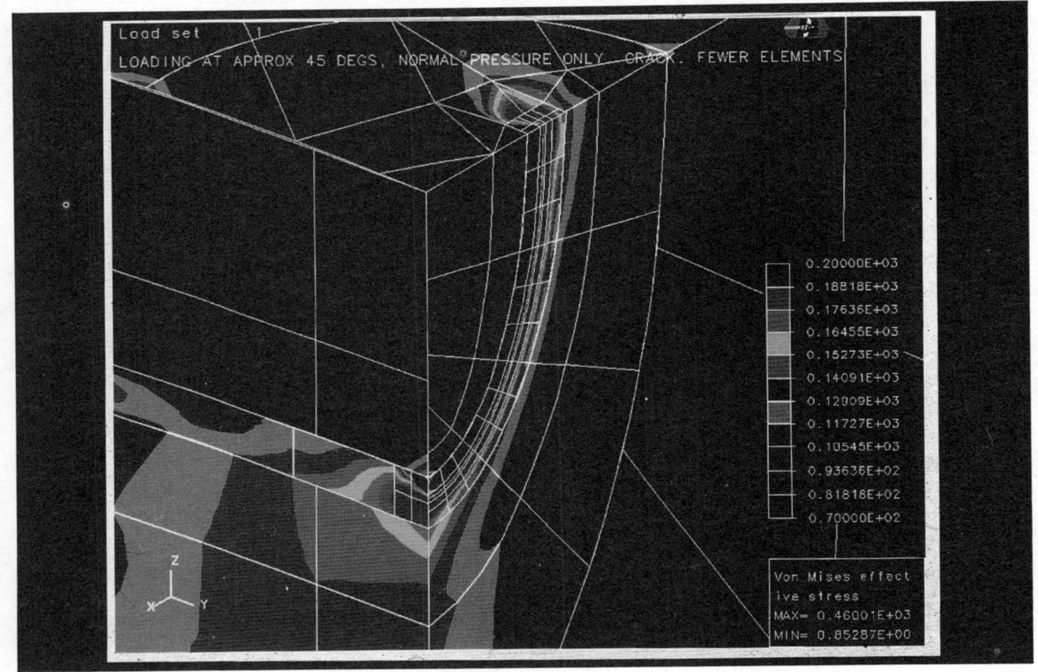

Figure 18: Contour plot of the Von Mises stresses near the crack

6 Conclusions

This chapter provides an introduction to the Boundary Element Method stressing its advantages in fracture mechanics. The technique not only requires a very simple surface discretization but also produces very accurate results. Furthermore the use of discontinuous elements, a property which is unique to boundary elements, permits sudden changes in mesh grading. Discontinuous elements are also an essential tool for crack propagation studies as will be seen in other chapters in this book. Modern techniques allow for following the crack trajectory simply by adding a series of new discontinuous elements without the need to carry out a complete new discretization.

References

1. Brebbia, C.A. *The Boundary Element Method for Engineers*, Pentech Press, London, 1978.

2. Brebbia, C.A., Telles, J.C.F. & Wrobel, L. *Boundary Element Techniques*, Computational Mechanics Publications, Southampton and Boston, 1984.

3. Brebbia, C.A. & Nardini, D. Dynamic analysis of solid mechanics by an alternative boundary element formulation, *Int. Jnl. Soil Dynamics and Earthquake Engineering*, 1983, **2**.

4. Partridge, P., Brebbia, C.A. & Wrobel, L. *The Dual Reciprocity Boundary Element Method*, Computational Mechanics Publications, Southampton and Boston, 1991.

5. Nowak, A.J. & Brebbia, C.A. The multiple reciprocity method. A new approach for transforming BEM domain integrals to the boundary, *Eng. Analysis*, 1989, **6**(3).

6. Trevelyan, J. *BEASY: Self-Teaching Guide*, Computational Mechanics Publications, Southampton and Boston, 1991.

7. Rigby, R.H. & Aliabadi, M.H. Mixed mode *J*-Integral method for analysis of 3D fracture problems using BEM, *Engineering Analysis with Boundary Elements* (to be published).

Chapter 3

Application of the boundary element method to fracture mechanics

M.H. Aliabadi
Wessex Institute of Technology, Ashurst Lodge, Ashurst Southampton SO4 2AA, UK

Introduction

Analytical solutions to problems in fracture mechanics are limited to a small number of idealized situations in which the domain is considered to be infinite, homogeneous and isotropic and the loading relatively simple. However, in practical situations the problems of fracture mechanics occur in flawed structures of finite size with complex loadings. This has frequently led to numerical approaches such as the Finite Element Method (FEM) and the Boundary Element Method (BEM) being used for the solution of these problems.

The boundary element method is now a well established numerical technique for the solution of crack problems in fracture mechanics.[2] In this Chapter the application of the boundary element method to crack problems in fracture mechanics is discussed, with particular emphasis on methods for evaluating stress intensity factors and crack growth analysis.

In section 1 difficulties in modelling co-planar crack surfaces in boundary element method is discussed and special techniques to overcome these difficulties are described. Section 2 presents an overview of some of the techniques used in BEM for evaluating the stress intensity factors. In section 3 the application of the BEM for evaluating weight functions is described. Several examples are presented to demonstrate the efficiency of the weight functions for crack problems.

Section 4 presents the application of the BEM to problems involving non-linear contact, and the effect of the contact forces on the behaviour of cracks. In the final section the application of the dual boundary element method to crack growth analyses in two- and three-dimensional structures are presented.

1 The boundary element method

In this section the fundamental equations of the boundary element method are briefly reviewed, more comprehensive details can be found in Chapter two of this book or in Brebbia & Domínguez[1] and Aliabadi & Rooke.[2]

The boundary integral representation of the displacement components u_i, at an internal source point $\mathbf{X}'$, is given by

$$u_i(\mathbf{X}') = \int_\Gamma U_{ij}(\mathbf{X}', \mathbf{x}) t_j(\mathbf{x}) d\Gamma(\mathbf{x}) - \int_\Gamma T_{ij}(\mathbf{X}', \mathbf{x}) u_j d\Gamma(\mathbf{x}) \tag{1}$$

where $\mathbf{x}', \mathbf{x} \in \Gamma$ and $\mathbf{X}', \mathbf{X} \in \boldsymbol{\Omega}$, and $U_{ij}(\mathbf{X}', \mathbf{x})$ and $T_{ij}(\mathbf{X}', \mathbf{x})$ are the Kelvin fundamental solutions for displacements and tractions respectively. Equation (1) is the well known Somigliana's identity relating the displacements in Ω to the displacements u_j and tractions t_j on boundary Γ. The strain field throughout the body may be obtained by differentiating eqn (1), which leads to the equation

$$\frac{\partial u_i(\mathbf{X}')}{\partial \mathbf{X}'_k} = \int_\Gamma \frac{\partial U_{ij}(\mathbf{X}', \mathbf{x})}{\partial \mathbf{X}'_k} t_j(\mathbf{x}) d\Gamma(\mathbf{x}) - \int_\Gamma \frac{\partial T_{ij}(\mathbf{X}', \mathbf{x})}{\partial \mathbf{X}'_k} u_j(\mathbf{x}) d\Gamma(\mathbf{x}) \tag{2}$$

Substituting eqn (2) into Hooke's law, that is

$$\sigma_{ij} = \lambda \delta_{ij} u_{k,k} + \mu(u_{i,j} + u_{j,i})$$

yields the Somigliana's identity for stresses at an interior point $\mathbf{X}'$:

$$\begin{aligned} \sigma_{ij}(\mathbf{X}') = & \int_\Gamma [\lambda \delta_{ij} U_{mk,m} + \mu(U_{ik,j} + U_{jk,i})] t_k(\mathbf{x}) d\Gamma(\mathbf{x}) + \\ & \int_\Gamma [\lambda \delta_{ij} T_{mk,m} + \mu(T_{ik,j} + T_{jk,i})] u_k(\mathbf{x}) d\Gamma(\mathbf{x}) + \\ & \int_\Omega [\lambda \delta_{ij} U_{mk,m} + \mu(U_{ik,j} + U_{jk,i})] b_k(\mathbf{X}) d\Omega(\mathbf{X}) \end{aligned} \tag{3}$$

where μ is the shear modulus of elasticity and λ is the Lamè constant. Equation (3) may be written in a more compact form as

$$\sigma_{ij} = \int_\Gamma D_{kij}(\mathbf{X}', \mathbf{x}) t_k(\mathbf{x}) d\Gamma(\mathbf{x}) - \int_\Gamma S_{kij}(\mathbf{X}', \mathbf{x}) u_k(\mathbf{x}) d\Gamma(\mathbf{x}) \tag{4}$$

where D_{kij} contains several derivatives of U_{ij}, and S_{kij} several derivatives of T_{ij} together with elastic constants.

The direct boundary element formulation relating the boundary displacements to boundary tractions can be obtained from eqn (1) by considering the limiting process as an internal point goes to the boundary (i.e. $\mathbf{X}' \rightarrow \mathbf{x}'$). It can be written as

$$C_{ij}(\mathbf{x}')u_j(\mathbf{x}') = \int_\Gamma U_{ij}(\mathbf{x}',\mathbf{x})\mathbf{t}_j(\mathbf{x})\mathrm{d}\Gamma(\mathbf{x}) - \int_\Gamma T_{ij}(\mathbf{x}',\mathbf{x})u_j(\mathbf{x})\mathrm{d}\Gamma(\mathbf{x}) \qquad (5)$$

where C_{ij} can be obtained from consideration of rigid body translation.

1.1 Numerical discretization

To solve the boundary element formulation in eqn (5) analytically is not generally possible. Therefore a numerical solution is required. The boundary Γ is divided into N elements, so that eqn (5) becomes

$$C_{ij}(\mathbf{x}')u_j(\mathbf{x}') = \sum_{n=1}^{N}\int_{\Gamma_n} U_{ij}(\mathbf{x}',\mathbf{x})\mathbf{t}_j(\mathbf{x})\mathrm{d}\Gamma_n(\mathbf{x}) - \sum_{n=1}^{N}\int_{\Gamma_n} T_{ij}(\mathbf{x}',\mathbf{x})u_j(\mathbf{x})\mathrm{d}\Gamma_n(\mathbf{x}) \qquad (6)$$

where

$$\Gamma = \sum_{n=1}^{N}\Gamma_n$$

On each element, the boundary parameter $\mathbf{x}$ (components x_j), the unknown displacement field $u_j(\mathbf{x})$ and the traction field $t_j(\mathbf{x})$ are approximated using interpolation functions, in the following manner:

$$\mathbf{x} = \sum_{\alpha=1}^{m} M^\alpha \mathbf{x}^\alpha \qquad (7)$$

$$\mathbf{u} = \sum_{\alpha=1}^{m} M^\alpha \mathbf{u}^\alpha$$

$$\mathbf{t} = \sum_{\alpha=1}^{m} M^\alpha \mathbf{t}^\alpha \qquad (8)$$

Shape functions M^α are polynomials of degree $m-1$ and have the property that they are equal to 1 at node α and 0 at all other nodes; $\mathbf{x}^\alpha, \mathbf{u}^\alpha$ and $\mathbf{t}^\alpha$ are the values of the function at node α.

For two-dimensional boundary element formulation the quadratic continuous shape functions are given as

$$\begin{aligned} M^1 &= \frac{1}{2}\xi(\xi-1) \\ M^2 &= (1-\xi^2) \\ M^3 &= \frac{1}{2}\xi(\xi+1) \end{aligned} \qquad (9)$$

where $-1 \leq \xi \leq 1$. For three-dimensional boundary element formulation, the shape functions for the quadratic 8-noded element are given as

$$
\begin{aligned}
M^\alpha &= \frac{1}{4}(1+\xi_\alpha\xi)(1+\eta_\alpha\eta)(\xi_\alpha\xi+\eta_\alpha\eta-1), \text{ for nodes at } \xi_\alpha=\eta_\alpha=\pm 1\\
M^\alpha &= \frac{1}{2}(1-\xi^2)(1+\eta_\alpha\eta), \text{ for nodes at } \xi_\alpha=0, \eta_\alpha=\pm 1\\
M^\alpha &= \frac{1}{2}(1+\xi_\alpha\xi)(1-\eta^2), \text{ for nodes at } \xi_\alpha=\pm 1, \eta_\alpha=0
\end{aligned} \tag{10}
$$

A discretized boundary element formulation can be obtained by substituting the expression in (9) or (10), into the integral eqn (6), to obtain

$$C_{ij}(\mathbf{x}')u_j(\mathbf{x}') + \sum_{n=1}^{N}\sum_{\alpha=1}^{m} P_{ij}^{n\alpha}u_j^{n\alpha} = \sum_{n=1}^{N}\sum_{\alpha=1}^{m} Q_{ij}^{n\alpha}t_j^{n\alpha} \tag{11}$$

The coefficients $P_{ij}^{n\alpha}$ and $Q_{ij}^{n\alpha}$ are defined in terms of integrals over Γ_n where $\mathrm{d}\Gamma_n(\xi)$ becomes $J^n(\xi)\mathrm{d}\xi$; that is

$$P_{ij}^{n\alpha} = \int_{-1}^{+1} M^\alpha(\xi)T_{ij}[\mathbf{x}',\mathbf{x}(\xi)]J^n(\xi)\mathrm{d}\xi$$

and

$$Q_{ij}^{n\alpha} = \int_{-1}^{+1} M^\alpha(\xi)U_{ij}[\mathbf{x}',\mathbf{x}(\xi)]J^n(\xi)\mathrm{d}\xi \tag{12}$$

for two-dimensional problems. Similarly for three-dimensional problems, $\mathrm{d}\Gamma_n(\xi,\eta)$ becomes $J^n(\xi,\eta)\mathrm{d}\xi\mathrm{d}\eta$; that is

$$P_{ij}^{n\alpha} = \int_{-1}^{+1} M^\alpha(\xi,\eta)T_{ij}[\mathbf{x}',\mathbf{x}(\xi,\eta)]J^n(\xi,\eta)\mathrm{d}\xi\mathrm{d}\eta$$

and

$$Q_{ij}^{n\alpha} = \int_{-1}^{+1} M^\alpha(\xi,\eta)U_{ij}[\mathbf{x}',\mathbf{x}(\xi,\eta)]J^n(\xi,\eta)\mathrm{d}\xi\mathrm{d}\eta \tag{13}$$

The most straightforward method of solution for integral eqn (11) is point collocation; eqn (11) is evaluated at nodal points $\mathbf{x}^\beta, \beta = 1, M$ (where M is the total number of nodes) to give

$$C_{ij}(\mathbf{x}^\beta)u_j(\mathbf{x}^\beta) + \sum_{n=1}^{N}\sum_{\alpha=1}^{m} P_{ij}^{n\alpha}(\mathbf{x}^\beta)u_j^{n\alpha} = \sum_{n=1}^{N}\sum_{\alpha=1}^{m} Q_{ij}^{n\alpha}(\mathbf{x}^\beta)t_j^{n\alpha}, \qquad \beta = 1, M \tag{14}$$

The double sum in (14) must be evaluated bearing in mind that some nodes are shared between elements and since the displacement values $u_j^{n\alpha}$ are uniquely defined at these nodes, they can be combined to give a sum over all nodes; so eqn (14) can be written as

$$C_{ij}(\mathbf{x}^\beta)u_j(\mathbf{x}^\beta) + \sum_{\gamma=1}^{M} H_{ij}^{\beta\gamma} u_j^\gamma = \sum_{n=1}^{N}\sum_{\alpha=1}^{m} G_{ij}^{\beta n\alpha} t_j^{n\alpha} \qquad \beta = 1, M \tag{15}$$

where $H_{ij}^{\beta\gamma}$ is made up from $P_{ij}^{n\alpha}(\mathbf{x}^\beta)$ and $G_{ij}^{\beta n\alpha}$ is equal to $Q_{ij}^{n\alpha}(\mathbf{x}^\beta)$. Collecting all the displacement unknown in (15) gives

$$\sum_{\gamma=1}^{M} \bar{H}_{ij}^{\beta\gamma} u_j^\gamma = \sum_{n=1}^{N}\sum_{\alpha=1}^{m} G_{ij}^{\beta n\alpha} t_j^{n\alpha} \tag{16}$$

where $\bar{H}_{ij}^{\beta\gamma} = C_{ij}(\mathbf{x}^\beta)\delta_{\beta\gamma} + H_{ij}^{\beta\gamma}$ and $\delta_{\beta\gamma}$ is the Kronecker delta function. The discretized boundary element equation may now be written in a matrix form as

$$[H]\{u\} = [G]\{t\} \tag{17}$$

where $[H]$ is a $2M \times 2M$ matrix and $[G]$ is $2M \times 2Nm$ matrix containing known integrals of the product of the shape functions, the Jacobian and the fundamental fields T_{ij} and U_{ij} respectively. The vector $\{u\}$ has $2M$ components and $\{t\}$ has $2Nm$ components; both contain field unknowns and prescribed boundary conditions. From (15) it can be seen that the diagonal terms which contain C_{ij} can be evaluated directly from the consideration of the rigid body motion; this leads to

$$C_{ij}(\mathbf{x}^\beta) = -\sum_{\gamma=1}^{M} H_{ij}^{\beta\gamma} \tag{18}$$

Thus the diagonal terms in $H_{ij}^{\beta\gamma}$ can be evaluated since eqn (18) can be rewritten as

$$C_{ij}(\mathbf{x}^\beta) + H_{ij}^{\beta\beta} = -\sum_{\gamma=1(\beta\neq\gamma)}^{M} H_{ij}^{\beta\gamma}$$

that is

$$\bar{H}_{ij}^{\beta\beta} = -\sum_{\gamma=1(\beta\neq\gamma)}^{M} \bar{H}_{ij}^{\beta\gamma} \tag{19}$$

After substitution of the prescribed boundary conditions, the resulting system of algebraic equations may be written as

$$[A]\{X\} = [B]\{Y\} = \{F\} \tag{20}$$

the vector $\{X\}$ contains all the unknown boundary displacements or tractions, $[A]$ is the coefficient matrix which is usually non-symmetric and densely populated and $[B]$ is a matrix which contains the coefficients corresponding to the prescribed boundary conditions $\{Y\}$.

1.2 Difficulties in crack modelling

The mathematical difficulties resulting from the application of the standard boundary element formulation (5) to modelling crack problems have been outlined by Cruse[3] and Aliabadi and Rooke.[2] It is recalled that Somigliana's identity for displacements at an interior point $\mathbf{X}'$, is given by

$$u_i(\mathbf{x}') = \int_{\Gamma+\Gamma_c^+ +\Gamma_c^-} U_{ij}(\mathbf{X}',\mathbf{x})t_j(\mathbf{x})d\Gamma(\mathbf{x}) - \int_{\Gamma+\Gamma_c^+ +\Gamma_c^-} T_{ij}(\mathbf{X}',\mathbf{x})u_j(\mathbf{x})d\Gamma(\mathbf{x}) \qquad (21)$$

where Γ_c^+ and Γ_c^- represent the upper and lower crack surfaces and Γ represents the remaining boundary, as shown in Fig. 1. The displacement kernel U_{ij} and traction kernel T_{ij} for points on the upper $\mathbf{x}^+$ and lower $\mathbf{x}^-$ crack surfaces have the property that

$$U_{ij}(\mathbf{X}',\mathbf{x}^+) = +U_{ij}(\mathbf{X}',\mathbf{x}^-)$$

and

$$T_{ij}(\mathbf{X}',\mathbf{x}^+) = -T_{ij}(\mathbf{X}',\mathbf{x}^-) \qquad (22)$$

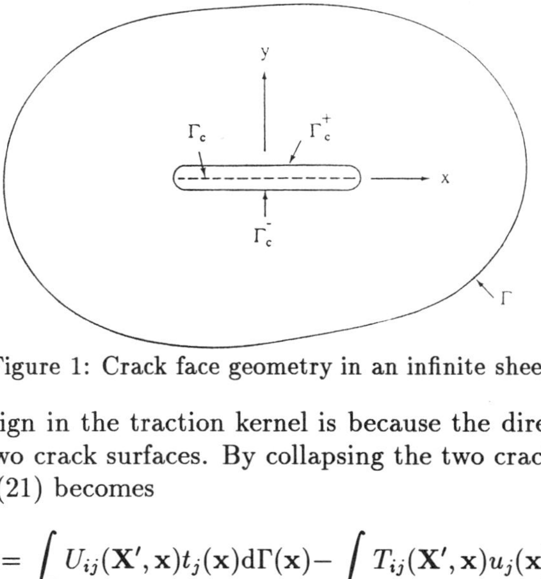

Figure 1: Crack face geometry in an infinite sheet

The change in sign in the traction kernel is because the direction of the normal is opposite on the two crack surfaces. By collapsing the two crack surfaces such that $\Gamma_c = \Gamma_c^+ = \Gamma_c^-$, eqn (21) becomes

$$u_i(\mathbf{X}') = \int_{\Gamma} U_{ij}(\mathbf{X}',\mathbf{x})t_j(\mathbf{x})d\Gamma(\mathbf{x}) - \int_{\Gamma} T_{ij}(\mathbf{X}',\mathbf{x})u_j(\mathbf{x})d\Gamma(\mathbf{x})$$

$$+ \int_{\Gamma_c} U_{ij}(\mathbf{X}',\mathbf{x}) \sum t_j(\mathbf{x})d\Gamma(\mathbf{x}) - \int_{\Gamma_c} T_{ij}(\mathbf{X}',\mathbf{x})\Delta u_j(\mathbf{x})d\Gamma(\mathbf{x}) \qquad (23)$$

where

$$\Delta u_j = u_j^+ - u_j^-$$

and

$$\sum t_j = t_j^+ + t_j^-$$

For traction free cracks or when the crack is loaded by equal and opposite tractions $\sum t_j = 0$.

Following the usual BEM derivation, let $\mathbf{X}' \rightarrow \mathbf{x}'$ on the boundary Γ_c, therefore eqn (23) for traction-free crack becomes

$$u_j(\mathbf{x}') - \frac{\Delta u_j(\mathbf{x}')}{2} = \int_\Gamma U_{ij}(\mathbf{x}', \mathbf{x}) t_j(\mathbf{x}) d\Gamma(\mathbf{x}) - \int_\Gamma T_{ij}(\mathbf{x}', \mathbf{x}) u_j(\mathbf{x}) d\Gamma(\mathbf{x}) + \int_{\Gamma_c} T_{ij}(\mathbf{x}', \mathbf{x}) \Delta u_j(\mathbf{x}) d\Gamma(\mathbf{x}) \quad (24)$$

Equation (24) has two serious deficiencies: i)any set of equal and opposite tractions on the crack surface Γ_c will result in the same equation since $\sum t_j = 0$; and ii) there are now two unknown displacement variables on Γ_c, namely Δu_j and u_j.

1.3 Multi-domain formulation

To overcome these problems Blandford *et al.*[4] divided the cracked body into two regions, as shown in Fig. 2, with each region containing one crack surface. The two regions are joined together such that equilibrium of tractions and compatibility of displacements are enforced.

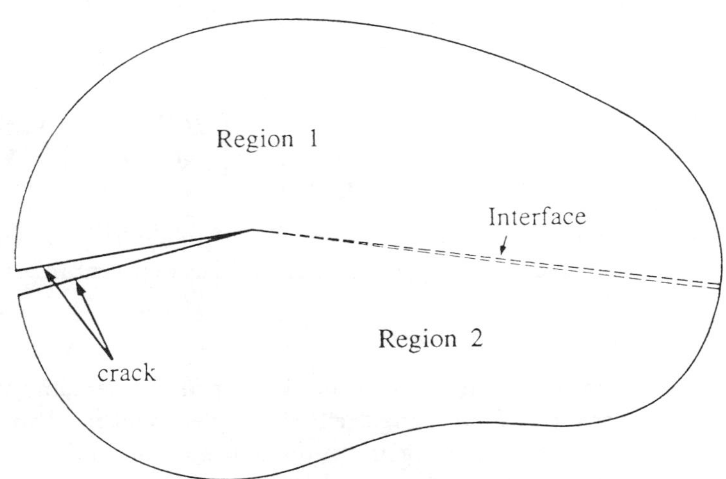

Figure 2: Partitioning of crack geometry

Consider a body partitioned into two sub-domains which have boundaries $\Gamma^{(1)}$ and $\Gamma^{(2)}$, each containing a crack surface. If $\mathbf{u}_z^{(i)}$ and $\mathbf{t}_z^{(i)}$ are displacements and tractions respectively on the interfacial boundaries $\Gamma_z^{(i)}$ for $i = 1, 2$, they must satisfy the following continuity and equilibrium conditions:

$$\mathbf{u}_z^{(1)} = \mathbf{u}_z^{(2)} \qquad (\equiv \mathbf{u}_z)$$

$$\mathbf{t}_z^{(1)} = -\mathbf{t}_z^{(2)} \qquad (\equiv \mathbf{t}_z) \tag{25}$$

where $\mathbf{u}_z$ and $\mathbf{t}_z$ are both unknowns on the interface z. The boundary element representation (17) is modified as follows:
for **region 1**

$$\begin{bmatrix} H^{(1)} & H_z^{(1)} \end{bmatrix} \begin{Bmatrix} \mathbf{u}^{(1)} \\ \mathbf{u}_z^{(1)} \end{Bmatrix} = \begin{bmatrix} G^{(1)} & G_z^{(1)} \end{bmatrix} \begin{Bmatrix} \mathbf{t}^{(1)} \\ \mathbf{t}_z^{(1)} \end{Bmatrix} \tag{26}$$

and for **region 2**

$$\begin{bmatrix} H^{(2)} & H_z^{(2)} \end{bmatrix} \begin{Bmatrix} \mathbf{u}^{(2)} \\ \mathbf{u}_z^{(2)} \end{Bmatrix} = \begin{bmatrix} G^{(2)} & G_z^{(2)} \end{bmatrix} \begin{Bmatrix} \mathbf{t}^{(2)} \\ \mathbf{t}_z^{(2)} \end{Bmatrix} \tag{27}$$

where $H_z^{(i)}$ and $G_z^{(i)}$ are the matrix coefficient entries corresponding to nodes on the interface z. Combining the coefficient matrices for region (1) and region (2) and enforcing the boundary conditions in (25), gives

$$\begin{bmatrix} H^{(1)} & H_z^{(1)} & -G_z^{(1)} & 0 \\ 0 & H_z^{(2)} & -G_z^{(2)} & H^{(2)} \end{bmatrix} \begin{Bmatrix} \mathbf{u}^{(1)} \\ \mathbf{u}_z \\ \mathbf{t}_z \\ \mathbf{u}^{(2)} \end{Bmatrix} = \begin{bmatrix} G^{(1)} & 0 \\ 0 & G^{(2)} \end{bmatrix} \begin{Bmatrix} \mathbf{t}^{(1)} \\ \mathbf{t}^{(2)} \end{Bmatrix} \tag{28}$$

Finally after substitution of the boundary conditions, the resulting system of equations can be written as

$$\begin{bmatrix} A^{(1)} & H_z^{(1)} & -G_z^{(1)} & 0 \\ 0 & H_z^{(2)} & -G_z^{(2)} & A^{(2)} \end{bmatrix} \begin{Bmatrix} \mathbf{X}^{(1)} \\ \mathbf{u}_z \\ \mathbf{t}_z \\ \mathbf{X}^{(2)} \end{Bmatrix} = \begin{bmatrix} B^{(1)} & 0 \\ 0 & B^{(2)} \end{bmatrix} \begin{Bmatrix} \mathbf{Y}^{(1)} \\ \mathbf{Y}^{(2)} \end{Bmatrix} \tag{29}$$

The above procedure overcomes the difficulty of modelling co-planar crack surfaces in BEM, and has been extensively used in the BEM literature.

1.4 The dual boundary element method

As stated above special techniques have been developed to overcome the difficulty of modelling co-planar crack surfaces using the boundary element method. Among these, the most general are the subregion or multi-domain method and the dual boundary element method. The multi-domain method introduces artificial boundaries into the structure, which connect the cracks to the boundary, in such a way that the domain is divided into subregions without cracks. The main disadvantage of this method is that the introduction of artificial boundaries is not unique, and this cannot be easily implemented as an automatic procedure in an incremental analysis of crack extension problems. In addition, the method generates a larger system of algebraic equations than is strictly required.

The dual boundary element method (DBEM) incorporates two independent boundary integral equations, with the displacement integral eqn (5) applied for collocation on one of the crack surfaces and the traction equation applied for collocation on the other. As a consequence, general mixed-mode crack problems can be solved in a single region formulation. Although, the integration path is still the same for coincident points on the crack surfaces, the respective boundary integral equations are not distinct.

The dual equations, on which DBEM is based, are the displacement

$$C_{ij}(\mathbf{x}')u_j(\mathbf{x}') = \int_\Gamma U_{ij}(\mathbf{x}',\mathbf{x})t_j(\mathbf{x})\mathrm{d}\Gamma(\mathbf{x}) - \int_\Gamma T_{ij}(\mathbf{x}',\mathbf{x})u_j(\mathbf{x})\mathrm{d}\Gamma(\mathbf{x}) \tag{30}$$

and the traction equation

$$\frac{1}{2}t_j(\mathbf{x}') = n_i(\mathbf{x}')\int_\Gamma D_{ijk}(\mathbf{x}',\mathbf{x})t_k(\mathbf{x})\mathrm{d}\Gamma(\mathbf{x}) - n_i(\mathbf{x}')\int_\Gamma S_{ijk}(\mathbf{x}',\mathbf{x})u_k(\mathbf{x})\mathrm{d}\Gamma(\mathbf{x}) \tag{31}$$

An efficient implementation of the dual boundary element method has been developed by Portela, Aliabadi and Rooke[5] for two-dimensional problems and Mi and Aliabadi[6] for three-dimensional problems. A more detailed description of the dual boundary element formulation will be presented in section 5.

2 Methods for evaluating the stress intensity factors

There have been several methods developed for evalauting the stress intensity factors using BEM analysis. The most important ones are the quarter-point elements, the J-integral method and the singularity subtraction technique. In this section these methods are described and test examples are presented to demonstrate their accuracy.

Quarter-point crack-tip elements

Williams[9] has shown that the stress fields are singular $O(\frac{1}{\sqrt{r}})$ at the crack tip and the displacement fields varies as $O(\sqrt{r})$. Therefore modelling crack regions with the usual shape functions, which may allow for polynomial variation only, does not lead to accurate solutions.

Henshell and Shaw[7] and Barsoum[8] have shown that by moving the mid-side node of a quadratic element to a quarter point position as shown in Fig. 3, the desired $\sqrt{r}$ variation for displacement can be achieved.

Let $x^1 = 0, x^2 = \frac{l}{4}$ and $x^3 = l$ (for $y = 0$) where l is the length of the element; from representation (9), with the Lagrangian shape functions, the coordinate x becomes

$$x = \frac{1}{2}\xi(1+\xi)l + (1-\xi^2)\frac{l}{4}$$

so that

$$\xi = -1 + 2\sqrt{\frac{x}{l}} = -1 + 2\sqrt{\frac{r}{l}} \tag{32}$$

Substitution for ξ in (8) gives the following displacement dependence on r,

$$\mathbf{u} = \mathbf{u}^1 + (-3\mathbf{u}^1 + 4\mathbf{u}^2 - \mathbf{u}^3)\sqrt{\frac{r}{l}} + 2(\mathbf{u}^1 - 2\mathbf{u}^2 + \mathbf{u}^3)\left(\frac{r}{l}\right) \tag{33}$$

The Jacobian of the transformation is $J(\xi) = \sqrt{lr}$, which vanishes as $r \to 0$. For a singular traction element the shape functions are divided by $\sqrt{\frac{r}{l}}$, to give

$$\mathbf{t} = \mathbf{t}^1\sqrt{\frac{l}{r}} + (-3\mathbf{t}^1 + 4\mathbf{t}^2 - \mathbf{t}^3) + 2(\mathbf{t}^1 - 2\mathbf{t}^2 + \mathbf{t}^3)\sqrt{\frac{r}{l}} \tag{34}$$

where $\mathbf{t}^1, \mathbf{t}^2$ and $\mathbf{t}^3$ correspond to the value of the traction $\mathbf{t}$ at the nodes, with $\mathbf{t}^1$ representing the values of the traction at the crack tip.

Stress intensity factor computation Stress intensity factors may be computed from conventional and/or special crack tip elements in a number of different ways, by equating the boundary element solutions to the theoretical expressions for displacements or stresses near the crack tip. Although, in this section the one-dimensional element representation is considered, the stress intensity factor formulae which will

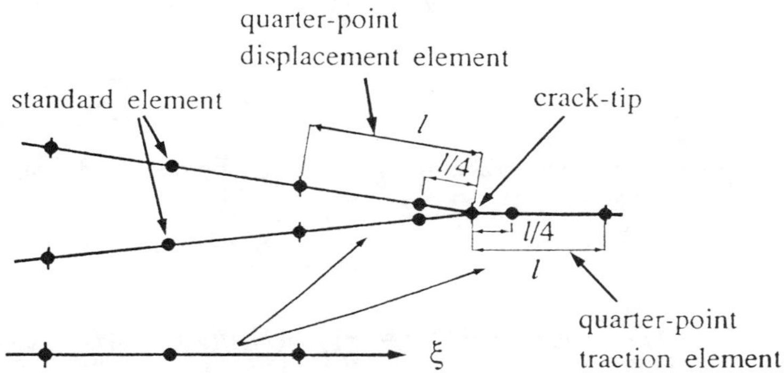

Figure 3: Quarter-point crack-tip elements

be represented are also valid for two-dimensional elements described previously, by simply considering displacement or stress values on planes normal to the crack front.

The displacement fields near the crack tip are given by Williams[9] as

$$
\begin{aligned}
u_1 &= \frac{K_I}{\mu}\sqrt{\frac{r}{2\pi}}\cos\frac{\theta}{2}\left[\frac{1}{2}(\kappa-1)+\sin^2\frac{\theta}{2}\right]+ \\
&+\frac{K_{II}}{\mu}\sqrt{\frac{r}{2\pi}}\sin\frac{\theta}{2}\left[\frac{1}{2}(\kappa+1)+\cos^2\frac{\theta}{2}\right] \\
u_2 &= \frac{K_I}{\mu}\sqrt{\frac{r}{2\pi}}\sin\frac{\theta}{2}\left[\frac{1}{2}(\kappa+1)-\cos^2\frac{\theta}{2}\right]+ \\
&+\frac{K_{II}}{\mu}\sqrt{\frac{r}{2\pi}}\cos\frac{\theta}{2}\left[\frac{1}{2}(1-\kappa)+\sin^2\frac{\theta}{2}\right]
\end{aligned}
\tag{35}
$$

where (r,θ) are the polar coordinates centred at the crack tip, $\mu = E/2(1+\nu)$ is the shear modulus of elasticity and $\kappa = 3-4\nu$ for plane strain and $\kappa = (3-\nu)/(1+\nu)$ for plane stress.

Similarly the near tip stress fields are given as

$$
\begin{aligned}
\sigma_{11} &= \frac{K_I}{\sqrt{2\pi r}}\cos\frac{\theta}{2}\left(1-\sin\frac{\theta}{2}\sin\frac{3\theta}{2}\right)-\frac{K_{II}}{\sqrt{2\pi r}}\sin\frac{\theta}{2}\left(2+\cos\frac{\theta}{2}\cos\frac{3\theta}{2}\right) \\
\sigma_{22} &= \frac{K_I}{\sqrt{2\pi r}}\cos\frac{\theta}{2}\left(1+\sin\frac{\theta}{2}\sin\frac{3\theta}{2}\right)+\frac{K_{II}}{\sqrt{2\pi r}}\sin\frac{\theta}{2}\cos\frac{\theta}{2}\cos\frac{3\theta}{2} \\
\sigma_{12} &= \frac{K_I}{\sqrt{2\pi r}}\cos\frac{\theta}{2}\sin\frac{\theta}{2}\cos\frac{3\theta}{2}+\frac{K_{II}}{\sqrt{2\pi r}}\sin\frac{\theta}{2}\left(1-\sin\frac{\theta}{2}\sin\frac{3\theta}{2}\right)
\end{aligned}
\tag{36}
$$

The displacement fields on the crack surfaces can be written from (35) as

$$u_2(\theta = \pi) - u_2(\theta = -\pi) = \frac{\kappa + 1}{\mu} K_I \sqrt{\frac{r}{2\pi}}$$

$$u_1(\theta = \pi) - u_1(\theta = -\pi) = \frac{\kappa + 1}{\mu} K_{II} \sqrt{\frac{r}{2\pi}} \tag{37}$$

The tractions ahead of the crack along $\theta = 0$ can be obtained from (36) as

$$t_1 = (\sigma_{11} n_1 + \sigma_{12} n_2) = \frac{1}{\sqrt{2\pi r}} (K_I n_1 + K_{II} n_2)$$

$$t_2 = (\sigma_{12} n_1 + \sigma_{22} n_2) = \frac{1}{\sqrt{2\pi r}} (K_{II} n_1 + K_I n_2) \tag{38}$$

Different techniques commonly used to evaluate the stress intensity factors are a) one-point and two-point displacement formulae, and b) traction formulae. These techniques are described below for a quarter point element shown in Fig. 4. The point A denotes the crack tip and the points B and D, are on the opposite faces of the crack to C and E (i.e. $\theta = \pm\pi$); the points F and G are located ahead of the crack-tip along $\theta = 0$. For quarter-point elements the points B,C and F are placed at quarter-point position near the crack-tip (i.e. $r = \frac{l}{4}$).

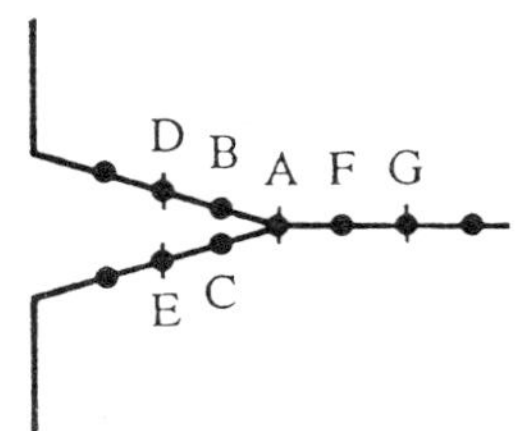

Figure 4: Nodes in the vicinity of the crack-tip

a) One-point and two-point displacement formulae The so-called one-point formula, is simply taking the value of the nodes closest to the crack tip and equating the displacements to the theoretical expressions in (37), to give

$$K_I = \frac{2\mu}{\kappa + 1} \sqrt{\frac{2\pi}{l}} (u_2^B - u_2^C)$$

and

$$K_{II} = \frac{2\mu}{\kappa + 1} \sqrt{\frac{2\pi}{l}} (u_1^B - u_1^C) \tag{39}$$

since $r = \frac{l}{4}$.

The stress intensity factors can also be obtained by considering the displacements over the whole element. This procedure was first suggested by Blandford *et al.*[4]

and is called a two-point formula by Smith and Mason.[10] For the shape function approximation in (33), K_I and K_{II} become

$$K_I = \frac{\mu}{\kappa+1}\sqrt{\frac{2\pi}{l}}\,[4(u_2^B - u_2^C) - (u_2^D - u_2^E)]$$

and

$$K_{II} = \frac{\mu}{\kappa+1}\sqrt{\frac{2\pi}{l}}\,[4(u_1^B - u_1^C) - (u_1^D - u_1^E)] \tag{40}$$

b) Traction formulae An alternative way for computing the stress intensity factors has been suggested by Martínez and Domínguez,[11] who used the nodal values of 'tractions' at the tip. The stress distribution ahead of the crack along $\theta = 0$, can be written as

$$\sigma_{11} = \sigma_{22} = \frac{K_I}{\sqrt{2\pi r}} \tag{41}$$

for the normal stresses, and

$$\sigma_{12} = \frac{K_{II}}{\sqrt{2\pi r}} \tag{42}$$

for the shear stress. Martínez and Domínguez[11] showed that for the singular traction representation in (34), the 'traction' at A is given by

$$t_1^A = \lim_{r\to 0} t_1\sqrt{\frac{r}{l}} = \lim_{r\to 0}(\sigma_{11}n_1 + \sigma_{12}n_2)\sqrt{\frac{r}{l}} \tag{43}$$

which from (38) becomes

$$t_1^A = \frac{1}{\sqrt{2\pi l}}(K_I n_1 + K_{II} n_2) \tag{44}$$

similarly

$$t_2^A = \lim_{r\to 0} t_2\sqrt{\frac{r}{l}} = \lim_{r\to 0}(\sigma_{12}n_1 + \sigma_{22}n_2)\sqrt{\frac{r}{l}} =$$

$$= \frac{1}{\sqrt{2\pi l}}(K_{II} n_1 + K_I n_2) \tag{45}$$

where n_i are the unit normals to the line $\theta = 0$ ahead of the crack. For symmetric crack configuration where the crack lies along $y = 0$, the normals $n_1 = 0$ and $n_2 = 1$;

$$K_I = t_2^A\sqrt{2\pi l} \quad \text{and} \quad K_{II} = t_1^A\sqrt{2\pi l} \tag{46}$$

In a recent study by Smith,[60] a comparison was made between methods of evaluating both mode I and mode II stress intensity factors using the one-point and two-point displacement formulae, and the traction formula. In this study Smith[60] concluded that the traction formula gave the most accurate results and was the less sensitive of the three methods, to the size of the crack-tip elements.

2.1 J-integral

The stress intensity factor can also be related to a path independent integral, termed the J-integral. This integral is independent of the actual path chosen, provided that the starting and end points of the contour Γ are on opposite faces of the crack and that the contour includes the crack tip.

The J-integral is related to the stress intensity factor, under plane strain conditions, the relationship is

$$J = \frac{1-\nu^2}{E}(K_I^2 + K_{II}^2) \tag{47}$$

J can be defined as a path independent integral which, if the crack lies along $y = 0$, is defined as

$$J = \int_C \left(W n_x - t_i \frac{\partial u_i}{\partial x} \right) \mathrm{d}C \tag{48}$$

where W is the strain energy per unit volume, n_x is the component in x-direction of the outward normal to the path $C, t_i (= \sigma_{ij} n_j)$ and u_i are the components of the interior tractions and displacements along C. The integration path C may be discretized into segments as follows:

$$C = \sum_{i=1}^{N} C_i$$

The J-integral can now be written in Cartesian components as

$$J = \sum_{i=1}^{N} \int_{C_i} \left\{ \frac{1}{2} \begin{bmatrix} \sigma_{xx} & \sigma_{yy} & \sigma_{xy} \end{bmatrix} \left\{ \begin{array}{c} \varepsilon_{xx} \\ \varepsilon_{yy} \\ 2\varepsilon_{xy} \end{array} \right\} n_x \right.$$
$$\left. - \begin{bmatrix} \sigma_{xx} & \sigma_{yy} & \sigma_{xy} \end{bmatrix} \begin{bmatrix} n_x & o \\ 0 & n_y \\ n_y & n_x \end{bmatrix} \left\{ \begin{array}{c} \frac{\partial u_x}{\partial x} \\ \frac{\partial u_y}{\partial x} \end{array} \right\} \right\} \mathrm{d}C \tag{49}$$

where the stresses (σ_{ij}) and the strains (ε_{ij}) are represented as 3-component vectors. For mixed mode problems Ishikawa, Kitagawa and Okamura[12] suggested that it is possible to decouple the J-integral into mode I and mode II components. They showed that the stress, strain, displacement and traction fields could be separated analytically into mode I and mode II components within a symmetric mesh region in the neighborhood of the crack tip. They considered two points $P(x, y)$ and $P'(x, -y)$ symmetric about the crack line. When the point $P(x, y)$ has field parameters $\sigma_{ij}, \varepsilon_{ij}, u_j$ and t_j and the point $P'(x, -y)$ has field parameters $\sigma'_{ij}, \varepsilon'_{ij}, u'_j$ and t'_j , they showed that

$$\sigma_{ij} = \sigma_{ij}^I + \sigma_{ij}^{II}$$

$$\varepsilon_{ij} = \varepsilon_{ij}^I + \varepsilon_{ij}^{II}$$

$$u_j = u_j^I + u_j^{II}$$

and

$$t_j = t_j^I + t_j^{II} \tag{50}$$

where

$$\left\{\begin{matrix} \sigma_{11}^I \\ \sigma_{22}^I \\ \sigma_{12}^I \end{matrix}\right\} = \frac{1}{2}\left\{\begin{matrix} \sigma_{11} & +\sigma_{11}' \\ \sigma_{22} & +\sigma_{22}' \\ \sigma_{12} & -\sigma_{12}' \end{matrix}\right\}, \left\{\begin{matrix} \sigma_{11}^{II} \\ \sigma_{22}^{II} \\ \sigma_{12}^{II} \end{matrix}\right\} = \frac{1}{2}\left\{\begin{matrix} \sigma_{11} & -\sigma_{11}' \\ \sigma_{22} & -\sigma_{22}' \\ \sigma_{12} & +\sigma_{12}' \end{matrix}\right\}$$

$$\left\{\begin{matrix} \varepsilon_{11}^I \\ \varepsilon_{22}^I \\ \varepsilon_{12}^I \end{matrix}\right\} = \frac{1}{2}\left\{\begin{matrix} \varepsilon_{11} & +\varepsilon_{11}' \\ \varepsilon_{22} & +\varepsilon_{22}' \\ \varepsilon_{12} & -\varepsilon_{12}' \end{matrix}\right\}, \left\{\begin{matrix} \varepsilon_{11}^{II} \\ \varepsilon_{22}^{II} \\ \varepsilon_{12}^{II} \end{matrix}\right\} = \frac{1}{2}\left\{\begin{matrix} \varepsilon_{11} & -\varepsilon_{11}' \\ \varepsilon_{22} & -\varepsilon_{22}' \\ \varepsilon_{12} & +\varepsilon_{12}' \end{matrix}\right\}$$

$$\left\{\begin{matrix} u_1^I \\ u_2^I \end{matrix}\right\} = \frac{1}{2}\left\{\begin{matrix} u_1 & +u_1' \\ u_2 & -u_2 \end{matrix}\right\}, \left\{\begin{matrix} u_1^{II} \\ u_2^{II} \end{matrix}\right\} = \frac{1}{2}\left\{\begin{matrix} u_1 & -u_1' \\ u_2 & +u_2 \end{matrix}\right\}$$

$$\left\{\begin{matrix} t_1^I \\ t_2^I \end{matrix}\right\} = \frac{1}{2}\left\{\begin{matrix} t_1 & +t_1' \\ t_2 & -t_2 \end{matrix}\right\}, \left\{\begin{matrix} t_1^{II} \\ t_2^{II} \end{matrix}\right\} = \frac{1}{2}\left\{\begin{matrix} t_1 & -t_1' \\ t_2 & +t_2 \end{matrix}\right\}$$

Substituting the above relationships into (49) gives

$$J = \sum_{i=1}^{N} \int_{\Gamma_i} \left\{ \frac{1}{2}[\sigma_{xx}^I \; \sigma_{yy}^I \; \sigma_{xy}^I] \left\{\begin{matrix} \varepsilon_{xx}^I \\ \varepsilon_{yy}^I \\ 2\varepsilon_{xy}^I \end{matrix}\right\} n_x - \right.$$

$$[\; \sigma_{xx}^I \quad \sigma_{yy}^I \quad \sigma_{xy}^I \;] \begin{bmatrix} n_x & o \\ 0 & n_y \\ n_y & n_x \end{bmatrix} \left\{\begin{matrix} \frac{\partial u_x^I}{\partial x} \\ \frac{\partial u_y^I}{\partial x} \end{matrix}\right\} +$$

$$\frac{1}{2}[\; \sigma_{xx}^I \quad \sigma_{yy}^I \quad \sigma_{xy}^I \;] \left\{\begin{matrix} \varepsilon^{II} \\ \varepsilon^{II} \\ 2\varepsilon^{II} \end{matrix}\right\} n_x - [\; \sigma_{xx}^I \quad \sigma_{yy}^I \quad \sigma_{xy}^I \;] \begin{bmatrix} n_x & o \\ 0 & n_y \\ n_y & n_x \end{bmatrix} \left\{\begin{matrix} \frac{\partial u_x^{II}}{\partial x} \\ \frac{\partial u_y^{II}}{\partial x} \end{matrix}\right\} +$$

$$\frac{1}{2}[\; \sigma_{xx}^{II} \quad \sigma_{yy}^{II} \quad \sigma_{xy}^{II} \;] \left\{\begin{matrix} \varepsilon_{xx}^I \\ \varepsilon_{yy}^I \\ 2\varepsilon_{xy}^I \end{matrix}\right\} n_x - [\; \sigma_{xx}^{II} \quad \sigma_{yy}^{II} \quad \sigma_{xy}^{II} \;] \begin{bmatrix} n_x & o \\ 0 & n_y \\ n_y & n_x \end{bmatrix} \left\{\begin{matrix} \frac{\partial u_x^I}{\partial x} \\ \frac{\partial u_y^I}{\partial x} \end{matrix}\right\} +$$

$$\left. \frac{1}{2}[\; \sigma^{II} \quad \sigma^{II} \quad \sigma^{II} \;] \left\{\begin{matrix} \varepsilon^{II} \\ \varepsilon^{II} \\ 2\varepsilon^{II} \end{matrix}\right\} n_x - [\; \sigma^{II} \quad \sigma^{II} \quad \sigma^{II} \;] \begin{bmatrix} n_x & o \\ 0 & n_y \\ n_y & n_x \end{bmatrix} \left\{\begin{matrix} \frac{\partial u^{II}}{\partial x} \\ \frac{\partial u^{II}}{\partial x} \end{matrix}\right\} \right\} \mathrm{d}\Gamma$$

$$= J_I + J_{I,II} + J_{II,I} + J_{II} \tag{51}$$

Now, since the integration path is taken to be symmetrical with respect to the crack along the x-axis ($y = 0$), the outward normal components (n_x, n_y) at points $P(x, y)$ and $P'(x, y)$ have the following relationships

$$(n'_x, n'_y) = (n_x, -n_y) \tag{52}$$

By using this relationship, the traction fields in (50) at point $P(x, y)$ can be related to those at $P'(x, y)$, that is

$$\left\{ \begin{array}{c} t_1^{I\prime} \\ t_2^{I\prime} \end{array} \right\} = \left\{ \begin{array}{c} t_1^{I} \\ -t_2^{I} \end{array} \right\} \text{ and } \left\{ \begin{array}{c} t_1^{II\prime} \\ t_2^{II\prime} \end{array} \right\} = \left\{ \begin{array}{c} -t_1^{II} \\ t_2^{II} \end{array} \right\} \tag{53}$$

Other field parameters are also related as follows:

$$\left\{ \begin{array}{c} \sigma_{11}^{I\prime} \\ \sigma_{22}^{I\prime} \\ \sigma_{12}^{I\prime} \end{array} \right\} = \left\{ \begin{array}{c} +\sigma_{11}^{I} \\ +\sigma_{22}^{I} \\ -\sigma_{12}^{I} \end{array} \right\}, \left\{ \begin{array}{c} \sigma_{11}^{II\prime} \\ \sigma_{22}^{II\prime} \\ \sigma_{12}^{II\prime} \end{array} \right\} = \left\{ \begin{array}{c} -\sigma_{11}^{II} \\ -\sigma_{22}^{II} \\ +\sigma_{12}^{II} \end{array} \right\} \tag{54}$$

$$\left\{ \begin{array}{c} \varepsilon_{11}^{I\prime} \\ \varepsilon_{22}^{I\prime} \\ \varepsilon_{12}^{I\prime} \end{array} \right\} = \left\{ \begin{array}{c} +\varepsilon_{11}^{I} \\ +\varepsilon_{22}^{I} \\ -\varepsilon_{12}^{I} \end{array} \right\}, \left\{ \begin{array}{c} \varepsilon_{11}^{II\prime} \\ \varepsilon_{22}^{II\prime} \\ \varepsilon_{12}^{II\prime} \end{array} \right\} = \left\{ \begin{array}{c} -\varepsilon_{11}^{II} \\ -\varepsilon_{22}^{II} \\ +\varepsilon_{12}^{II} \end{array} \right\} \tag{55}$$

$$\left\{ \begin{array}{c} u_1^{I\prime} \\ u_2^{I\prime} \end{array} \right\} = \left\{ \begin{array}{c} +u_1^{I} \\ -u_2^{I} \end{array} \right\}, \left\{ \begin{array}{c} u_1^{II\prime} \\ u_2^{II\prime} \end{array} \right\} = \left\{ \begin{array}{c} -u_1^{II} \\ +u_2^{II} \end{array} \right\} \tag{56}$$

Using (52) and (53)-(56) with the J-integral in (51) gives

$$J'_I = J_I, J'_{II} = J_{II}, J'_{I,II} = -J_{I,II} \text{ and } J'_{II,I} = -J_{II,I} \tag{57}$$

Substituting (57) into (51) will result in

$$J = J_I + J_{II} \tag{58}$$

Therefore using the above relationships the J-integral can be decoupled into mode I and mode II components, hence K_I and K_{II} can be evaluated separately from (47).

The J-integral being a energy approach has the advantage that elaborate representation of the crack tip singular fields is not necessary. This is due to the relatively small contribution that the crack-tip fields make to the total J (i.e. strain energy) of the body. Further advantage of the J-integral approach is that it can be used as a post-processing procedure for the evaluation of stress intensity factors with the internal stress components σ_{ij} and the partial derivatives of the displacement $\frac{\partial u_i}{\partial x_j}$ are evaluated using (4) and (2) respectively.

In order to asses the accuracy of the J-integral method, Portela, Aliabadi and Rooke,[5] studied several crack geometries, including an edge crack, centre crack and kinked crack problem. They considered a square plate with a single edge crack, as represented in Fig. 5. The crack length is denoted by a and the width of the plate is defined by w. The plate is subjected to the action of a uniform traction $\bar{t}$, applied symmetrically at the ends in the direction perpendicular to the crack.

Results have been obtained and compared with those published by Civelek and Erdogan.[13] Five cases were considered $a/w = 0.2, 0.3, 0.4, 0.5$ and 0.6. A convergence

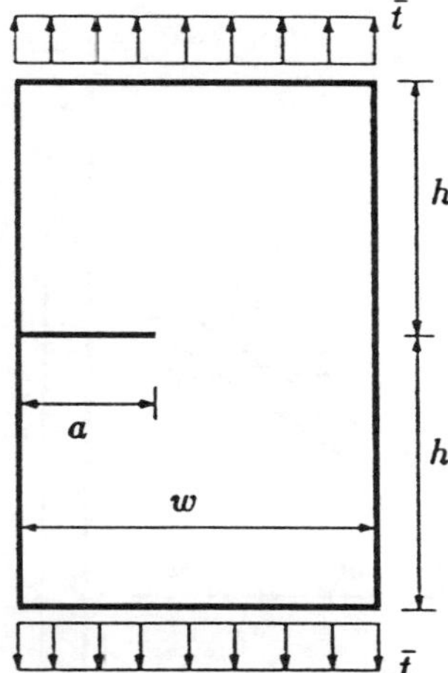

Figure 5: Rectangular plate with a single edge crack ($h/w = 0.5$)

Table 1: Stress intensity factors for a single edge crack in a square plate

a/w	$K_I/(\bar{t}\sqrt{\pi a})$					
	J-integral contour Path					Ref. 13
	2	3	4	5	8	
0.2	1.496	1.495	1.495	1.494	1.495	1.488
0.3	1.860	1.859	1.858	1.857	1.858	1.848
0.4	2.340	2.338	2.338	2.336	2.335	2.324
0.5	3.032	3.029	3.028	3.025	3.021	3.010
0.6	4.188	4.185	4.184	4.179	4.168	4.152

study was carried out with three different meshes 32, 40 and 64 quadratic elements, in which the crack was discretized with 4, 5 and 8 quadratic discontinuous elements on each surface. The results, obtained with the mesh of 32 elements, in which the crack discretization was graded towards the tip with the ratios 0.4, 0.3, 0.2 and 0.1, are presented in Table 1. They show a high level of accuracy when compared with those of Ref. 13.

For the J-integral computation, 10 internal points and the trapezoidal rule were considered. The stability of the J-integral results, for any contour path, is noticeable.

Consider, now the analysis of a central slant crack in a rectangular plate, represented in Fig. 6. The ratio between the height and the width of the plate is given by $h/w = 2$. The crack has a length 2a and makes an angle $\theta = 45°$ with the direction perpendicular to the applied traction. The plate is loaded with a uniform traction $\bar{t}$, applied symmetrically at the ends.

To solve this problem, Portela, Aliabadi and Rooke,[5] used a mesh of 36 quadratic

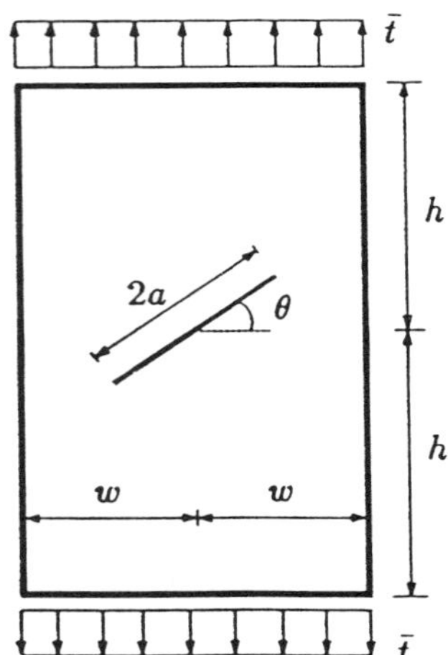

Figure 6: Retangular plate with a central slant crack ($h/w = 2, \theta = 45^o$)

Table 2: Mode I stress intensity factors for a central slant crack in a rectangular plate

a/w	$K_I/(\bar{t}\sqrt{\pi a})$					
	J-integral contour Path					Ref. 14
	2	3	4	5	8	
0.2	0.521	0.519	0.521	0.521	0.521	0.518
0.3	0.544	0.542	0.544	0.544	0.544	0.541
0.4	0.575	0.574	0.576	0.576	0.576	0.572
0.5	0.616	0.614	0.617	0.617	0.616	0.612
0.6	0.666	0.665	0.667	0.667	0.666	0.661

boundary elements, in which 6 discontinuous elements were used on each face of the crack, graded from the centre towards the tips, with the ratios 0.25, 0.15 and 0.1. Accurate results for this problem were published by Murakami.[14] The results obtained in Ref. 5, are presented in Tables 2 and 3. They show an excellent accuracy when compared with the results of Ref. 14.

The J-integral computation were carried out with 30 internal points and the trapezoidal integration rule.

Recently Rigby and Aliabadi[61] presented the application of BEM to the J-integral method for three-dimensional problems. In their analysis, the J-integral was decoupled into mode I, II and III and presented as

$$J^M = \int_C (W^M n_1 - \sigma_{ij}^M \frac{\partial u_i^M}{\partial x_1} n_j) \mathrm{d}C + \int_A (\frac{\partial \sigma_{i1}^M}{\partial x_1} + \frac{\partial \sigma_{i2}^M}{\partial x_2}) \frac{\partial u_i^M}{\partial x_1} \mathrm{d}A - \int_A \sigma_{i3}^M \frac{\partial^2 u_i^M}{\partial x_1 \partial x_3} \mathrm{d}A$$

Table 3: Mode I stress intensity factors for a central slant crack in a rectangular plate

a/w	$K_I/(\bar{t}\sqrt{\pi a})$					
	J-integral contour Path					Ref. 14
	2	3	4	5	8	
0.2	0.499	0.499	0.501	0.503	0.508	0.507
0.3	0.508	0.508	0.511	0.512	0.517	0.516
0.4	0.521	0.521	0.523	0.525	0.529	0.529
0.5	0.538	0.538	0.541	0.542	0.547	0.546
0.6	0.560	0.560	0.562	0.564	0.569	0.567

where $M = I, II$ and III; and C represents a contour normal to the crack front and A is the area enclosed by it. The stress intensity factors can be related to the J-integral as follows

$$J = J^I + J^{II} + J^{III}$$

$$= \frac{1}{E^*}(K_I^2 + K_{II}^2) + \frac{1}{2\mu}K_{III}^2$$

where E^* is equal to the Young's modulus E for equivalent plane stress ($E^* = E/(1-\nu^2)$) for equivalent plane strain.

2.2 Subtraction of singularity method

Most techniques for the evaluation of the stress intensity factors, attempt to model the singular stress behaviour at the crack tip. In contrast, the subtraction of the singularity technique avoids the need for this difficult task; it removes the singular fields completely. This leaves a non-singular field to be modelled numerically - much simpler task. This approach was first introduced into the two-dimensional boundary element formulation of potential problems by Symm[15] and Papamichel and Symm[16] who used constant elements to model a symmetrical slit. Later Xanthis *et al.*[17] used this formulation to solve the same problem of a symmetrical slit using quadratic isoparametric elements. The extension of the formulation to two-dimensional crack problems in elasticity was introduced by Aliabadi *et al.*,[18,19] who obtained both mode I and mode II stress intensity factors. This formulation was later extended to V-notch plates[20] and three-dimensional problems.[21]

In general, the displacement and traction fields in a given crack problem can be represented as

$$\mathbf{u} = \mathbf{u}^R + \mathbf{u}^S \tag{59}$$

$$\mathbf{t} = \mathbf{t}^R + \mathbf{t}^S \tag{60}$$

where $\mathbf{u}^R$ and $\mathbf{t}^R$ denote a regular field and $\mathbf{u}^S$ and $\mathbf{t}^S$ denote singular fields. The stress and displacement fields (i.e. $\mathbf{u}^S, \sigma_{ij}^S$), in the vicinity of the crack tip where the singularities occur, are known and given in eqns (35) and (36).

These expressions can be utilized to modify the boundary element formulation so that the traction singularity is removed from the numerical analysis. The new functions u_j^R and t_j^R are given by

$$u_j^R = u_j - u_j^S \tag{61}$$

and

$$t_j^R = t_j - t_j^S \tag{62}$$

are regular in the sense that the fields described by them contain no singularities. Since the singularity in t_j has the same functional form as that in t_j^S, then the left hand side of eqn (62) can be made nonsingular by a suitable choice of the unknown coefficient K_I; the variables u_j^R and t_j^R will now be smoothly varying (non-singular) functions everywhere. By substituting eqns (61) and (62) into the boundary element formulation and bearing in mind that, since each term in the series expansion of u_j^S and t_j^S satisfies the governing differential equation the integrals involving the singular functions u_j^S and t_j^S will cancel out exactly, the boundary integral equation becomes

$$C_{ij}(\mathbf{x}')u_j^R(\mathbf{x}') = \int_\Gamma U_{ij}(\mathbf{x}',\mathbf{x})\mathbf{t}_j^R(\mathbf{x})\mathrm{d}\Gamma(\mathbf{x}) - \int_\Gamma T_{ij}(\mathbf{x}',\mathbf{x})u_j^R(\mathbf{x})\mathrm{d}\Gamma(\mathbf{x}) \tag{63}$$

which contains only regular fields. This equation is now solved subject to the modified boundary conditions which do contain the singular functions, since

$$\bar{u}_j^R = \bar{u}_j - \bar{u}_j^S \tag{64}$$

$$\bar{t}_j^R = \bar{t}_j - \bar{t}_j^S \tag{65}$$

For example consider the central crack configuration shown in Fig. 7; the symmetry means that only one quarter of the problem needs to be modelled, if boundary conditions $\bar{u}_2 = 0$ and $\bar{u}_1 = 0$ are prescribed along the symmetry lines. The boundary conditions for this problem are:

$$\begin{aligned}
\bar{u}_2 = 0, \bar{t}_1 = 0 \quad & \text{along side } 1-2;\\
\bar{t}_j = 0, j = 1,2 \quad & \text{along side } 2-3;\\
\bar{t}_1 = 0, \bar{t}_2 = \sigma \quad & \text{along side } 3-4;\\
\bar{u}_1 = 0, \bar{t}_2 = 0 \quad & \text{along side } 4-5;\\
\bar{t}_j = 0, j = 1,2 \quad & \text{along side } 5-1
\end{aligned} \tag{66}$$

The modified boundary conditions from (64) and (65) are:

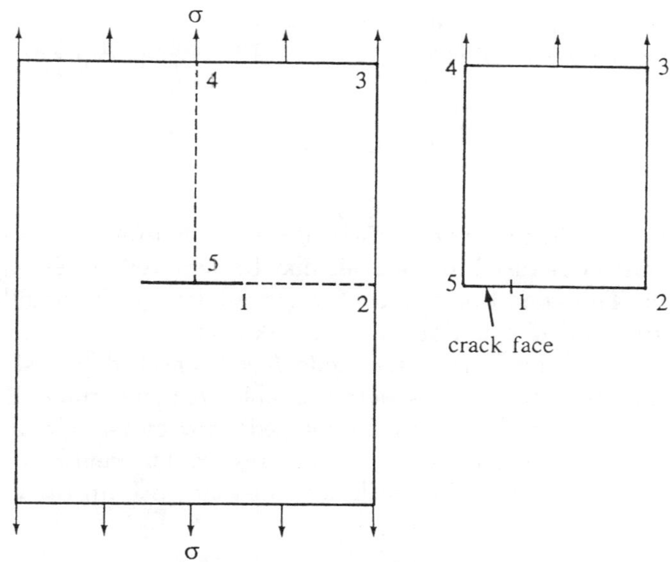

Figure 7: Central crack in a rectangular sheet

$$
\begin{array}{rl}
\bar{u}_2 = -\bar{u}_2^S, \bar{t}_1 = -\bar{t}_1^S & \text{along side } 1-2; \\
\bar{t}_j = -\bar{t}_j^S, j = 1,2 & \text{along side } 2-3; \\
\bar{t}_1 = -\bar{t}_1^S, \bar{t}_2 = \sigma - \bar{t}_2^S & \text{along side } 3-4; \\
\bar{u}_1 = -\bar{u}_1^S, \bar{t}_2 = -\bar{t}_2^S & \text{along side } 4-5; \\
\bar{t}_j = --\bar{t}_j^S, j = 1,2 & \text{along side } 5-1
\end{array}
\tag{67}
$$

The coefficient K_I which arises from the new boundary conditions is now an unknown in the numerical problem. Following the same discretization procedure as described in section 1, the matrix form of (63) can be written as

$$[H]\{\mathbf{u}^R\} = [G]\{\mathbf{t}^R\} \tag{68}$$

After substitution of the modified boundary conditions, the resulting system of equations can be written as

$$[A]\{X\} = [B]\{Y\} - [B]\{C\}K_I \tag{69}$$

where, if M collocation points are used then, $[A]$ and $[B]$ are $2M \times 2M$ matrices containing the appropriate coefficients from $[H]$ and $[G]$,$\{X\}$ is vector containing $2M$ components of the unknown variable u_j^R or t_j^R, $\{\mathbf{Y}\}$ is a vector containing $2M$ components of the prescribed boundary conditions $\bar{u}_j$ or $\bar{t}_j$; $\{C\}$ is a vector containing $2M$ components of $\bar{u}_j$ or $\bar{t}_j$ and K_I is the stress intensity factor.

Since the stress intensity factor is an unknown in the solution, there are $2M$ equations with $2M+1$ unknowns. Rearranging eqns (69), so that all the unknowns are collected on the left-hand side gives

$$\left[\ [A]\ \ \{D\}\ \right]\left\{\begin{array}{c} X \\ K_I \end{array}\right\} = \{F\} \tag{70}$$

where $\{D\} = [B]\{C\}$ and $\{F\} = [B]\{Y\}$.

In order to solve the matrix eqn (70), an extra equation is required or the number of unknowns must be reduced by one; this may be achieved by setting t_j^R equal to zero at the crack-tip. This condition ensures the cancellation of the singularity in eqn (61). Hence there are now $2M$ equations for $2M$ unknowns.

The procedure outlined above for mode I, was checked[18] by studying a configuration for which alternative results were available. A rectangular sheet of width $2W$, and a crack length of $2a$, symmetrically located between two circular holes of radius R, is subjected to a remote uniaxial tensile stress σ. The centres of the holes are on the perpendicular bisector of the crack, at a distance h from the crack, as shown in Fig. 8.

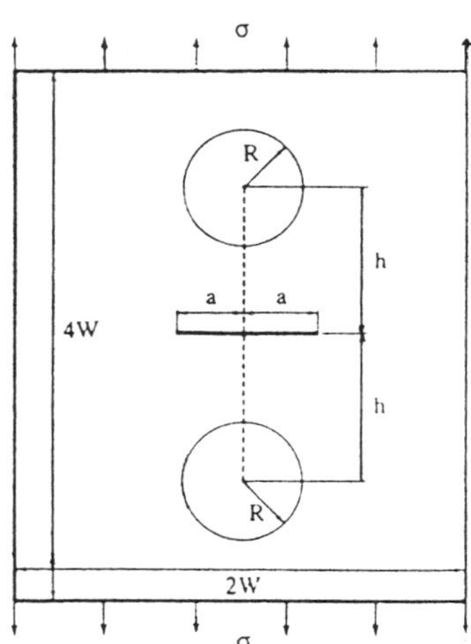

Figure 8: Central crack between two circular holes in a rectangular sheet

Since this configuration is symmetric with respect to the crack line, only the opening mode stress intensity factors K_I/K_o (K_o is the stress intensity factor for a crack, of length $2a$, in a large sheet subjected to remote uniform tensile stress σ perpendicular to crack direction and is given by $K_o = \sigma\sqrt{\pi a}$) in a large sheet ($W \gg a, R$) are presented in Fig. 9, for different ratios of h/R and a/R.

These values agree well (<1% difference) with those obtained by Newman,[22] who used boundary collocation of complex functions. In the limit as the crack length approaches zero, the normalized stress intensity factors tend to within 1% of the local stress concentration factor. The values of local stress concentration factors are obtained by using the same standard boundary element method; in this case (large sheet) the results agreed well (< 0.5% difference) with those found using the analytical formulation, derived by Ching-Bing Ling[23] for an infinite sheet.

In a paper by Smith and Aliabadi,[24] comparisons were made between the traction singular quarter-point element (TSQP) and the subtraction of singularity technique (SST). The configuration studied was that of a central slant crack in a large sheet. For optimal performance of quarter-point elements the element at the crack tip were taken to be 0.2 of the crack length. Figure 10 shows the accuracy of the results.

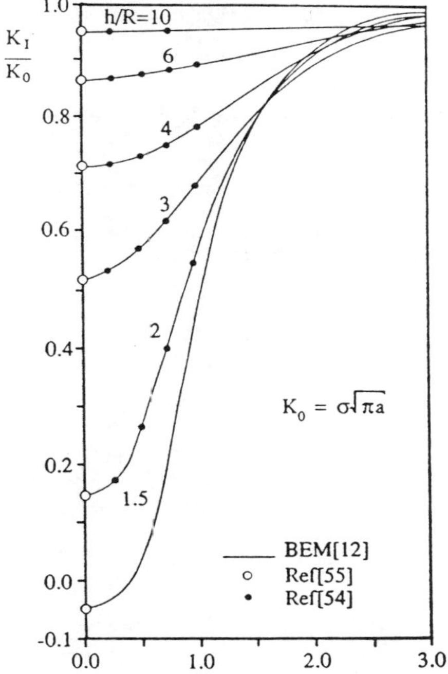

Figure 9: Stress intensity factors for a central crack between two circular holes

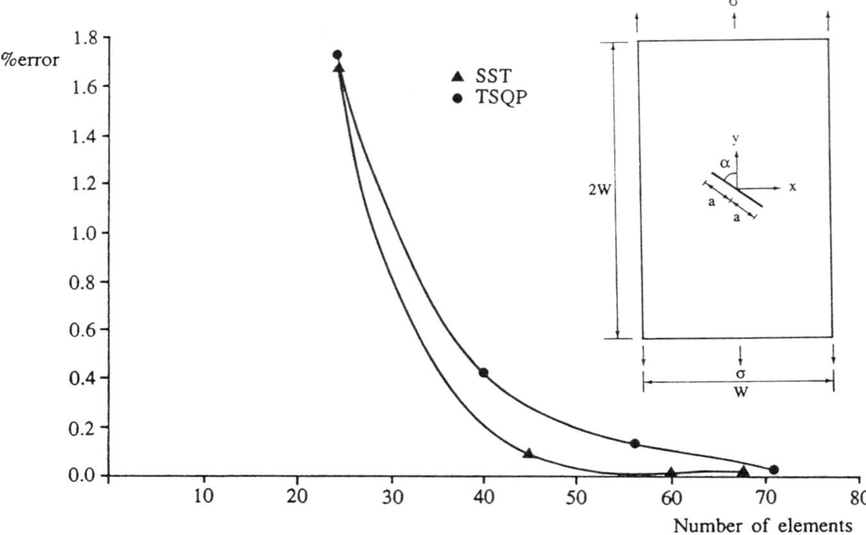

Figure 10: Comparison of singularity subtration technique with traction singular quarter-point element.

3 Weight function method

Bueckner[25] introduced the concept of 'weight functions' for two-dimensional elastic crack analysis in 1970. His weight functions satisfy the equations of linear elasticity, but they have a singularity at the crack tip which normally would not be admissible as a physical field. He refers to them as 'fundamental fields'. Rice[26] showed that the weight functions could equally well be determined by differentiating known elastic solutions for displacement fields with respect to the crack length. It was also shown by Rice that the knowledge of the two-dimensional elastic solution of a crack problem for any one loading, enable the determination of the solution for the same cracked body under any other loading system. Rice also presented the derivation of three-dimensional weight functions in the Appendix of Ref. 26. The three-dimensional derivation is based on the displacement field variation associated, to first order, with an arbitrary variation in the position of the crack front. Bueckner[27,28] also presented the formulation for weight functions in three-dimensional weight functions based on fundamental fields. Other theoretical advances in three-dimensional weight functions have been made by Rice,[29,30] Gao and Rice[31,32] and Bueckner.[33,34]

3.1 Two-dimensional weight functions

Rice[26] has derived through the concept of strain energy release rate a generalized integral expression for the stress intensity factor. He showed that if the displacement field $\mathbf{u}^{(2)}$ and the stress intensity factor $K_I^{(2)}$ are known for any one symmetrical load on a given geometry with a crack of total length l, then the stress intensity factor $K_I^{(1)}$ for any other symmetrical loading can be obtained from

$$K_I^{(1)} = \int_\Gamma \mathbf{t}^{(1)}(\mathbf{x}').\mathbf{H}_I(\mathbf{x}',l)\mathrm{d}\Gamma + \int_\Omega \mathbf{F}^{(1)}(\mathbf{X}').\mathbf{H}_I(\mathbf{X}',l)\mathrm{d}\Omega,\ \mathbf{x}' \in \Gamma, \mathbf{X}' \in \Omega \qquad (71)$$

where Γ is the boundary on which the surface tractions $\mathbf{t}^{(1)}$ are applied and $\mathbf{F}^{(1)}$ are the body forces within the region Ω. The function $\mathbf{H}_I$ is called the weight function and is given by

$$\mathbf{H}_I(\mathbf{x}',l) = H_{I1}\hat{\imath} + H_{I2}\hat{\jmath} = \frac{E'}{2K_I^{(2)}}\left(\frac{\partial u_1^{(2)}}{\partial l}\hat{\imath} + \frac{\partial u_2^{(2)}}{\partial l}\hat{\jmath}\right) \qquad (72)$$

where $E' = E/(1-\nu^2)$ for plane strain and $E' = E$ for plane stress. The unit vector $\hat{\imath}$ and $\hat{\jmath}$ lie in the x' and y' directions respectively where the origin of coordinates is at one end of the crack such that l is measured along $y' = 0$ in the positive x' direction.

The general extension of Rice's formula to mixed-mode crack problems has been proposed by a number of researchers, for example, Paris, McMeeking and Tada,[35] Bortmann and Banks-Sills[36] and Chen.[37]

Paris, MacMeeking and Tada[35] proposed an alternative derivation of the weight functions based on reciprocity relationships and Bueckner's fundamental fields. For an elastic body, Betti's reciprocal theorem can be written as

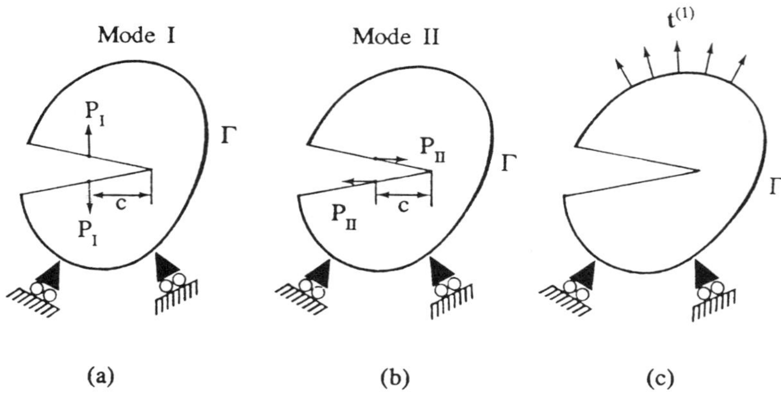

Figure 11: Arbitrary two-dimensional elastic body containing a crack

$$\int_{\Gamma} \mathbf{u}^{(1)}.\mathbf{t}^{(2)} d\Gamma = \int_{\Gamma} \mathbf{u}^{(2)}.\mathbf{t}^{(1)} d\Gamma \tag{73}$$

where it is assumed that the state (1) corresponds to the required solution of a crack in the body, see Fig. 11c, subjected to a set of known self-equilibrated tractions $\mathbf{t}^{(1)}$. The leading term in the series expansion of displacement fields near the crack tip are, given by

$$\begin{aligned} u_1^{(1)} &= \frac{K_I}{\mu}\sqrt{\frac{r}{2\pi}}\cos\frac{\theta}{2}\left[\frac{1}{2}(\kappa-1)+\sin^2\frac{\theta}{2}\right]+ \\ &+\frac{K_{II}}{\mu}\sqrt{\frac{r}{2\pi}}\sin\frac{\theta}{2}\left[\frac{1}{2}(\kappa+1)+\cos^2\frac{\theta}{2}\right] \\ u_2^{(1)} &= \frac{K_I}{\mu}\sqrt{\frac{r}{2\pi}}\sin\frac{\theta}{2}\left[\frac{1}{2}(\kappa+1)-\cos^2\frac{\theta}{2}\right]+ \\ &+\frac{K_{II}}{\mu}\sqrt{\frac{r}{2\pi}}\cos\frac{\theta}{2}\left[\frac{1}{2}(1-\kappa)+\sin^2\frac{\theta}{2}\right] \end{aligned} \tag{74}$$

where $\kappa = (3-4\nu)$ for plane strain and $\kappa = (3-\nu)/(1+\nu)$ for plane stress. From the above equations, the displacement components on the crack surfaces $\theta = \pm\pi$ at a small distance c behind the crack front are given by

$$u_1^{(1)} = \pm\frac{(\kappa+1)}{2\mu}\sqrt{\frac{c}{2\pi}}K_{II}$$

and

$$u_2^{(1)} = \pm\frac{(\kappa+1)}{2\mu}\sqrt{\frac{c}{2\pi}}K_I \tag{75}$$

The state 2 is chosen to correspond to the solution for pairs of symmetrical forces located on the crack surfaces a distance c from the crack tip, as shown in Fig. 11. The traction $\mathbf{t}^{(2)}$ can be written in terms of the point forces $\mathbf{P}$ on the crack surface, which are given by

$$\mathbf{P} = P_{II}\,\hat{\imath} + P_I\,\hat{\jmath} \tag{76}$$

where P_I and P_{II} are the magnitudes of the point forces in the normal and tangential directions respectively. For point forces a distance c from the crack tip, the traction vector $\mathbf{t}^{(2)}$ can be written as

$$\mathbf{t}^{(2)} = \delta(x_1 + c)\mathbf{P} \tag{77}$$

where δ is the Dirac delta function. Substitution of $\mathbf{t}^{(2)}$ from eqn (77), and $\mathbf{u}^{(1)}$, from (75), into eqn (73) followed by integration over both crack surfaces gives

$$P_I K_I^{(1)} + P_{II} K_{II}^{(1)} = \frac{E\sqrt{\pi}}{4\sqrt{2}(1-\nu^2)\sqrt{c}} \int_\Gamma \mathbf{t}^{(1)}.\mathbf{u}^{(2)} \mathrm{d}\Gamma \tag{78}$$

for plane strain conditions.

Equation (78) can be rewritten in a more compact form as

$$K_N^{(1)} = \frac{E'}{4}\sqrt{\frac{\pi}{2}}\frac{1}{P_N\sqrt{c}} \int_\Gamma \mathbf{t}^{(1)}.\mathbf{u}_N^{(2)} \mathrm{d}\Gamma\ ,\ N = I, II \tag{79}$$

As c tends to zero, a singularity is introduced into the fields at the crack tip due to the point forces; the strength of these fields is defined as of magnitude B_N, which is given by

$$B_N = \lim_{c \to 0}\left(\frac{P_N\sqrt{c}}{\pi}\right) \tag{80}$$

Substitution of eqn (80) into eqn (79) gives

$$K_N^{(1)} = \frac{E'}{4\sqrt{2\pi}B_N} \int_\Gamma \mathbf{t}^{(1)}.\mathbf{u}_N^{(2)} \mathrm{d}\Gamma\ ,\ N = I, II \tag{81}$$

where $\mathbf{u}_N^{(2)}$ now represents the displacement field on the boundary Γ, which results from the singular field of strength B_N at the crack tip. From the comparison of eqn (81) and (72) it can be seen that the weight function $\mathbf{H}_N(\mathbf{x},a)$ is given by

$$\mathbf{H}_N(\mathbf{x},a) = \frac{E'}{4\sqrt{2\pi}B_N}\mathbf{u}_N^{(2)}\ ,\ N = I, II \tag{82}$$

The weight function for mixed boundary conditions can be derived in a similar manner. Assuming that Γ_u and Γ_t respectively represent the portion of the boundary Γ where the displacement and traction fields are prescribed for state (1), the expression for $K_N^{(1)}$ in eqn (81) can be rewritten as

$$K_N^{(1)} = \frac{E'}{4\sqrt{2\pi}B_N}\left\{\int_{\Gamma_t} \mathbf{t}^{(1)}.\mathbf{u}_N^{(2)} - \int_{\Gamma_u} \mathbf{t}^{(2)}.\mathbf{u}_N^{(1)}\right\} \mathrm{d}\Gamma \tag{83}$$

where $\mathbf{u}_N^{(1)}$ are the prescribed displacement boundary conditions of state (1) and $\mathbf{t}^{(2)}$ are the reactions on Γ_u that result from displacement constraints when the point forces act at the crack tip.

For three-dimensional problems, as in two-dimensional cases, a formulation is possible through the analysis of Bueckner's fundamental (point force) fields.

Following Rooke, Cartwright and Aliabadi,[38] it is assumed that state (1) corresponds to the desired solution of a crack in the body whose surface is subjected to a self-equilibriated traction $\mathbf{t}^{(1)}$. The leading terms in the general expansion of the displacement on the crack surface $\theta = \pm\pi$ can be expressed in terms of a local coordinate system centred at a general point s. If the crack lies in the y=0 plane, then the displacement at a point $(-c, 0, 0)$ behind the crack front can be obtained as

$$u_1^{(1)}(s) = \pm\frac{2(1-\nu^2)}{E}\sqrt{\left(\frac{2c}{\pi}\right)}K_{II}(s)$$

$$u_2^{(1)}(s) = \pm\frac{2(1-\nu^2)}{E}\sqrt{\left(\frac{2c}{\pi}\right)}K_{I}(s)$$

and

$$u_3^{(1)}(s) = \pm\frac{2(1+\nu)}{E}\sqrt{\left(\frac{2c}{\pi}\right)}K_{III}(s) \qquad (84)$$

where 1,2, and 3 refer to the orthogonal directions which are tangential, normal and binormal to the crack front at s.

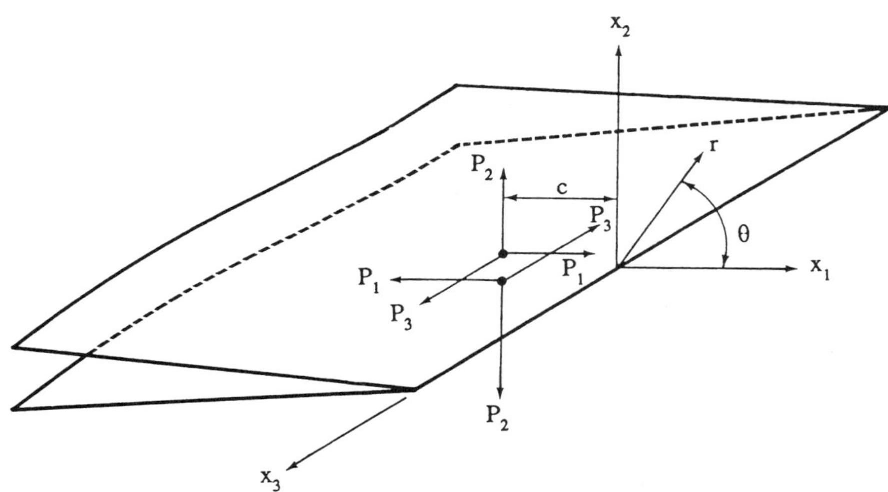

Figure 12: Concentrated forces near the tip of planar crack

The state (2) is taken to be that corresponding to a solution of a body containing a crack which is subjected to a pair of concentrated symmetrical forces located on the crack surfaces near the crack tip as shown in Fig. 12. The traction $\mathbf{t}^{(2)}$ is represented in terms of a point force in each of the three orthogonal directions which are given by the vector

$$\mathbf{P} = P_1\hat{\imath} + P_2\hat{\jmath} + P_3\hat{k} \tag{85}$$

where P_1, P_2 and P_3 are the magnitudes of the point forces in the x_1, x_2 and x_3 directions respectively. For a point force located on the crack surface at $x_1 = -c$, $x_3 = 0$, the traction vector $\mathbf{t}^{(2)}$ can be written as

$$\mathbf{t}^{(2)} = \delta(x_1 + c)\delta(x_3)\mathbf{P} \tag{86}$$

where δ represents the Dirac delta function.

Substituting eqn (84) and (86) into Betti's reciprocal eqn (83) gives

$$P_I K_{II}^{(1)} + P_{II} K_I^{(1)} + \frac{1}{1-\nu} P_3 K_{III}^{(1)} = \frac{E\sqrt{\pi}}{4\sqrt{2}(1-\nu^2)\sqrt{c}} \int_\Gamma \mathbf{t}^{(1)}.\mathbf{u}^{(2)} \mathrm{d}\Gamma \tag{87}$$

Equation (87) can be rewritten in a more general form as

$$K_N = \frac{\Lambda E\sqrt{\pi}}{4\sqrt{2}(1-\nu^2)\sqrt{c}} \int_\Gamma \mathbf{t}^{(1)}.\mathbf{u}^{(2)} \mathrm{d}\Gamma$$

$$= \int_\Gamma \mathbf{t}^{(1)}.\mathbf{W}^{(2)} \mathrm{d}\Gamma \tag{88}$$

where $\Lambda = 1$ for $N = I, II$ and $\Lambda = (1-\nu)$ for $N = III$; $W^{(2)}$ can be defined as the weight function. Thus each stress intensity factor K_N $(N = I, II, III)$, for a given position along the crack front, can be obtained independently by making P_j corresponding to that mode non-zero, and all other P_j zero.

3.2 Application of the BEM to weight functions

A boundary element formulation for obtaining weight functions, based on Bueckner's fundamental fields, has been developed for mode I and mixed-mode two-dimensional problems,[39,40] and mode I three-dimensional problems.[41] These boundary element formulations, which follow the derivation of Paris *et al.*[35] require only a single solution to a numerical problem to obtain either mode I or mode II weight functions. This weight function approach, requires the solution of the problem that results from two equal and opposite localized forces acting on opposite faces of the crack at the tip. The stress and displacement fields resulting from a pair of point forces (either normal $\pm\mathbf{P}_I$ or shear $\pm\mathbf{P}_{II}$) at the tip of a crack in an infinite sheet can be derived from the stress function Z^F given by Irwin.[42] In general, the stress function can be interpreted as a series stress functions,[42] as follows:

$$Z^F(z) = Z_o^F(z) + \sum_N \sum_{m=0}^{\infty} Z_m^F(z)\ , \ N = I, II \tag{89}$$

where $Z_o^F(z) = B_o^F z^{m-\frac{3}{2}}$ describes the highly singular field at the crack tip due to the point forces only ($F = P_I$ or P_{II}), and $Z_m^N = B_m^N z^{m-\frac{3}{2}}$ ($m \neq 0$) describes the fields due to the presence of boundaries. It is convenient to write $B_m^N = B_o^F b_m^N$, with $b_o^N = 1$. In a general non-symmetric stress problem, stress functions for both opening- and sliding-mode deformation must be considered; because of the lack of symmetry a pair of normal forces ($\pm P_I$) on the crack may cause some sliding-mode deformation and shear forces ($\pm P_{II}$) may cause opening-mode deformation.

For each value of m there exists a displacement field of the form

$$u_j^m(r,\theta) = B_m^N r^{m-\frac{1}{2}} f_j^{Nm}(\theta) \tag{90}$$

where the summation over N is assumed, and (r,θ) are polar coordinate centred at the crack tip; and there exists a stress field of the form

$$\sigma_{ij}^m(r,\theta) = B_m^N r^{m-\frac{3}{2}} g_{ij}^{Nm}(\theta) \tag{91}$$

These fields can be derived from the general relationships for Westergaard stress function $Z(z)$; the functions $f_j^{Nm}(\theta)$ and $g_{ij}^{Nm}(\theta)$ for $N = I$ are listed below:

$$f_1^{I0} = \frac{1}{\mu}\cos\frac{\theta}{2}\left[\frac{1}{2}(1-\kappa) + \sin\frac{\theta}{2}\sin\frac{3\theta}{2}\right]$$

$$f_2^{I0} = \frac{1}{\mu}\sin\frac{\theta}{2}\left[\frac{1}{2}(1+\kappa) - \cos\frac{\theta}{2}\cos\frac{3\theta}{2}\right]$$

$$g_{11}^{I0} = \cos\frac{3\theta}{2} - \frac{3}{2}\sin\theta\sin\frac{5\theta}{2}$$

$$g_{22}^{I0} = \cos\frac{3\theta}{2} + \frac{3}{2}\sin\theta\sin\frac{5\theta}{2}$$

and

$$g_{12}^{I0} = \frac{3}{2}\sin\theta\cos\frac{5\theta}{2}$$

for $m = 0$; and

$$f_1^{I1} = \frac{1}{\mu}\cos\frac{\theta}{2}\left[\frac{1}{2}(\kappa-1) + \sin^2\frac{\theta}{2}\right]$$

$$f_2^{I1} = \frac{1}{\mu}\sin\frac{\theta}{2}\left[\frac{1}{2}(1+\kappa) - \cos^2\frac{\theta}{2}\right]$$

$$g_{11}^{I1} = \cos\frac{\theta}{2}\left[1 - \sin\theta\sin\frac{3\theta}{2}\right]$$

$$g_{22}^{I1} = \cos\frac{\theta}{2}\left[1 + \sin\theta\sin\frac{3\theta}{2}\right]$$

and

$$g_{12}^{I1} = \frac{1}{2}\sin\theta\cos\frac{3\theta}{2}$$

for $m = 1$. For $N = II$ and $m = 0$,

$$f_1^{II0} = \frac{1}{\mu}\sin\frac{\theta}{2}\left[\frac{1}{2}(1+\kappa) + \cos\frac{\theta}{2}\cos\frac{3\theta}{2}\right]$$

$$f_2^{II0} = \frac{1}{\mu}\cos\frac{\theta}{2}\left[\frac{1}{2}(1-\kappa) + \sin\frac{\theta}{2}\sin\frac{3\theta}{2}\right]$$

$$g_{11}^{II0} = -2\sin\frac{3\theta}{2} - \frac{3}{2}\sin\theta\cos\frac{5\theta}{2}$$

$$g_{22}^{II0} = \frac{3}{2}\sin\theta\cos\frac{5\theta}{2}$$

and

$$g_{12}^{II0} = \cos\frac{3\theta}{2} - \frac{3}{2}\sin\theta\sin\frac{5\theta}{2}$$

and for $m = 1$,

$$f_1^{II1} = \frac{1}{\mu}\sin\frac{\theta}{2}\left[\frac{1}{2}(\kappa+1) + \cos^2\frac{\theta}{2}\right]$$

$$f_2^{II1} = \frac{1}{\mu}\cos\frac{\theta}{2}\left[\frac{1}{2}(1-\kappa) + \sin^2\frac{\theta}{2}\right]$$

$$g_{11}^{II1} = -\sin\frac{\theta}{2}\left[2 + \cos\frac{\theta}{2}\cos\frac{3\theta}{2}\right]$$

$$g_{22}^{II1} = \frac{1}{2}\sin\theta\cos\frac{3\theta}{2}$$

and

$$g_{12}^{II1} = \cos\frac{\theta}{2}\left[1 - \sin\frac{\theta}{2}\sin\frac{3\theta}{2}\right]$$

It can be seen from eqn (90) that the displacement fields u_j^0 become singular $O(r^{-\frac{1}{2}})$ at the crack tip ($r \to 0$), and from eqn (91) that σ_{ij}^0 and σ_{ij}^1 become singular at the crack tip $O(r^{-\frac{3}{2}})$ and $O(r^{-\frac{1}{2}})$ respectively. By using the above general displacement and stress fields it is possible to define new displacement and traction fields u_j^{RM} and t_j^{RM} as follows

$$u_j^{RM} = u_j - U_j^M \quad , \quad U_j^M = \sum_{m=0}^{M} u_j^m \tag{92}$$

and

$$t_j^{RM} = t_j - T_j^M \quad , \quad T_j^M = \sum_{m=0}^{M} t_j^m \tag{93}$$

where u_j and t_j are the desired solution for the original problem of a crack subjected to a pair of point forces ($\pm P_I$ or $\pm P_{II}$). Substitution of u_j and t_j, from (92) and (93), into the boundary element equation, gives

$$C_{ij}(\mathbf{x}')u_j^{RM}(\mathbf{x}') + \int_\Gamma T_{ij}(\mathbf{x}',\mathbf{x})u_j^{RM}\mathrm{d}\Gamma(\mathbf{x}) = \int_\Gamma U_{ij}(\mathbf{x}',\mathbf{x})t_j^{RM}\mathrm{d}\Gamma(\mathbf{x}) \tag{94}$$

since u_j^m and t_j^m automatically satisfy the equations of elasticity, eqn (94) is to be solved subjected to the following boundary conditions:

$$\bar{u}_j^{RM} = \bar{u}_j - \bar{U}_j^M \tag{95}$$

and

$$\bar{t}_j^{RM} = \bar{t}_j - \bar{T}_j^M \tag{96}$$

For $m = 0$, the procedure for solving the boundary element eqn (94) is straightforward. The fields chosen to be subtracted for $M = 0$ are not mixed-mode, since they are the fields for a pair of forces at the tip of the crack in an infinite sheet and $B_m^N = B_0^F$. The boundary values of U_j^0 and T_j^0 needed will, in general, be non-zero and can be calculated from eqns (90) to (93) and the functions f_j^{N0} and g_{ij}^{N0} ($N = I, II$).

The original boundary conditions for the weight function problem are given by

$$\bar{t}_j = 0 \text{ on } \Gamma_e \quad ; \quad \bar{t}_j = \lim_{c\to 0}\{F_j\delta(z+c)\} \text{ on } \Gamma_c \tag{97}$$

where $F_1 = P_{II}$ and $F_2 = P_I$, the forces on the crack surfaces Γ_c. On the crack surfaces Γ_c the resultant tractions from all the stress functions Z_m^N must equal the applied traction; that is

$$T_j^M = \sum_{m=0}^{\infty} t_j^m = \lim_{c\to 0}\{F_j\delta(z+c)\} \tag{98}$$

Since at the crack tip, where the forces act, $| t_j^{m+1} | / | t_j^m | \to 0$, it follows that

$$T_j^0 = \sum_{m=0}^{\infty} t_j^m = \lim_{c\to 0}\{F_j\delta(z+c)\} \tag{99}$$

Therefore from (93), (97) and (99), the boundary conditions $\bar{T}_j^0$ are proportional to B_o^F, therefore the solution (u_j^{R0} and t_j^{R0}) is proportional to B_o^F. Thus in the numerical problem the constant B_o^F can be put equal to unity without any loss of generality.

Once the boundary element eqn (94) is solved, for either $\pm P_I$ or $\pm P_{II}$ forces, the weight function u_j can be obtained (for that mode) by substituting u_j^{R0} into (92), since U_j^0 is known. It is worth noting that the above analysis does not introduce any modification into the BEM procedure other than the modified boundary conditions, hence it can be readily used with standard BEM programs. However, the solution (u_j^{R0} and t_j^{R0}) is not regular, because the fundamental fields resulting from stress function (89) have higher order singularities than the Williams field. For greater accuracy the singularity in t_j^1 should be removed before the numerical solution of eqn (94) is attempted.

The traction T_j^0 acting on the boundary will, in general, produce a K–like field ($t_j^{R0} \propto r^{-\frac{1}{2}}$) with both K_I and K_{II} components for each pair of forces ($\pm P_I$ or $\pm P_{II}$). These fields can be described by

$$\sum_N Z_1^N = Z_1^I + Z_1^{II} = (b_1^I + b_1^{II})z^{-\frac{1}{2}} \quad , \quad B_0^F = 1 \tag{100}$$

In order to regularize the problem the displacements and tractions for Z_1^N must be subtracted, that is $M = 1$ in eqns (92) and (93). Thus from (90) and (91)

$$U_j^1 = u_j^0 + u_j^1 = r^{-\frac{1}{2}} f_j^{I0} + r^{-\frac{1}{2}} \left[b_1^I f_j^{I1} + b_1^{II} f_j^{II1} \right] \tag{101}$$

and

$$T_j^1 = t_j^0 + t_j^1 = n_i \{ r^{-\frac{3}{2}} g_{ij}^{I0} + r^{-\frac{1}{2}} \left[b_1^I g_{ij}^{I1} + b_1^{II} g_{ij}^{II1} \right] \} \tag{102}$$

for the determination of mode I weight functions; and

$$U_j^1 = u_j^0 + u_j^1 = r^{-\frac{1}{2}} f_j^{II0} + r^{-\frac{1}{2}} \left[b_1^I f_j^{I1} + b_1^{II} f_j^{II1} \right] \tag{103}$$

and

$$T_j^1 = t_j^0 + t_j^1 = n_i \{ r^{-\frac{3}{2}} g_{ij}^{II0} + r^{-\frac{1}{2}} \left[b_1^I g_{ij}^{I1} + b_1^{II} g_{ij}^{II1} \right] \} \tag{104}$$

for the determination of mode II weight functions. Thus in each analysis there are two unknown coefficients b_1^I and b_1^{II}; these coefficients will in general be different in the two cases. It is clear that eqns (101) to (104) can be extended to any value of ($M > 1$) as follows:

$$U_j^M = u_j^0 + u_j^1 = r^{-\frac{1}{2}} f_j^{N0} + \sum_{m=1}^{M} r^{m-\frac{1}{2}} \left[b_m^I f_j^{Im} + b_m^{II} f_j^{IIm} \right] \tag{105}$$

and

$$T_j^M = t_j^0 + t_j^1 = n_i \{ r^{-\frac{3}{2}} g_{ij}^{N0} + \sum_{m=1}^{M} r^{m-\frac{3}{2}} \left[b_m^I g_{ij}^{Im} + b_m^{II} g_{ij}^{IIm} \right] \} \tag{106}$$

where $N = I$ for the opening-mode weight functions and $N = II$ for sliding-mode weight functions.

It can be seen from (105) and (106) that the subtraction of U_j^1 and T_j^1 in (92) and (93) can be used to remove all field singularities from the boundary element formulation of (94). Higher order values of $M (> 1)$ can be used, but will not in general be necessary. Thus for $M = 1$ eqn (92) and (93) become

$$u_j^{R1} = u_j - u_j^0 - u_j^{I1} - u_j^{II1} \tag{107}$$

and

$$t_j^{R1} = t_j - t_j^0 - t_j^{I1} - t_j^{II1} \tag{108}$$

where

$$u_j^{N1} = b_1^N r^{\frac{1}{2}} f_j^{N1} \quad \text{and} \quad t_j^{N1} = b_1^N r^{-\frac{1}{2}} g_{ij}^{N1} n_i \tag{109}$$

for $N = I$ or $N = II$. Equation (94) must now be solved for u_j^{R1} and t_j^{R1} subjected to the following boundary conditions:

$$\bar{u}_j^{R1} = \bar{u}_j - \bar{u}_j^0 - \bar{u}_j^{I1} - \bar{u}_j^{II1} \tag{110}$$

and

$$\bar{t}_j^{R1} = \bar{t}_j - \bar{t}_j^0 - \bar{t}_j^{I1} - \bar{t}_j^{II1} \tag{111}$$

where $\bar{u}_j^{N1}$ and $\bar{t}_j^{N1}$ contain unknown coefficients b_1^I and b_1^{II} which must be evaluated as part of the solution. After discretization the matrix form of (94) will be

$$[A]\{X\} = [B]\{Y\} - [B]\{C_0\} - [B]\{C_1^I\}b_1^I - [B]\{C_1^{II}\}b_1^{II} \tag{112}$$

where matrices $[A]$, $[B]$ contain the coefficients H_{ij} and/or G_{ij}; the vector $\{X\}$ contains the unknown boundary values u_j^{R1} and t_j^{R1}; the vector $\{Y\}$ contains the original boundary conditions $\bar{u}_j$ and $\bar{t}_j$; the vector $\{C_0\}$ contains the functions $r^{-\frac{1}{2}} f_j^{N0}$ and/or $r^{-\frac{3}{2}} g_{ij}^{N0} n_i$ (for $N = I$ or $N = II$); the vectors $\{C_1^I\}$ and $\{C_1^{II}\}$ contain the coefficients of b_1^N respectively, evaluated on the boundary from eqn (109). Rearranging (112) so that all the unknowns are on the left-hand side, gives

$$\left[\, [A] \quad [D_I] \quad [D_{II}] \,\right] \left\{ \begin{array}{c} X \\ b_1^I \\ b_1^{II} \end{array} \right\} = \{R\} \tag{113}$$

where $\{D_N\} = [B]\{C_1^N\}$, $N = I, II$ and $\{R\} = [b]\{Y\} - [B]\{C_0\}$.

It can be seen that there are two more unknowns than equations in (113); thus two more conditions are required. These two are obtained by setting the modified traction to zero at the crack tip; that is $t_j^{R1} = 0$ for $j = 1, 2$; with these conditions, eqn (113) can now be solved for u_j^{R1}. Substitution of u_j^{R1} into eqn (92) yields u_j, the desired weight functions.

Some numerical values of the stress intensity factor derived from the weight functions for $M = 0$ and $M = 1$ are shown in Table 4 for an edge crack specimen of width $w = b$ and total height of $2h = 6b$ subjected to a uniform tensile stress $\bar{t}$ on the ends (see Fig. 13). Four crack lengths $a/b = 0.3, 0.4, 0.5$ and 0.6 were considered, and the specimen was modelled with 30 quadratic elements. The results for both $M = 0$ and 1 are compared in Table 4 with accurate results of Keer and Freedman.[43] The accuracy

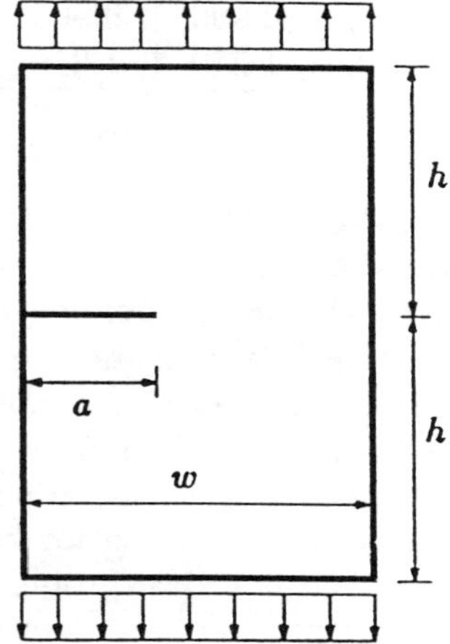

Figure 13: Rectangular plate with a single edge crack

Table 4: Comparison of stress intensity factors for $M = 0$ and 1 with Keer and Freedman[43]

	$K_I/(\sigma\sqrt{\pi a})$		
a/b	$M = 0$	$M = 1$	Ref. 43
0.3	1.632	1.656	1.660
0.4	2.070	2.109	2.112
0.5	2.767	2.829	2.826
0.6	3.956	4.065	4.039

Table 5: Comparison of Mode I stress intensity factors for $M = 0$ and 1

	$K_I/(\sigma\sqrt{\pi a})$		
a/w	$M=0$	$M=1$	Ref. 14
0.1	0.761	0.757	0.756
0.5	0.906	0.905	0.905
0.8	1.249	1.248	1.245

for $M = 0$ is about 1-3% worse than that for $M = 1$; but by increasing the number of elements to 38 for $M = 0$ the difference reduces to within 1% of $M = 1$.

Rooke and Aliabadi[40] studied the problem of central crack of length $2a$ in a rectangular sheet of height to width ratio $h/w = 2$ and the angle of inclination $\theta = 30^o$ (as shown in Fig. 14). The calculated values of stress intensity factors due to a uniform tensile stress σ at the ends of the sheet for different ratios of a/w are shown in Tables 5 and 6 for both mode I and mode II where they are compared with accurate results given by Murakami.[14] The results were obtained for both $M = 0$ and $M = 1$, which remove all the singularities from the numerical problem.

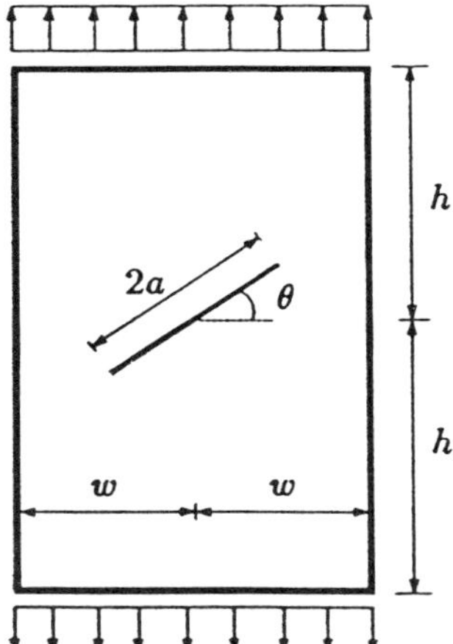

Figure 14: Rectangular plate with a central slant crack ($h/w = 2, \theta = 30^o$)

As can be seen from Tables and the results for $M = 0$ are close ($< 1\%$) to the reference values, and the results for $M = 1$ are even closer. The differences are small in this configuration because the crack tips are not very close to any external boundaries and the perturbation traction boundary conditions $\bar{t}_j^{N1}$ are small; hence the coefficients b_1^N are small.

Bains, Aliabadi and Rooke[41,44] have extended the subtraction of fundamental field formulation to three-dimensional problems in which the highly singular fields were subtracted reducing the problem to the same level as that of the standard crack problems.

Table 6: Comparison of Mode II stress intensity factors for $M = 0$ and 1

	$K_{II}/(\sigma\sqrt{\pi a})$		
a/w	$M=0$	$M=1$	Ref. 14
0.1	0.436	0.434	0.434
0.5	0.463	0.462	0.462
0.8	0.551	0.55	0.550

3.3 Application of weight functions to strip yield cracks

Another important application of the weight functions is that they provide an efficient means of solving for the plastic zone size in the Dugdale[45] elastic-plastic model of a crack of length l with a strip yield-zone of length s. The formulation consists of arbitrary applied tractions $\mathbf{t}^{(2)}$ on boundaries other than the crack, and a constant traction equivalent to the yield stress σ_Y on the strip yield-zone which extends a distance s ahead of the original crack tip. If the arbitrary traction, $\mathbf{t}^{(2)}$ is denoted by $\mathbf{t}^{(2)} = \sigma\mathbf{t}$, then the stress intensity factor K^σ that results is given, from (81), by

$$K^\sigma = \frac{E'\sigma}{4\sqrt{(2\pi)}B_N}\int_\Gamma \mathbf{t}.\mathbf{u}^{(2)}\mathrm{d}\Gamma \tag{114}$$

where $\mathbf{u}^{(2)}$ is the displacement due to point forces applied at the tip of a crack of length $c = (l+s)$; σ is a suitable normalizing factor which depends on the loading. The stress intensity factor K^Y for the yield-zone tractions, is obtained by letting $\mathbf{t}^{(2)} = \sigma_Y\hat{\jmath}$ in eqn (81) to give

$$K^\sigma = \frac{E'\sigma_Y}{4\sqrt{(2\pi)}B_N}\int_{\Gamma_s} u_2^{(2)}\mathrm{d}\Gamma \tag{115}$$

where Γ_s is the boundary over which the yield zone tractions are distributed.

The condition that the stresses are bounded in the yield zone is given by superimposing the stress intensity factors for the crack of length c, K^σ for the applied traction and K^Y for the strip yield traction, and equating them to zero; that is

$$K^\sigma + K^Y = 0 \tag{116}$$

Substitution of (114) and (115) into (116) gives

$$\frac{\sigma}{\sigma_Y}\int_{\Gamma-\Gamma_s} \mathbf{t}.\mathbf{u}^{(2)}\mathrm{d}\Gamma + \int_{\Gamma_s} \mathbf{u}_2^{(2)}\mathrm{d}\Gamma = 0 \tag{117}$$

The stress ratio σ/σ_Y, for each specific zone length s, can be obtained by carrying out the integration, in eqn (117).

Application of the boundary element formulation coupled with Bueckner singular fields to strip yield cracks was developed by Aliabadi and Cartwright.[46] In that study the ratios of applied stress to yield stress were obtained for various plastic zone to crack length ratios (s/l) by a single boundary element analysis. Once the length of the strip-yield is obtained, the calculation of the corresponding crack-opening displacement δ can be achieved through further integration of the weight function times the stress intensity factor.[47]

Aliabadi and Cartwright[46] studied the problem of a crack of length l emanating from a hole of radius R in an infinite sheet. The crack has a strip yield zone of length s starting a distance l from the hole boundary and ending a distance c from the boundary $(c = l + s)$. The infinite sheet was modelled as a rectangular sheet of height $2H$ and width W containing a central hole of radius R with $H/R = W/(R+c) = 20$. The normalizing stress σ is that due to the uniform stress on the sheet perpendicular to crack. The crack-line Green's functions obtained using the boundary element formulation are in agreement with those found by Shivakumar and Forman[48] to within 0.4%. The stress intensity factors, derived from the weight functions, for a remote uniaxial tensile stress perpendicular to the crack line are in agreement with those obtained by Tweed and Rooke[49] to within 1%. The stress ratio σ/σ_Y for various values of s/c are presented in Fig. 15. The results presented for the stresses and those by Rich and Roberts[50] differ by a maximum of 4% for $s/c = 0.5$ and $l/R = 4$.

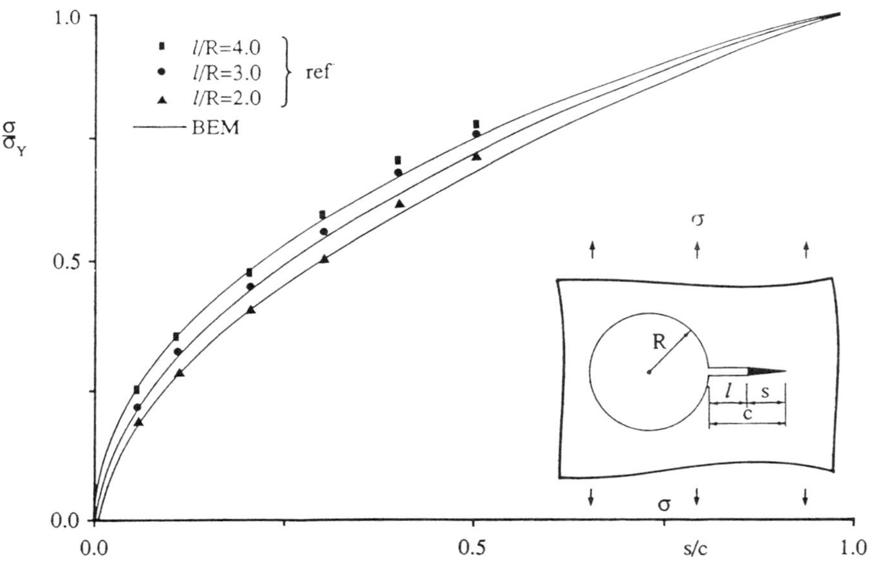

Figure 15: Comparison of normalized strip yield crack lengths for a crack at a hole

3.4 Application of the weight functions in residual stress field

The presence of a residual stress field may cause the value of the effective stress intensity factor K_{eff} to be significantly different from the value of the stress intensity factor due to external loads. The approach commonly used to account for the effect of the residual stresses on the stress intensity factor in linear elastic analysis involves the superposition of the stress intensity factor for residual and applied stress fields. The stress intensity factor for residual stress field K_{res} is obtained by applying to the crack faces the residual stresses which exist in the uncracked body; K_{res} is added to the stress intensity factor resulting from external load K_{ext}; that is

$$K_{eff} = K_{ext} + K_{res}$$

Recently Leitao, Aliabadi, Cook and Rooke,[51] have used the above procedure to study the effect of prestress on crack growth. In their analysis of the configuration shown in Fig. 16, a prestressing technique was used whereby a single uniaxial tensile load is applied to cause local plasticity, and then completely unloaded. The magnitude of the average net section stress used to produce the residual stress fields was 60%,70% and 80% of the yield stress, corresponding to applied loads of 237, 277 and 317 KN respectively. The test specimens were then subjected to uniaxial fatigue loading; the loading was constant amplitude sinusoidal with a mean stress of $110Nmm^{-2}$ and alternating stress of $\pm 96.5Nmm^{-2}$ on the net section. This corresponds to minimum and maximum gross section stresses of $3.03Nmm^{-2}$ and $46.34Nmm^{-2}$ at the free end.

The residual stress fields were evaluated using an elastoplastic boundary element method, and are presented in Fig. 17. Using a boundary element weight function formulation they evaluated the stress intensity factors due to residual stress fields and the applied fatigue load (see Figs 18,19). The values of K_{eff} are shown in Fig. 20.

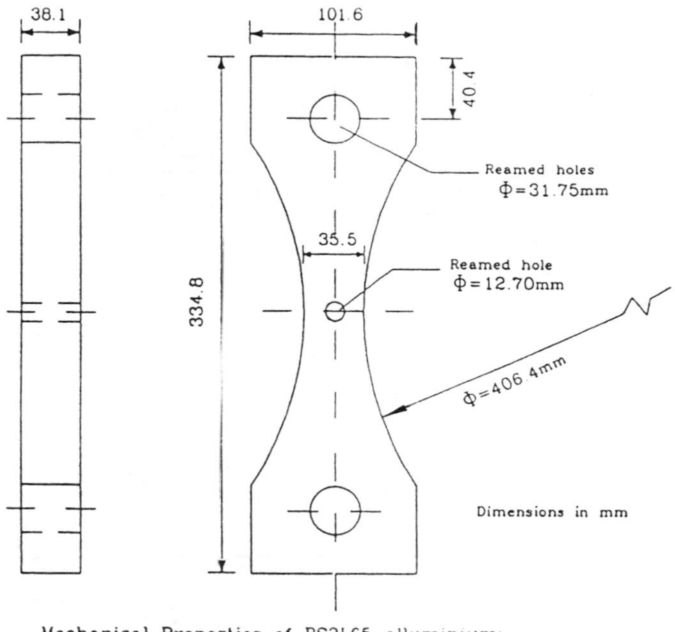

Figure 16: Fatigue test specimen

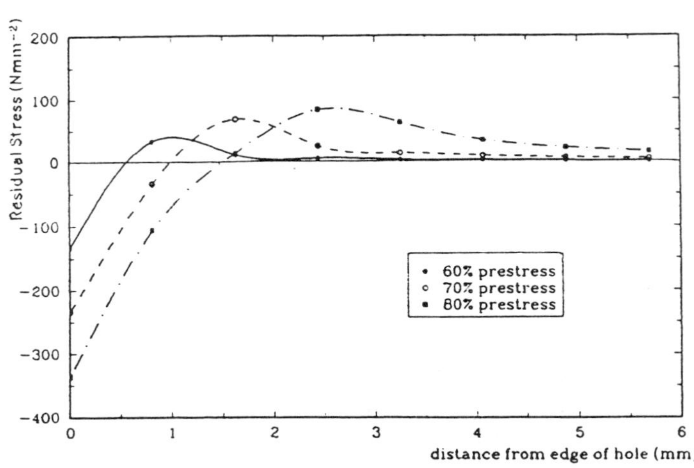

Figure 17: Residual stress field distribution

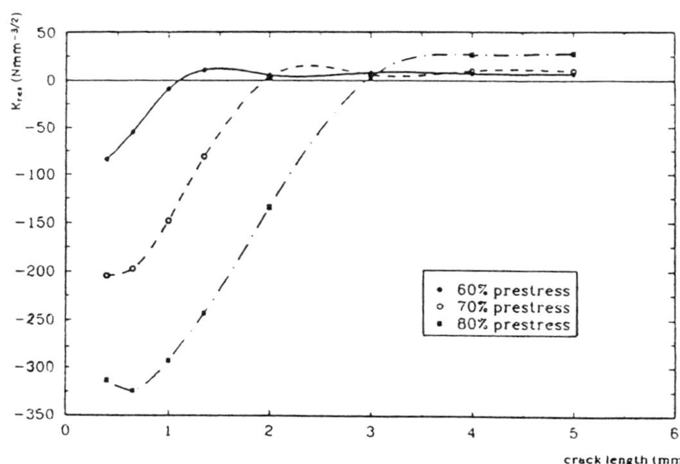

Figure 18: K due to residual stresses

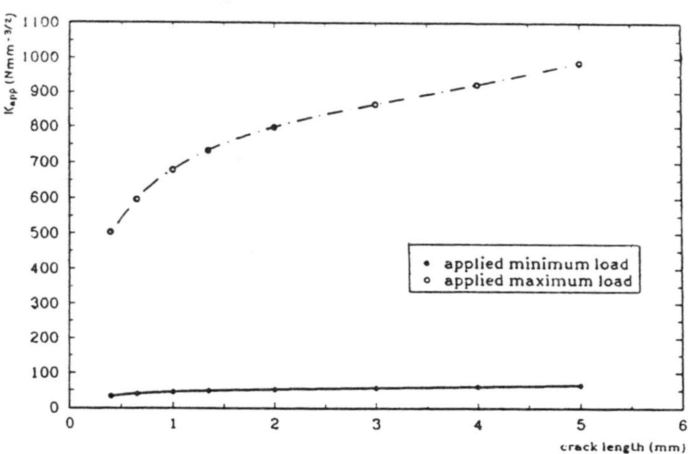

Figure 19: K due to the applied remote load

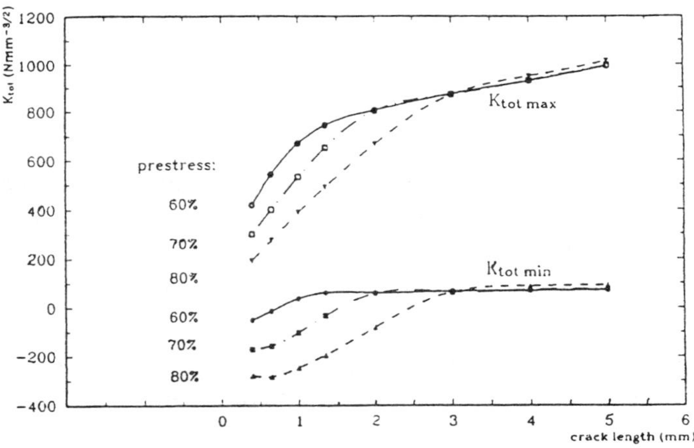

Figure 20: K due to applied and residual stresses

4 A nonlinear contact analysis for cracked structures

Dangerous cracks often occur at holes or notches which act as initiation sites. These may appear early in the life of the structure, particularly when fretting is present near stress concentration, e.g. in pin-loaded lugs and bolted or riveted joints. Therefore, it is important to consider the contact fretting forces when solving a contact problem.

The formulation presented in this section is based on the two-dimensional boundary element formulation developed by Man, Aliabadi and Rooke,[52,53,54] to calculate the load distribution at contact region in a frictional situation. Their technique is based on an iterative and fully incremental loading technique and deals with both conforming and non-conforming types of contact.

Basic contact definitions

Contact modes may be viewed as boundary constraints: they are required to be prescribed in those regions where the surfaces are in contact or potentially coming into contact. For a given contact state, the contact conditions of a contact node-pair (a and b say) may be represented by any one of three modes listed below:

Separation

$$t_{\mathrm{t}}^{a} - t_{\mathrm{t}}^{b} = 0, \quad t_{\mathrm{t}}^{a} - t_{\mathrm{n}}^{b} = 0, \quad t_{\mathrm{t}}^{a} = 0 \;\; t_{\mathrm{n}}^{a} = 0$$

Slip

$$t_{\mathrm{t}}^{a} - t_{\mathrm{t}}^{b} = 0, \quad t_{\mathrm{t}}^{a} - t_{\mathrm{n}}^{b} = 0, \quad t_{\mathrm{t}}^{a} \pm \mu t_{\mathrm{n}}^{a} = 0 \;\; u_{\mathrm{n}}^{a} + u_{\mathrm{n}}^{b} = gap^{ab}$$

Stick

$$t_{\mathrm{t}}^{a} - t_{\mathrm{t}}^{b} = 0, \quad t_{\mathrm{t}}^{a} - t_{\mathrm{n}}^{b} = 0, \quad u_{\mathrm{t}}^{a} + u_{\mathrm{t}}^{a} = 0 \;\; u_{\mathrm{n}}^{a} + u_{\mathrm{n}}^{b} = gap^{ab}$$

where t_{t} and t_{n} are the tangential and normal tractions, and u_{t} and u_{n} are the tangential and normal displacement respectively, express in local coordinates. gap^{ab} denotes the original gap between nodes a and b.

4.1 Application of BEM to contact problems

Let the boundary of two homogeneous isotropic linearly elastic bodies A and B be represented by Γ^{A} and Γ^{B}. Generally, when the bodies come into contact only parts of each boundary are in contact with each other, therefore their total boundaries may be divided into regions of *contact boundary* Γ_{c} and *non-contact boundary* Γ_{nc}, as illustrated in Fig. 21; that is

$$\Gamma^{A} = \Gamma_{\mathrm{nc}}^{A} + \Gamma_{\mathrm{c}}^{A} \tag{118}$$

$$\Gamma^{B} = \Gamma_{\mathrm{nc}}^{B} + \Gamma_{\mathrm{c}}^{B} \tag{119}$$

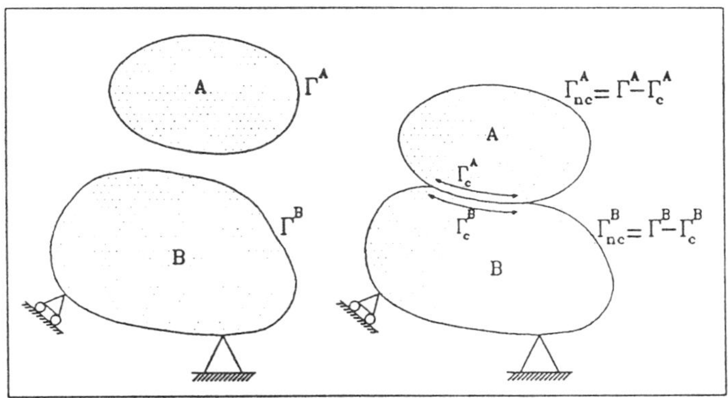

Figure 21: Contact region Γ_c and non-contact region Γ_{nc}

A contact boundary Γ_c will contain both actual contacting boundary sections and potential contacting boundary sections; it may therefore contain regions of STICK (st), regions of SLIP (sl) and regions of SEPARATION (sp). Hence Γ_c for bodies A, B may be expressed as

$$\Gamma_c^{A,B} = \Gamma_{st}^{A,B} + \Gamma_{sl}^{A,B} + \Gamma_{sp}^{A,B} \tag{120}$$

The numerical solution of contact problems require a mathematical model to represent the bodies in contact. In this work, the mathematical model is the Boundary Element Method (BEM).

If two bodies subjected to an external load are in contact over an area Γ_c, then the deformation can be described by two coupled integral equations, one for each body; they are

$$C_{ij}^A u_j^A + \int_{\Gamma_{nc}^A} T_{ij}^A u_j^A \mathrm{d}\Gamma^A + \int_{\Gamma_n^A} T_{ij}^A u_j^A \mathrm{d}\Gamma^A = \int_{\Gamma_{nc}^A} U_{ij}^A t_j^A \mathrm{d}\Gamma^A + \int_{\Gamma_n^A} U_{ij}^A t_j^A \mathrm{d}\Gamma^A \tag{121}$$

and

$$C_{ij}^B u_j^B + \int_{\Gamma_{nc}^B} T_{ij}^B u_j^B \mathrm{d}\Gamma^B + \int_{\Gamma_n^B} T_{ij}^B u_j^B \mathrm{d}\Gamma^B = \int_{\Gamma_{nc}^B} U_{ij}^B t_j^B \mathrm{d}\Gamma^B + \int_{\Gamma_n^B} U_{ij}^B t_j^B \mathrm{d}\Gamma^B \tag{122}$$

where u_j and t_j denote components of the boundary displacement vector and traction vector respectively, and U_{ij} and T_{ij} are Kelvien's fundamental solutions.

For a numerical solution to the problem the boundary integral eqns (121) and (122), the boundaries of body A and body B are discretized separately. This produces two sets of equations (one for A and one for B) given by,

$$C_{ij}^A u_j^A + \sum_{n=1}^{N} H_{ij}^{n^A} u_j^{n^A} = \sum_{n=1}^{N} G_{ij}^{n^A} t_j^{n^A} \tag{123}$$

$$C_{ij}^B u_j^B + \sum_{n=1}^{M} H_{ij}^{n^B} u_j^{n^B} = \sum_{n=1}^{M} G_{ij}^{n^B} t_j^{n^B} \tag{124}$$

where N and M are the total number of nodes for bodies A and B respectively. Two sets of simultaneous linear equations are obtained which can be expressed in matrix form as

$$[H]^A\{u\}^A = [G]^A\{t\}^A \text{ and } [H]^B\{u\}^B = [G]^B\{t\}^B \tag{125}$$

The vectors $\{u\}^{A,B}$ and $\{t\}^{A,B}$ represent boundary values of displacements and tractions. At the contact region, the two systems of equations share the boundary variables of the problem; that is the equations are coupled and must be solved simultaneously for any given combination of external load contact conditions. If the boundary conditions inside and outside any region of contact are implemented, then eqn (125) can be reduced to the single unified system of equations of the form

$$[A]\{X\} = \{F\} \tag{126}$$

For linear problems, once this system of equations has been solved, the final solution for the displacements and tractions everywhere on the boundaries can be obtained.

However, contact problems may be nonlinear and the extend of the contact region may not be known *a priori,* but must be determined as part of the solution. This means some contact problems require an iterative solution procedure. During the iterative process, coefficients in $[A]$ and $\{F\}$, derived form inside the contact zone only may change from one iteration to the next; the number of changes in matrix $[A]$ is small, because the number of elements whose contact conditions change is usually a small fraction of the total. Under normal procedures the entire system of equations would have to be reordered for the next iteration, in order to accommodate the changes in the contact zone, and then the matrix $[A]$ updated. Repetition of this procedure, until the final solution is found, would be inefficient and costly.

In order to solve the updated system of equations efficiently, without resources to a reformation of the entire system matrix, it is necessary to keep the unknowns in the potential contact zone separate from the unknowns outside it. The separation of unknowns is accomplished using a technique similar to the special bound elements used in the finite element method. This technique can substantially speed up the iterative process. Since the actual contact zone may not be known, *a priori*, it is essential to choose a potential contact zone larger than the likely final contact region. For potential contact zones, the equations obtained from the contact conditions have to be expressed explicitly, so that they can be separated from those outside the zone. In this way, a coefficient sub-matrix $\mathbf{A}_c$ can be set up for the variables of the contact (potential) region. This separation of unknowns enables eqn (126) to be arranged in such a way that the final matrix form of the merged system is as shown below:

$$\begin{bmatrix} H^A_{nc} - G^A_{nc} & H^A_c - G^A_c & 0 & 0 \\ 0 & 0 & H^B_c - G^B_c & H^B_{nc} - G^B_{nc} \\ 0 & & A_c & 0 \end{bmatrix} \{X\} = \{F\} \qquad (127)$$

The subscripts nc and c denote non-contact (potential or actual contact) zones respectively.

If the total number of nodes outside the potential contact zone for body A and body B is N^A_{nc} and M^B_{nc} respectively, this results in $2(N^A_{nc}+M^B_{nc})$ linear equations since there are two unknowns per node. Inside the contact zone at each boundary point both traction components and both displacement components are unknown. Hence, for a pair of contact nodes there are eight unknowns. To account for these eight unknowns, eight equations are provided by considering displacement compatibility and traction equilibrium at the contact interface. These compatibility and equilibrium equations can be derived explicitly for each potential contact node pair by considering the contact state of the node pair itself and its immediate neighboring node pairs, to give systems of contact equations.

Once these contact equations are defined they are grouped to form the coefficient sub-matrix called $\mathbf{A}_c$ as shown in eqn (127). The final matrix is now square and eqn (126) is in a standard form which is solvable; the vector $\{X\}$ contains all the unknowns, both inside and outside the contact region and the vector $\{F\}$ contains all the boundary conditions from outside the contact region. Solution of the matrix equation can be obtained by standard procedures and all boundary values determined. The advantage for such a structured formulation is that time is saved by avoiding re-assembly of the entire matrix when only a few changes are made to the contact region (i.e. $\mathbf{A}_c$). More importantly, a high speed matrix solver exploiting this feature can be employed (see Man, Aliabadi and Rooke,[53] such that the solution process of the system of equations is speeded up considerably. This technique has shown to be highly efficient for solving contact problems.

4.2 Incremental loading technique

In the numerical modelling of contact problems, the two bodies are brought into contact, at discrete points along the common boundary (contact boundary). In the initial unloaded state, node pairs may either be in contact or in separation. Under the influence of an external load the bodies will respond by undergoing a deformation which may lead to different possible events taking place in the potential contact region.

The final solution of a contact problem is usually governed by either the final maximum external load or the ultimate allowable contact area. But, since in most contact problems the contact area is a function of the external loads and the friction coefficient inside the contact region, the numerical solution of the problem has to be obtained iteratively; and if it is a non-conforming contact problem with friction, then the solution of the problem may have to be approximated by way of many small load steps. How the load steps are applied is highly critical in view of the nonlinear nature of the problem.

Consider, a discrete incremental load step ΔP^m_j applied to a system which is initially in equilibrium. The system will respond to the applied load by undergoing small perturbation in displacements and tractions everywhere on the boundary, to give a new equilibrium state. The new total external discrete load is defined as,

$$P_j^m = P_j^{m-1} + \Delta P_j^m \tag{128}$$

and the changes in displacements u_j and tractions t_j can be defined as,

$$u_j^m = u_j^{m-1} + \Delta u_j^m \quad \text{and} \quad u_j^m = u_j^{m-1} + \Delta u_j^m \tag{129}$$

where Δt_j^m and Δu_j^m are the incremental changes in tractions and displacements respectively, due to the incremental of load ΔP_j^m. Substituting eqn (129) into the boundary element equation, gives

$$C_{ij}(u_j^{m-1} + \Delta u_j^m) + \int_\Gamma T_{ij}(u_j^{m-1} + \Delta u_j^m)\mathrm{d}\Gamma = \int_\Gamma U_{ij}(t_j^{m-1} + \Delta t_j^m)\mathrm{d}\Gamma \tag{130}$$

From the principle of superposition an integral equation in the Δ(incremental) variable can be obtained;

$$C_{ij}\Delta u_j^m + \int_\Gamma T_{ij}\Delta u_j^m \mathrm{d}\Gamma = \int_\Gamma U_{ij}\Delta t_j^m \mathrm{d}\Gamma \tag{131}$$

The nodes on the discretized boundaries have to be organized at the contact region in such a way that they form contact node-pairs, representing the contact interface. Each contact node-pair has to be coupled and treated as an independent contact system where both the incremental traction Δt and the incremental displacement Δu must satisfy both equilibrium and compatibility conditions. Discretization of two bodies (A and B) produce two individual systems of equations; they are given, in incremental terms, by

$$[H]^A\{\Delta u\}^A = [G]^A\{\Delta t\}^A \quad \text{and} \quad [H]^B\{\Delta u\}^B = [G]^B\{\Delta t\}^B \tag{132}$$

The vectors $\{\Delta u\}^{A,B}$ and $\{\Delta t\}^{A,B}$ contain the boundary values of incremental displacements and tractions. This incremental problem can be reduced to a form of eqn (126) and solved in the usual way, once the conditions outside the contact region and the contact constraints inside are applied.

4.2.1 Implementation of contact constraints

In the incremental load (ΔP) approach presented in eqn (131) above, contact conditions of a node pair a and b have to be expressed in a form such that the incremental response controls the total response. That is, as each ΔP^m is applied, the incremental quantities ($\Delta u, \Delta t$) inside the contact region are obtained subject to the total traction equilibrium and the total displacement compatibility at the node-pair. The constraints for the different contact conditions are as follows:

1. **Stick** mode

$$\Delta(t_t^a)^m - \Delta(t_t^b)^m = -\left[(t_t^a)^{m-1} - (t_t^b)^{m-1}\right]$$

$$\Delta(t_n^a)^m - \Delta(t_n^b)^m = -\left[(t_n^a)^{m-1} - (t_n^b)^{m-1}\right]$$

$$\Delta(u_t^a)^m + \Delta(u_t^b)^m = 0$$

$$\Delta(u_n^a)^m + \Delta(u_n^b)^m = g_o - \left[(u_n^a)^{m-1} + (u_n^b)^{m-1}\right] \equiv g_o^m \tag{133}$$

2. **Slip** mode

$$\Delta(t_t^a)^m - \Delta(t_t^b)^m = -\left[(t_t^a)^{m-1} - (t_t^b)^{m-1}\right]$$

$$\Delta(t_n^a)^m - \Delta(t_n^b)^m = -\left[(t_n^a)^{m-1} - (t_n^b)^{m-1}\right]$$

$$\Delta(t_t^a)^m \pm \mu\Delta(t_n^b)^m = -\left[(t_t^a)^m \pm \mu(t_n^b)^{m-1}\right]$$

$$\Delta(u_n^a)^m + \Delta(u_n^b)^m = g_o - \left[(u_n^a)^{m-1} + (u_n^b)^{m-1}\right] \equiv g_o^m \tag{134}$$

3. **Separation** mode

$$\Delta(t_t^a)^m - \Delta(t_t^b)^m = -\left[(t_t^a)^{m-1} - (t_t^b)^{m-1}\right]$$

$$\Delta(t_n^a)^m - \Delta(t_n^b)^m = -\left[(t_n^a)^{m-1} - (t_n^b)^{m-1}\right]$$

$$\Delta(t_t^a)^m = -\left[(t_t^b)^m\right]$$

$$\Delta(t_n^a)^m = -\left[(t_n^a)^{m-1}\right] \tag{135}$$

The previous total traction and total displacements (t^{m-1}, u^{m-1}) which are already known make up the right-hand-sides; they are updated from load-step to load-step. The unknown quantities are the (Δ) incremental values on the left-hand-side as shown in eqns (133) to (135); they are expressed in the local coordinate system.

4.3 Numerical examples

A pin-loaded plate is a common engineering connection in which load is transmitted through a pin to a plate by contact pressure on the bore of a hole. Small oscillatory, relative tangential movements of touching metallic surfaces often occur in clamped or fastened engineering structures. Repeated rubbing of these contacting surfaces causes damage to the surface and can lead to the initiation of cracks. Man *et al.*[54] simulated this problem using an Al-alloy plate and a stiff steel pin with Young's modulus E_p of $2.1 \times 10^5 Nmm^{-2}$; the modulus of the plate E_s is $E_p/3$. The same Poisson's ratio assumed $\nu = 0.3$ and plane strain is assumed. The dimensions of the problem have been chosen such that $H/W = 2, W/R = 8$ and $R = 10mm$, where H is the height and $W =$ is the width of the plate, and R is the radius of the pin. The same problem is also solved with a pair of radial cracks introduced at the edge of the hole bore. The crack length to radius ratio was chosen to be $a/R = 2$, where a is the distance from the centre of the hole to the crack tip. Here, the clearance-fit is simulated by having difference in the radius between the hole and the pin.

Stress intensity factors are calculated using J-integral technique. Numerical results for these problems are described below with solutions for the cracked and uncracked cases directly compared; for all the cases studied, the maximum total loads were approximately the same. In all cases linear boundary elements are used in the contact region and quadratic elements outside.

4.3.1 Tolerance-fit in a cracked or uncracked plate

A tolerance-fit pin configuration is a load dependent problem, non-conforming and is of a progressive contact nature. It is solved here using the load incremental technique.

The potential contact area is defined by, $-65^o \leq \theta \leq 65^o$, where θ is measured from the lowest point of the pin (centre of contact). The potential contact region is represented by a total of 26 equal size linear element pairs. The linear elements are evenly and symmetrically distributed along the potential contact region such that each half of the region contains a total of 13 node-pairs. The pin is loaded through a small cavity at its centre by the traction field derived from that of a line force P per unit thickness in an infinite plate. The load was increased step by step until the maximum allowable contact area was reached; that is when $| \theta_{\mathbf{max}} |= 65^o$. In Fig. 22, the load history is plotted for $-65^o \leq \theta \leq 0^o$ for the uncracked plate; and for $0^o \leq \theta \leq 65^o$ for the cracked plate. The variation of the stress intensity factors against contact angle and load are presented in Figs 23 and 24 respectively. In Figs 25 and 26, the variation of the contact tractions for $\mu = 0$ and 0.4 are shown respectively.

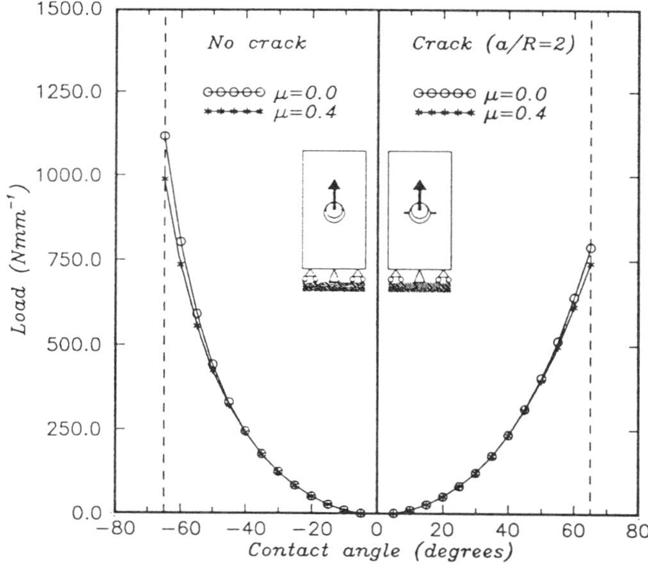

Figure 22: Load history vs. contact angle θ

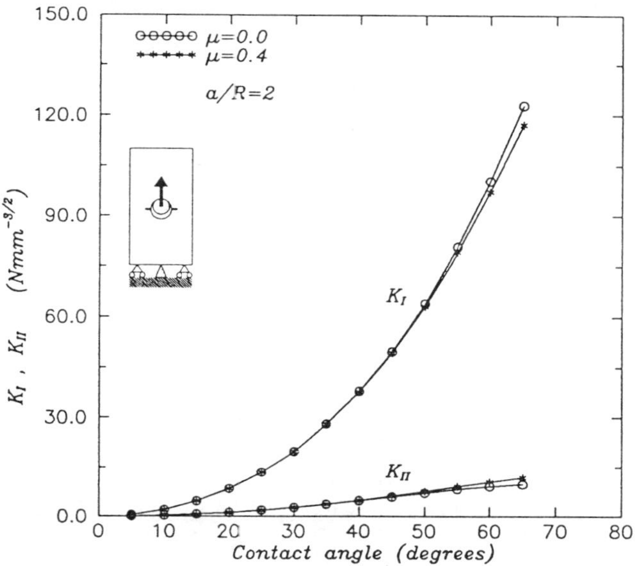

Figure 23: Stress intensity factor vs. contact angle

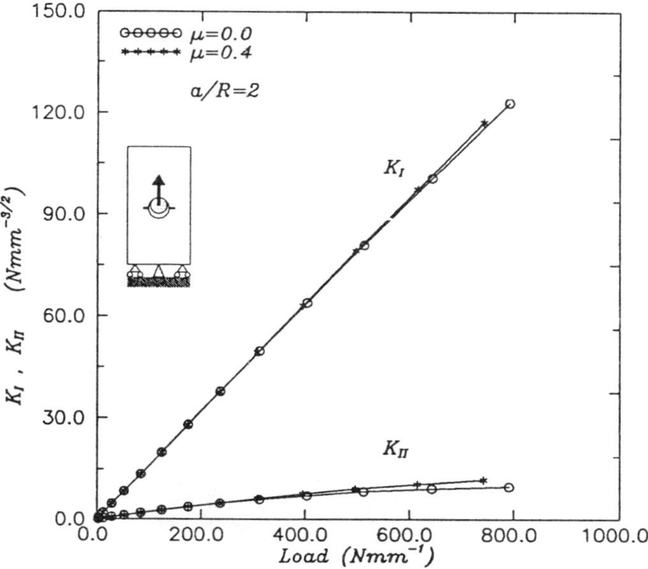

Figure 24: Stress intensity factor vs. load

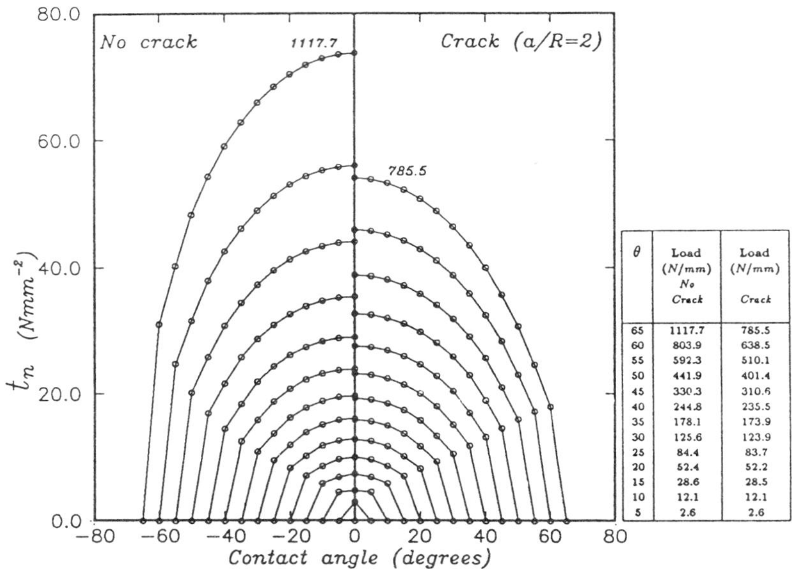

θ	Load (N/mm) No Crack	Load (N/mm) Crack
65	1117.7	785.5
60	803.9	638.5
55	592.3	510.1
50	441.9	401.4
45	330.3	310.6
40	244.8	235.5
35	178.1	173.9
30	125.6	123.9
25	84.4	83.7
20	52.4	52.2
15	28.6	28.5
10	12.1	12.1
5	2.6	2.6

Figure 25: Normal traction t_n distribution for $\mu = 0$

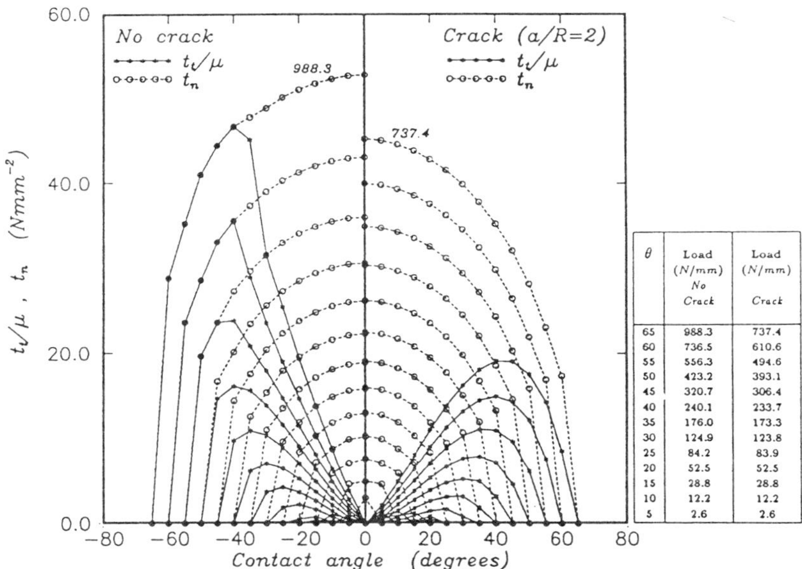

θ	Load (N/mm) No Crack	Load (N/mm) Crack
65	988.3	737.4
60	736.5	610.6
55	556.3	494.6
50	423.2	393.1
45	320.7	306.4
40	240.1	233.7
35	176.0	173.3
30	124.9	123.8
25	84.2	83.9
20	52.5	52.5
15	28.8	28.8
10	12.2	12.2
5	2.6	2.6

Figure 26: Normal t_n and tangential traction distribution for $\mu = 0.4$

5 The dual boundary element method

As described in section 1.2, the solution of general crack problems cannot be achieved with the direct application of the BEM, in a single-region analysis, because the coincidence of the crack surfaces gives rise to a singular system of algebraic equations.

Some special techniques have been devised to overcome this difficulty. Among these, the most general are the subregion method[4] and the dual boundary element method.[5,6,55,56] The subregion method introduces artificial boundaries into the structure, which connect the cracks to the boundary, in such a way that the domain is divided into the subregions without cracks. In an incremental crack extension analysis, these artificial boundaries must be repeatedly introduced for each increment of the crack extension. The main drawback of this method is that the introduction of artificial boundaries is not unique, and thus cannot be easily implemented as an automatic procedure in an incremental analysis of crack-extension problems. In addition, the method generates a larger system of algebraic equations than is strictly required.

The dual boundary element method (DBEM) incorporates two independent boundary integral equations, with the displacement equation applied for collocation on one of the crack surfaces and the traction equation on the other. As a consequence, general mixed-mode crack problems can now be solved in a single region formulation. Although the integration path is still the same for coincident points on the crack surfaces, their respective boundary integral equations are now distinct. The single-region analysis, characteristic of the DBEM, can eliminate the remeshing problems associated with crack growth problems which are typical of the finite element and subregion boundary element methods.

5.1 The dual boundary integral equations

The displacement boundary integral equation relating the boundary displacement components u_j and traction components t_j can be written as

$$C_{ij}(\mathbf{x}')u_j(\mathbf{x}') = \int_\Gamma U_{ij}(\mathbf{x}',\mathbf{x})t_j(\mathbf{x})d\Gamma(\mathbf{x}) - \int_\Gamma T_{ij}(\mathbf{x}',\mathbf{x})u_j(\mathbf{x})d\Gamma(\mathbf{x}) \qquad (136)$$

In absence of body forces and assuming continuity of both strains and tractions at $\mathbf{x}'$ on a smooth boundary, the stress components σ_{ij} are given by

$$\frac{1}{2}\sigma_{ij}(\mathbf{x}')= \int_\Gamma U_{ijk}(\mathbf{x}',\mathbf{x})t_k(\mathbf{x})d\Gamma(\mathbf{x}) - \int_\Gamma T_{ijk}(\mathbf{x}',\mathbf{x})u_k(\mathbf{x})d\Gamma(\mathbf{x}) \qquad (137)$$

and the traction integral equation can be obtained form (137), as

$$\frac{1}{2}t_j(\mathbf{x}') = n_i(\mathbf{x}')\int_\Gamma D_{ijk}(\mathbf{x}',\mathbf{x})t_k(\mathbf{x})d\Gamma(\mathbf{x}) - n_i(\mathbf{x}')\int_\Gamma S_{ijk}(\mathbf{x}',\mathbf{x})u_k(\mathbf{x})d\Gamma(\mathbf{x}) \qquad (138)$$

where n_i denotes the i component of the unit outward normal to the boundary at the point $\mathbf{x}'$. The boundary integral eqns (136) and (138) constitute the dual boundary element method.

The fundamental solution U_{ij} contains a weak singularity of order $\ln(\frac{1}{r})$ in 2D and singularity of order $(\frac{1}{r})$ in 3D; T_{ij} has a singularity of order $(\frac{1}{r})$ in 2D and $(\frac{1}{r^2})$ in 3D; D_{ijk} has a singularity of order $(\frac{1}{r})$ in 2D and $(\frac{1}{r^2})$ in 3D; S_{ijk} has a singularity of order $(\frac{1}{r^2})$ in 2D and $(\frac{1}{r^3})$ in 3D. Special techniques for dealing with this type of highly singular integrands are described in Refs 5 and 6.

5.1.1 Crack modelling strategy

For the sake of efficiency and to keep the simplicity of the standard boundary elements, the DBEM formulation uses discontinuous quadratic elements for the crack modelling as shown in Fig. 27. The general modelling strategy can be summarized as follows:

a) the crack boundaries are modelled with discontinuous quadratic elements

b) continuous quadratic elements are used along the remaining boundaries of the structure, except at an intersection between a crack and an edge, where 'edge discontinuous' elements are used

c) the displacement eqn (30) is applied for collocation on one of the crack surfaces

d) the traction integral eqn (31) is applied for collocation on other crack surface

e) the displacement integral eqn (30) is applied for collocation on all non-crack boundaries.

5.2 Crack growth analysis

In many practical situations structures are subjected to shear and torsional as well as tensile loadings, which will lead to mixed-mode cracking. Two criteria for mixed-mode loading that allow non-coplanar crack growth have been proposed; one based on the maximum principal stress by Erdogan and Sih[58] and the other on the strain energy density factor proposed by Sih.[59]

5.2.1 Maximum principal stress criterion

The maximum principal stress criterion postulates that a crack will grow in a direction perpendicular to the maximum principal stress. Considering two-dimensional combined mode I and mode II loading the stress $\sigma_{\theta\theta}$ and $\sigma_{r\theta}$ at the crack tip are given by

$$\sigma_{\theta\theta} = \frac{1}{\sqrt{2\pi r}} \cos\frac{\theta}{2}\left[K_I \cos^2\frac{\theta}{2} - \frac{3}{2}K_{II}\sin\theta\right] \tag{139}$$

and

$$\sigma_{r\theta} = \frac{1}{2\sqrt{2\pi r}} \cos\frac{\theta}{2}\left[K_I \sin\theta + K_{II}(3\cos\theta - 1)\right] \tag{140}$$

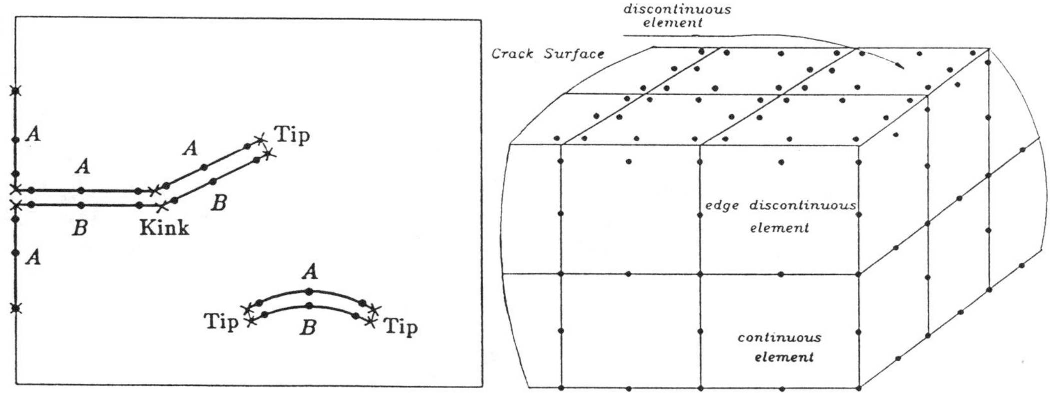

Figure 27: Crack modelling strategy

The stress $\sigma_{\theta\theta}$ will be the maximum principal stress at $\phi = \theta$, where ϕ is defined by $\sigma_{r\theta} = 0$, that is

$$K_I \sin\phi + K_{II}(3\cos\phi - 1) = 0 \tag{141}$$

Thus, the local crack growth direction ϕ is determined from (141); ϕ is measured from the crack axis ahead of the tip.

5.2.2 Strain energy density criterion

The strain energy density criterion states that crack growth takes place in the direction of minimum strain energy density factor S. With reference to the coordinate system (n, b, t) in Fig. 28, the strain energy stored in a volume element $dv = dndbdt$ is given as

$$\frac{dW}{dv} = \frac{1}{2E}(\sigma_n^2 + \sigma_b^2 + \sigma_t^2) - \frac{\nu}{E}(\sigma_n\sigma_b + \sigma_b\sigma_t + \sigma_t\sigma_n) + \frac{1+\nu}{E}(\tau_{nb}^2 + \tau_{bt}^2 + \tau_{tn}^2) \tag{142}$$

where ν is the Poisson's ratio and E is the Young's modulus. The stress field in the vicinity of the crack is given by

$$\sigma_n = \frac{K_I}{\sqrt{2\pi r}}\cos\frac{\theta}{2}\left[1 - \sin\frac{\theta}{2}\sin\frac{3\theta}{2}\right] - \frac{K_{II}}{\sqrt{2\pi r}}\sin\frac{\theta}{2}\left[2 + \cos\frac{\theta}{2}\cos\frac{3\theta}{2}\right]$$

$$\sigma_b = \frac{K_I}{\sqrt{2\pi r}}\cos\frac{\theta}{2}\left[1 + \sin\frac{\theta}{2}\sin\frac{3\theta}{2}\right] + \frac{K_{II}}{\sqrt{2\pi r}}\sin\frac{\theta}{2}\cos\frac{\theta}{2}\cos\frac{3\theta}{2}$$

$$\sigma_t = 2\nu \frac{K_I}{\sqrt{2\pi r}} \cos\frac{\theta}{2} - 2\nu \frac{K_{II}}{\sqrt{2\pi r}} \sin\frac{\theta}{2}$$

$$\tau_{nb} = \frac{K_I}{\sqrt{2\pi r}} \cos\frac{\theta}{2} \sin\frac{\theta}{2} \cos\frac{3\theta}{2} + \frac{K_{II}}{\sqrt{2\pi r}} \cos\frac{\theta}{2}\left[1 - \sin\frac{\theta}{2}\sin\frac{3\theta}{2}\right]$$

$$\tau_{nt} = -\frac{K_{III}}{\sqrt{2\pi r}} \sin\frac{\theta}{2}$$

$$\tau_{bt} = \frac{K_{III}}{\sqrt{2\pi r}} \cos\frac{\theta}{2} \tag{143}$$

where K_I, K_{II} and K_{III} are mode I, II and III stress intensity factors and (r, θ) are the polar coordinate system in the plane normal to the crack front. Substituting the above crack border stress fields in to (142), the strain energy per unit volume may be written as

$$\frac{dW}{dv} = \frac{S}{r} + \text{nonsingular terms} \tag{144}$$

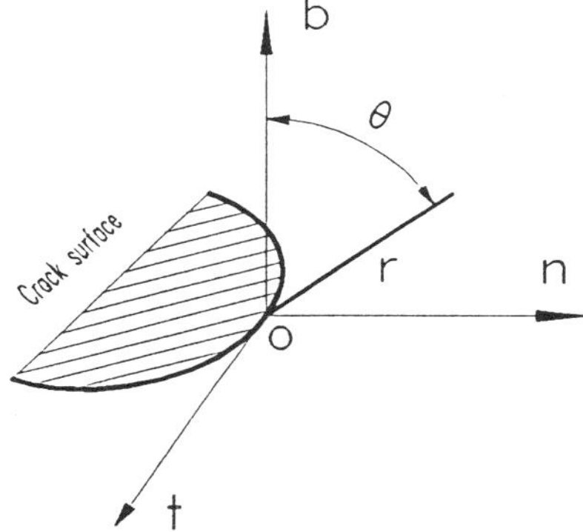

Figure 28: Crack coordinate system

The magnitude of S in eqn (144) is referred to as the strain energy density factor. It can be expressed in terms of stress intensity factor as

$$S = a_{11}K_I^2 + 2a_{12}K_IK_{II} + a_{22}K_{II}^2 + a_{33}K_{III}^2 \tag{145}$$

where

$$a_{11} = \frac{1}{16\mu}(3 - 4\nu - \cos\theta)(1 + \cos\theta)$$

$$a_{12} = \frac{1}{8\mu}\sin\theta(\cos\theta - 1 + 2\nu)$$

$$a_{22} = \frac{1}{16\mu}[4(1 - \nu)(1 - \cos\theta) + (3\cos\theta - 1)(1 + \cos\theta)]$$

$$a_{33} = \frac{1}{4\mu}$$

in which μ stands for the shear modulus and θ is the angle in the crack front coordinate system shown in Fig. 28.

The fundamental assumptions of strain energy density factor criterion for three-dimensional crack problem are as follows:

Hypothesis (1): *The direction of crack propagation at any point along the crack front is towards the region with the minimum value of strain energy density factor S as compared with other regions on the same spherical surface surrounding the point.*

Hypothesis (2): *Crack extension occurs when the strain energy density factor in the region determined by hypothesis (1), $S = S_{\min}$, reaches a critical value, say S_{cr}.*

Hypothesis (3): *The length, r_o, of the initial crack extension is assumed to be proportional to $S_{\min}$ such that $S_{\min}/r_o$ remains constant along the new crack front.*

The crack incremental direction (i.e. θ), at each point along the front is obtained through minimizing the strain energy density factor S with respect to θ. The crack incremental size r_o is decided by keeping $S_{\min}/r_o$ constant.

5.3 Crack extension application

The incremental crack extension analysis assumes a piece-wise linear discretization of the unknown crack path. For each increment of the crack extension, the DBEM is applied to carry out a stress analysis of cracked structure and the J-integral method is applied to evaluate the stress intensity factors. After each DBEM analysis, the direction of the next increment of crack extension is evaluated using the maximum principal stress criterion for two-dimensional problems and strain energy density criterion for three-dimensional problems.

The steps of the computational cycles are summarized as follows:

(i) Carry out a DBEM stress analysis of cracked structure;

(ii) Compute the stress intensity factors along the crack front;

(iii) Compute the direction of crack extension

(iv) Repeat all the above steps sequentially until a specified number of crack extension increment is reached.

5.3.1 Test example in 2D

Portela, Aliabadi and Rooke[56] considered a square plate of height h, with an edge crack of length $a = h/3$, subjected to either pure mode I or pure mode II loadings, as presented in Fig. 29. Results of the incremental crack extension analysis of these problems were obtained with maximum principal stress criterion and the maximum crack extension increment equal to three times the initial crack-tip element. Figures 30 and 31 present these results in terms of the crack paths and stress intensity factor diagrams, for each one of the problems.

5.3.2 Test example in 3D

Mi and Aliabadi[57] studied an elliptical crack of major semi-axis a and minor semi-axis b in a cylindrical bar of radius R and height h subjected to a tensile stress σ as shown in Fig. 32. The angle φ between y-axis and the crack surface is taken as 45°, resulting in a mixed-mode problem. The ratio $b/a = 0.5, R/a = 10$ and $h/R = 6$ are chosen and the Young's modulus is taken to be 1000 units and the Poissons ratio 0.3.

A total of five increments is performed. The crack surface after the last analysis, including five crack fronts are shown in Fig. 33. Mode I, II and III stress intensity factors corresponding to each analysis are plotted in Fig. 34.

The stress intensity factors for the original problem were found to be within 2% of the exact solution for an embedded elliptical crack in an infinite domain. As it can be seen the crack is growing into the shape which gives constant mode I stress intensity factors along the crack front with vanishing mode II stress intensity factors. The mode III stress intensity factors remain more or less unchanged with crack growth.

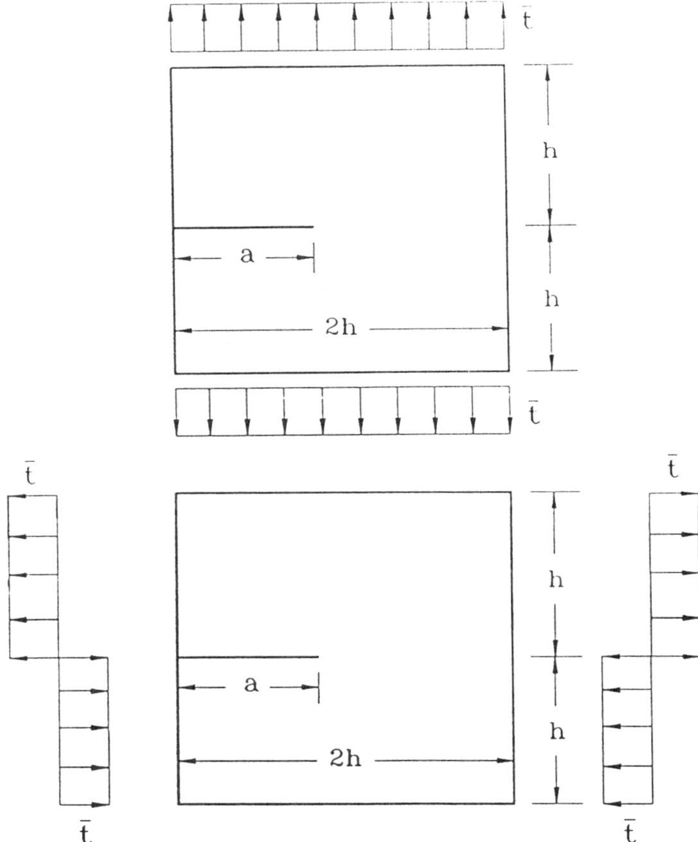

Figure 29:

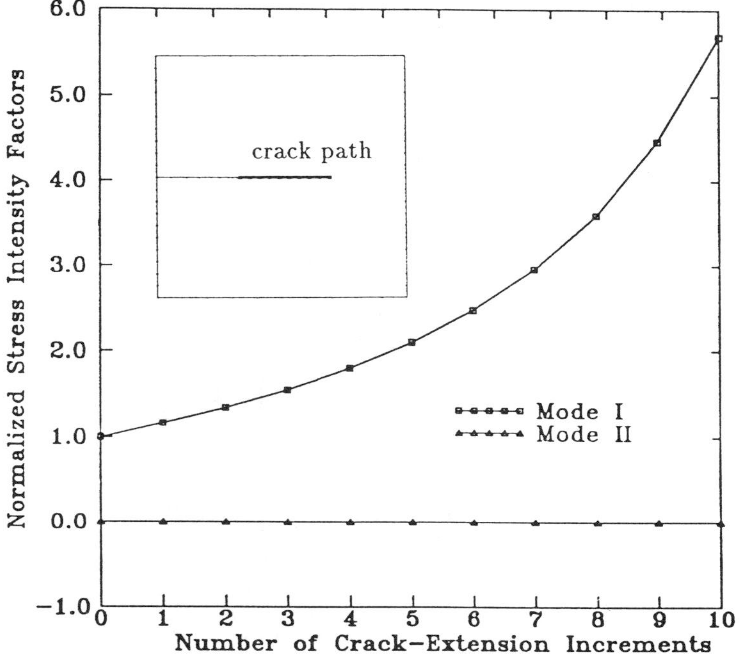

Figure 30:

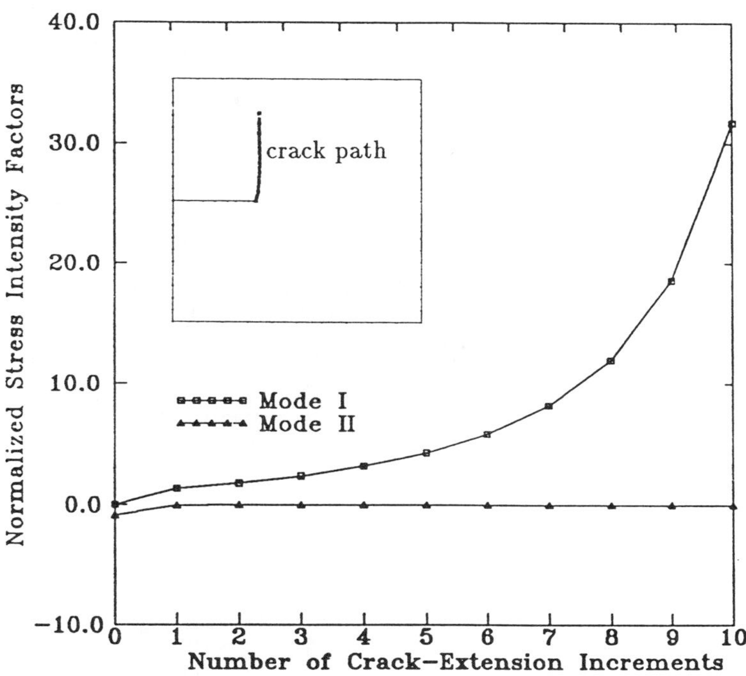

Figure 31:

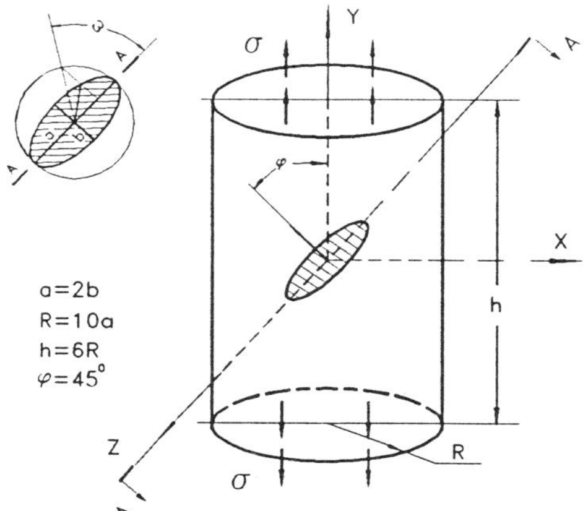

Figure 32:

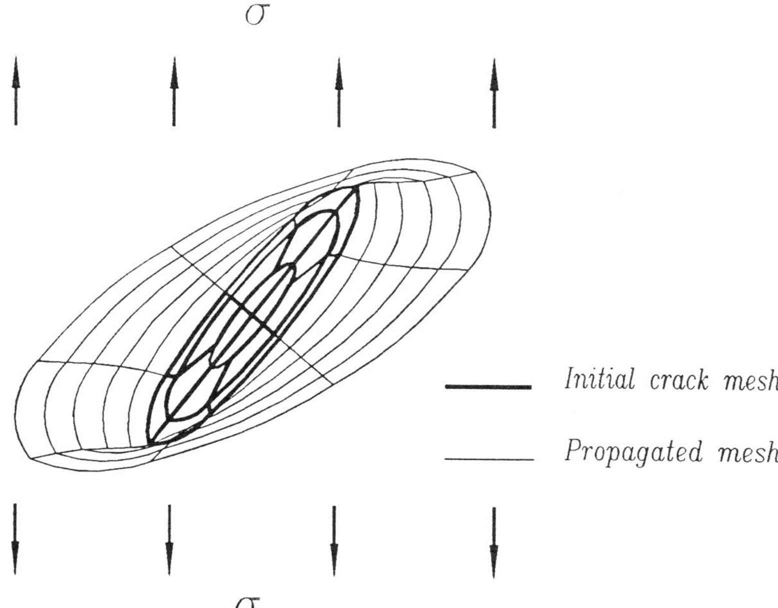

Figure 33:

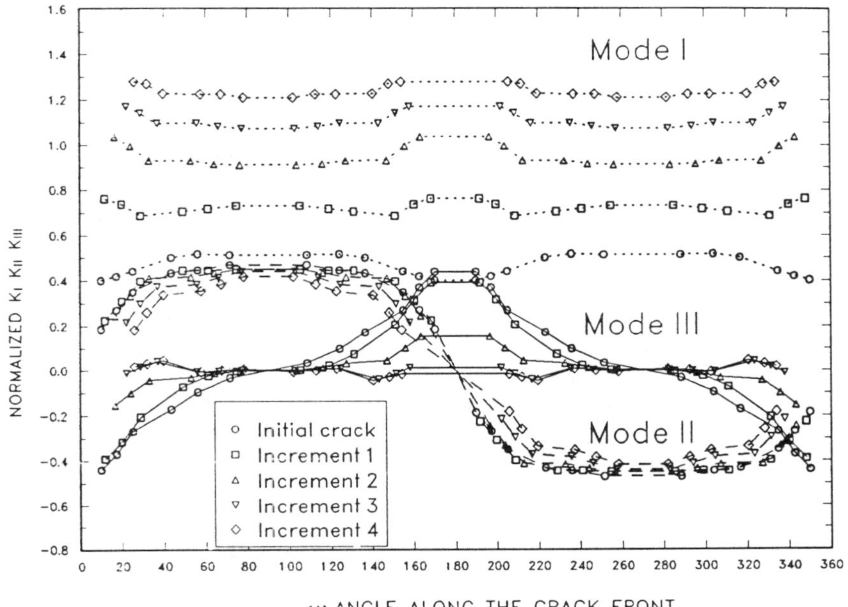
Mode I
Mode III
Mode II
Initial crack
Increment 1
Increment 2
Increment 3
Increment 4
NORMALIZED KI KII KIII
ω: ANGLE ALONG THE CRACK FRONT

Figure 34:

References

1. Brebbia, C.A & Domínguez, J. *Boundary Elements - An Introductory Course*, Computational Mechanics Publications, Southampton, 1992.

2. Aliabadi, M.H & Rooke, D.P. *Numerical Fracture Mechanics*, Kluwer Academic Publishers, Dordrecht and Computational Mechanics Publications, Southampton, 1991.

3. Cruse, T.A. Numerical evaluation of elastic stress intensity factors by the boundary integral equation method, in *The Surface Crack: Physical Problems and Computational Solutions*, J.L. Swedlow (ed), ASME, New York, 1972, pp. 153-170.

4. Balndford, G.E., Ingraffea, A.R. & Liggett, J.A. Two-dimensional stress intensity factor computation using the boundary element method, *Int. J. Num. Meth. Engng.*, 1981, **17**, 387-404.

5. Portela, A., Aliabadi, M.H. & Rooke, D.P. The dual boundary element method: effective implementation for crack problems, *Int. J. Num. Meth. Engng.*, 1992, **33**, 1269-1287.

6. Mi, Y. & Aliabadi, M.H. Dual boundary element method for three-dimensional fracture mechanics analysis, *Engng. Anal. with Bound. Elem.*, 1992, **10**, 161-171.

7. Henshell, R.D. & Shaw, K.G. Crack-tip finite elements are unnecessary, *Int. J. Num. Meth. Egng.*, 1975, **9**, 495-507.

8. Barsoum, R.S. On the use of isoparametric finite elements in linear fracture mechanics, *Int. J. Num. Meth. Engng.*, 1976, **10**, 25-37.

9. Williams, M.L. Stress singularities resulting from various boundary conditions, *J. Appl. Mech.*, 1952, **19**, 526-528.

10. Smith, R.N.L. & Mason, J.C. A boundary element method for curved crack problems in two-dimensions, in *Proc. of 4th Int. Seminar on BEM*, C.A.Brebbia (ed), 1987, Springer-Verlag, Berlin, pp. 472-484.

11. Martínez, J. & Domínguez, J. On the use of quarter-point boundary elements for stress intensity factor computations, *Int. J. Num. Meth. Engng.*, 1984, **20**, 1941-1950.

12. Ishikawa, H., Kitagawa, H. & Okamura, H. J-integral of a mixed mode crack and its application, *Proc. 3rd Int. Conf. on Mechanical Behaviour of Materials*, Vol. 3, 1980, Pergamon Press, Oxford, pp. 447-455.

13. Civelek, M.B. & Erdogan, F. Crack problems for a rectangular sheet and an infinite strip, *Int. J. Fracture*, 1982, **19**, 139-159.

14. Murakami, Y. *Stress Intensity Factors Handbook*, Pergamon Press, 1987.

15. Symm, G.T. Treatment of singularities in the solution of Laplace's equation by an integral equation method, National physical laboratory report NAC31, 1973.

16. Papamichel, N. & Symm, G.T. Numerical techniques for two-dimensional Laplacian problems, *Comp. Meth. Appl. Mech. Egng.*, 1975, **6**, 175-194.

17. Xanthis, L.S., Bernal, M.J.M. & Atkinson, C. The treatment of singularities in the calculation of stress intensity factors using the integral equation method, *Comp. Methods in Appl. Mechs. Engng.*, 1981, **26**, 285-304.

18. Aliabadi, M.H., Rooke, D.P. & Cartwright, D.J. An improved boundary element formulation for calculating stress intensity factors: application to aerospace structures, *J. Strain. Analysis*, 1987, **22**, 203-207.

19. Aliabadi, M.H. An enhanced boundary element method for determining fracture parameters, *Proc. 4th Int. Conf. on Numerical Methods in Fracture Mechanics*, 1987, San Antonio, Texas, Pineridge Press, 27-39.

20. Portela, A., Aliabadi, M.H. & Rooke, D.P. Efficient boundary element analysis of sharp notched plates, *Int. J. Num. Meth. Engng*, 1991, **32**, 445-470.

21. Aliabadi, M.H. & Rooke, D.P. A new procedure for calculating three-dimensional stress intensity factors using boundary elements, in *Advances in BEM*, Vol. 3, C.A. Brebbia (ed), 1989, Computational Mechanics Publication, Southampton, pp. 123-131.

22. Newman, J.C. An improved method of collocation for the stress analysis of cracked plates with various shaped boundaries, NASA, TN,D-6376, 1971.

23. Ching-Bing Ling. On the stresses in a plate containing two circular holes, *J. Appl. Phys.*, 1948, **19**, 77-82.

24. Smith, R.N.L. & Aliabadi, M.H. Boundary integral equation method for the solution of cracked problems, *Math. Comput. Modelling*, 1981, **15**, 285-293.

25. Bueckner, H.F. A novel principle for the computation of stress intensity factors, *Z. Agnew. Meth.*, 1970, **50**, 529-546.

26. Rice, J.R. Some remarks on elastic crack-tip stress fields, *Int. J. Solids and Strucutres*, 1972, **8**, 751-758.

27. Bueckner, H.F. Field singularities and related representations, in *Methods of Analysis and Solutions of Crack Problems, Mechanics of Fracture I*, G.C. Sih (ed), 1973, Noordhoff, Leyden, 239-314.

28. Bueckner, H.F. Weight function and fundamental fields for the penny shaped and the half-plane crack in three-space, *Int. J. Solids and Structures*, 1987, **23**, 57-93.

29. Rice, J.R. First order variation in elastic fields due to variation in location of a planar crack front, *J. Appl. Mech.*, 1985, **52**, 571-579.

30. Rice, J.R. Weight function theory for three-dimensional elastic crack analysis, in *Fracture Mechanics: Perspectives and Directives*, STP 1020, R.P.Wei & R.P.Gangloff (ed), 1989, ASTM, pp. 29-57.

31. Gao, H. & Rice, J.R. Somewhat circular tensile cracks, *Int. J. Fracture*, 1987, **33**, 155-174.

32. Gao, H. & Rice, J.R. Shear stress intensity factors for a planar crack with slightly curved front, *J. Appl. Mechanics*, 1986, **53**, 774-778.

33. Bueckner, H.F. The weight function of the configuration of collinear cracks, *Int. J. Fracture*, 1975, **11**, 71-83.

34. Bueckner, H.F. Observations of weight functions, *Engng. Analysis with Boundary Elements*, 1989, **6**, 3-18.

35. Paris, P.C., McMekking, R.M. & Tada, H. The weight function method for determining stress intensity factors, in *Cracks and Fracture*, J.L. Swedlow & M.L. Williams (ed), STP 601, 1976, ASTM, pp. 471-489.

36. Bortmann, Y. & Banks-Sills, L. An extended weight function method for mixed-mode elastic crack analysis, *J. Appl. Mech.*, 1983, **50**, 907-909.

37. Chen, Y.Z. Weight function technique in a more general case, *Engng. Frac. Mech.*, 1989, **33**, 983-986.

38. Rooke, D.P., Cartwright, D.J. & Aliabadi, M.H. Boundary elements combined with singular field for three-dimensional cracked solids, *Proc. of 4th Int. Conf. on Numerical Methods in Fracture Mechanics*, A.P. Luxmore *et al.* (ed), 1987, Pineridge Press, pp. 15-26.

39. Aliabadi, M.H., Cartwright, D.J. & Rooke, D.P. Fracture mechanics weight functions by the removal of singular fields using boundary element analysis, *Int. J. Fracture*, 1987, **34**, 131-147.

40. Rooke, D.P. & Aliabadi, M.H. The use of fundamental fields to obtain weight functions for mixed-mode cracks, to appear in *J. Engng. Sciences*.

41. Bains, R.S., Aliabadi, M.H. & Rooke, D.P. Fracture mechanics weight functions in three dimensions: subtraction of fundamental fields, *Int. J. Num. Meth. Engng.*, 1992, **35**, 179-202.

42. Irwin, G.R. Analysis of stresses and strains near the end of a crack traversing a plate, *J. Appl. Mech.*, 1957, **24**, 361-364.

43. Keer, L.M. & Freedman, J.M. Tensile strip with edge cracks, *Int. J. Egng. Sce.*, 1973, **11**, 1265-1275.

44. Bains, R.S., Aliabadi, M.H. & Rooke, D.P. Weight functions for curved crack fronts using BEM, *J. Strain Analysis*, 1993, **28**, 67-78.

45. Dugdale, D.S. Yielding of steel sheets containing slits, *J. Mech. Phys. Solids*, 1960, **8**, 100-104.

46. Aliabadi, M.H. & Cartwright, D.J. Boundary element analysis of strip yield cracks, *Engng. Analysis*, 1991, **8**, 9-12.

47. Lorenzo, J.M., Cartwright, D.J. & Aliabadi, M.H. Boundary element weight function analysis for crack surface displacements and strip yield cracks, *Proc. of BEM XV*, Vol. 2, C.A.Brebbia & J.J. Rencis (ed), 1993, Computational Mechanics Publications, Southampton and Elsevier Applied Science, London, pp. 315-330.

48. Shivakumar, V. & Forman, R.G. Green's function for a crack emanating from a circular hole in an infinte sheet, *Int. J. Fracture*, 1980, **16**, 305-316.

49. Rooke, D.P. & Tweed, J. Stress intensity factors for cracks at edge of a pressurized hole, *Int. J. Engng. Sci.*, 1980, **18**, 109-121.

50. Rich, T. & Roberts, R. Plastic enclave sizes for internal cracks emanating from circular cavities within elastic plates, *Engng. Fracture Mechanics*, 1968, **1**, 167-173.

51. Leitao, V., Aliabadi, M.H., Cook, R. & Rooke, D.P. Residual stress fields effect on fatigue crack growth, *Boundary Elements XIV*, C.A. Brebbia *et al.* (ed), 1992, Computational Mechanics Publication, Southmapton, Vol. 2, pp. 331-350.

52. Man, K., Aliabadi, M.H. & Rooke, D.P. BEM frictional analysis: load incremental technique, *Computers and Structures*, 1993, **47**, 893-905.

53. Man, K., Aliabadi, M.H. & Rooke, D.P. BEM frictional contact analysis: modelling consideration, *Egng. Anal. with Bound. Elem.*, 1993, **11**, 77-85.

54. Man, K., Aliabadi, M.H. & Rooke, D.P. Analysis of contact friction using the boundary element method, in *Computational Methods in Contact Mechanics*, M.H. Aliabadi & C.A.Brebbia (ed), 1993, Computational Mechanics Publications, Southampton and Elsevier Applied Sciences, London.

55. Man, K., Aliabadi, M.H. & Rooke, D.P. A fully incremental contact analysis for cracked structures, *Boundary Element XIV*, Vol. 2, Stress Analysis and Computational Aspects, C.A. Brebbia *et al.* (ed), 1992, Computational Mechanics Publication, Southampton, pp. 299-314.

56. Portela, A., Aliabadi, M.H. & Rooke, D.P. Dual boundary element incremental analysis of crack propagation, *Computers and Structures*, 1993, **46**, 237-247.

57. Mi, Y. & Aliabadi, M.H. Dual boundary element method for three-dimensional crack growth analysis, *Boundary Elements XV*, C.A. Brebbia & J.J. Rencis (ed), 1993, Computational Mechanics Publication, Southampton, Elsevier Aplied Science, London, pp. 249-260.

58. Erdogan, F. & Shi, G.C. On the crack extension in plates under plane loading and transverse shear, *J. Basic Engng.*, 1963, **85**, 519-527.

59. Sih, G.C. Strain energy density factor applied to mixed mode crack problems, *Int. J. Fracture*, 1974, **10**, 305-321.

60. Smith, R.N.L. The solution of mixed-mode fracture problems using the boundary element method, *Engng. Anal. with Bound. Elem.*, 1988, **5**, 75-80.

61. Rigby, R.H. & Aliabadi, M.H. Mixed-mode J-integral method for analysis of 3D fracture problems using BEM, *Engng. Anal. with Bound. Elem.*, 1993, **11**, 239-256.

Chapter 4

Dynamic fracture mechanics

V.Z. Parton

Laboratoire de Mécanique des Sols, Structures et Matériaux, École Centrale Paris, Grande Voie des Vignes, 92295 Châtenay-Malabry Cedex, France

Introduction

In spite of all the brilliant achievements in the field of fracture mechanics and its numerous applications, the formulation and solution of the dynamic problems of this theory remained unknown, until recently, on account of their extremely complicated nature. Only the latest elegant analytic solutions of certain model problems and the development of new effective numerical methods have helped in surmounting this obstacle. In order to understand the growing interest towards investigations in dynamic fracture mechanics, it is necessary to grasp the essence of the subject and its interaction with quasi-static fracture mechanics. Indeed, the process of fracture is characterized (at least in its final stage) by a rapid propagation of the arterial crack or a set of branched cracks and is therefore essentially a dynamic process.

A large number of problems still remain unsolved in the description of this process on microscopic and macroscopic levels. Hence, when we state that fracture mechanics is an essential tool for computing the strength of bodies and structures, we mean the quasi-static fracture mechanics which determines whether or not an arterial crack is stable. Indeed, the quasi-static mechanics of brittle fracture, which is based on the idealized model of a sharp arterial crack and the concept of the stress intensity factor (SIF) at its tip, has been developed quite extensively; however, it provides only the first approximation to the description of fracture and can simply indicate whether or not a catastrophic growth of the crack sets in.

As without first approximation all the others are impossible, the chapter begins with the basis of quasi-static dynamic fracture mechanics (Section 1), then it is followed by the construction of analytical and numerical solutions (finite element method), determination of SIF in different problems for bodies with cracks subjected to steady-state vibrations (Sections 1 and 2).

Section 3 is dedicated to the determination of dynamic SIF in limited and unlimited bodies in the case of impact loading. Under investigation is the influence of microcracks on the propagation of arterial cracks.

The two following sections treat a new field of fracture mechanics, the influence of electromagnetic fields on cracks propagation. It is noted that a deceleration (treatment) of cracks with the help of strong electromagnetic fields is possible.

Finally, section 6 compares theoretical and experimental results in dynamic fracture mechanics that lead to the conclusion of the necessity of construction of new models for an adequate description of essentially dynamic processes. In this case, it is necessary to take into consideration a microstructure of material which reacts in the first place to the propagation of waves passing ahead of the fracture.

Unlike the static fracture theory, the dynamic theory studies wave propagation. If the system under consideration contains a stationary or propagating defect, the wave field becomes very complicated and this has to be taken into account. Thus the temporal behavior of SIF for a specimen under shock loading with due regard for the reflected waves is characterized by strong oscillations. One more example - the tips of branching cracks themselves became the sources of new propagating waves. Even microdefects formed ahead of the tip of the main crack generate waves and interact with the main crack, which can by no means be neglected.

Thus at the present stage of its development, the theory of dynamic fracture mechanics is rather inconsistent. Today, the efforts of scientists working in fracture mechanics should be focused in this direction. But contradictions and inconsistencies in science always gave an impetus for new investigations and therefore the accumulated knowledge on fracture dynamics should inevitably lead to the formulation of a new rigorous and consistent theory. It is also worth noting that the recent decade has been characterized by a drastic increase of the number of investigations in the field. These include the creation of new models of fracture, analytical and numerical solutions of the problems of dynamic elasticity and plasticity theories for bodies with stationary or propagating cracks, and the development of new experimental methods.

1 Fundamentals of dynamic fracture mechanics

In this section we shall describe the general laws of stress and displacement field distribution in the vicinity of a crack tip. The stress intensity factors, which are the basic characteristics of the stress state of a cracked body, will be introduced.

The stress intensity factors are employed for determining the limiting equilibrium of a body with a crack (in the stationary state), and the nature of its propagation (in the nonstationary case). However, the equations of motion and equilibrium in the mechanics of a continuous medium cannot provide a solution to these basic problems in fracture mechanics. It will be shown that this difficulty can be overcome by the introduction of some additional conditions linking the stress intensity factors with some experimentally determined parameters which are constant for a given material under given conditions and characterize the ability and tendency of the material to crack formation.

1.1 Equations of elastodynamics

In the absence of body forces, the stress tensor components σ_{ij} satisfy the following equations of motion in the Cartesian system of coordinates:

$$\sum_j \frac{\partial \sigma_{ij}}{\partial x_j} = \rho_o \frac{\partial^2 u_i}{\partial t^2} (i = 1,2,3;\ j = 1,2,3) \tag{1}$$

Using the relation between the stress tensor and the displacement vector components (generalized Hook's law), viz.,

$$\sigma_{ij} = \mu\left(\frac{\partial u_i}{\partial x_j} + \frac{\partial u_j}{\partial x_i}\right) + \lambda \delta_{ij}\ \mathrm{div}\{u\} \tag{2}$$

we obtain the equation of motion in the following form:

$$(\lambda + 2\mu)\mathrm{grad\ div}\{u\} - \mu\ \mathrm{rot\ rot}\{u\} = \rho_o \frac{\partial^2 \{u\}}{\partial t^2} \tag{3}$$

For further analysis, we shall require the equations of motion for the plane problem (for cracks by mixed opening and inplane shear modes in planes), the antiplane problem (for cracks due to antiplane shear in plates), and the axisymmetric problem (for penny-shaped cracks in space).

It is well known that in the plane problem of the theory of elasticity, either $u_z = \sigma_{xz} = \sigma_{yz} = 0$ and $\sigma_{zz} = \lambda\ \mathrm{div}\{u\}$ (plane deformation), or $\sigma_{xz} = \sigma_{yz} = \sigma_{zz} = 0$ (plane stress). In both cases, the introduction of the wave potentials φ and ψ, i.e.,

$$\begin{aligned} u_x &= \frac{\partial \varphi}{\partial x} + \frac{\partial \psi}{\partial y}\ ;\ u_y = \frac{\partial \varphi}{\partial y} - \frac{\partial \psi}{\partial x} \\ \sigma_{xx} &= \lambda \nabla^2 \varphi + 2\mu\left(\frac{\partial^2 \varphi}{\partial x^2} + \frac{\partial^2 \psi}{\partial x \partial y}\right) \end{aligned}$$

$$\begin{aligned}\sigma_{yy} &= \lambda\nabla^2\varphi + 2\mu\left(\frac{\partial^2\varphi}{\partial y^2} - \frac{\partial^2\psi}{\partial x\partial y}\right)\\ \sigma_{xy} &= \mu\left(2\frac{\partial^2\varphi}{\partial x\partial y} + \frac{\partial^2\psi}{\partial y^2} - \frac{\partial^2\psi}{\partial x^2}\right)\end{aligned} \tag{4}$$

transforms the equations of motion into two Helmholtz equations for potentials:

$$\begin{aligned}\frac{\partial^2\varphi}{\partial x^2} + \frac{\partial^2\varphi}{\partial y^2} &= \frac{1}{c_1^2}\frac{\partial^2\varphi}{\partial t^2}\\ \frac{\partial^2\psi}{\partial x^2} + \frac{\partial^2\psi}{\partial y^2} &= \frac{1}{c_2^2}\frac{\partial^2\psi}{\partial t^2}\end{aligned} \tag{5}$$

where c_1 and c_2 are the velocities of propagation of the dilatational and shear waves, respectively. The potential φ describes the propagation of dilatational waves (in which the particles of the medium do not rotate), while the potential ψ characterizes the propagation of shear, or distortion, waves. In such a field, the volume expansion is equal to zero.

In the antiplane problem, $u_x = u_y = \sigma_{xx} = \sigma_{yy} = \sigma_{xy} = \sigma_{zz} = 0$, and hence $\sigma_{xz} = \mu\partial u_z/\partial x$, $\sigma_{yz} = \mu\partial u_z/\partial y$, while the component $u_z = w$ of the displacement vector satisfies the equation

$$\frac{\partial^2 w}{\partial x^2} + \frac{\partial^2 w}{\partial y^2} = \frac{1}{c_2^2}\frac{\partial^2 w}{\partial t^2} \tag{6}$$

Thus, the stress waves responsible for the longitudinal deformation of crack boundaries are distortion waves propagating with the velocity of the shear waves.

In the axisymmetric case, the displacement vector has two nonzero components u_r and u_z in cylindrical coordinates r, Θ, z. The third component u_Θ and all derivatives with respect to Θ are equal to zero. In this case, the displacements u_z and u_r are expressed in terms of the wave potentials, i.e.,

$$u_r = \frac{\partial\varphi}{\partial r} - \frac{\partial\psi}{\partial z};\quad u_z = \frac{\partial\varphi}{\partial z} + \frac{\partial\psi}{\partial r} - \frac{\psi}{z} \tag{7}$$

The potentials φ and ψ satisfy the wave equations

$$\begin{aligned}\frac{\partial^2\varphi}{\partial r^2} &+ \frac{1}{r}\frac{\partial\varphi}{\partial r} + \frac{\partial^2\varphi}{\partial z^2} = \frac{1}{c_1^2}\frac{\partial^2\varphi}{\partial t^2}\\ \frac{\partial^2\psi}{\partial r^2} &+ \frac{1}{r}\frac{\partial\psi}{\partial r} - \frac{\psi}{r^2} + \frac{\partial^2\psi}{\partial z^2} = \frac{1}{c_2^2}\frac{\partial^2\psi}{\partial t^2}\end{aligned} \tag{8}$$

It should be noted that in the case of harmonic oscillations, all quantities, including the potentials φ and ψ, have the same time dependence, which is characterized by the factor $e^{i\omega t}$, i.e., $\varphi = \varphi_* e^{-i\omega t}$ and $\psi = \psi_* e^{-i\omega t}$. The potentials φ_* and ψ_* satisfy the well-known Sommerfield radiation and finiteness conditions. For the plane problem, these potentials have the form

$$\varphi_* = 0\left(\frac{1}{\sqrt{r}}\right); \quad \frac{\partial \varphi_*}{\partial r} - i\alpha_1\varphi_* = 0\left(\frac{1}{\sqrt{r}}\right)$$
$$\psi_* = 0\left(\frac{1}{\sqrt{r}}\right); \quad \frac{\partial \psi_*}{\partial r} - i\alpha_2\psi_* = 0\left(\frac{1}{\sqrt{r}}\right) \tag{9}$$

Obviously, the peak values of the potentials φ_* and ψ_* satisfy the equations

$$\nabla^2\varphi_* + \alpha_1^2\varphi_* = 0; \quad \nabla^2\psi_* + \alpha_2^2\psi_* = 0 \tag{10}$$

1.2 Stress and displacement fields near the tip of a running crack

For a mathematical description of the propagation of a crack it is most important to reveal the general laws governing the distribution of the stress and displacement fields in the vicinity of the crack tip.

It is found that if a crack tip moves along a smooth curve at an arbitrary velocity, the angular distribution of stresses in a local system of coordinates fixed to the tip depends only on the instantaneous velocity of the tip. The components of the stress tensor can be written in the form

$$\begin{aligned}
\sigma_{xx} &= \frac{K_I(t)}{\sqrt{2\pi r}}\sum\nolimits_{xx}^{I}(\Theta, v) + \frac{K_{II}(t)}{\sqrt{2\pi r}}\sum\nolimits_{xx}^{II}(\Theta, v) + 0(1)\\
\sigma_{yy} &= \frac{K_I(t)}{\sqrt{2\pi r}}\sum\nolimits_{yy}^{I}(\Theta, v) + \frac{K_{II}(t)}{\sqrt{2\pi r}}\sum\nolimits_{yy}^{II}(\Theta, v) + 0(1)\\
\sigma_{xy} &= \frac{K_I(t)}{\sqrt{2\pi r}}\sum\nolimits_{xy}^{I}(\Theta, v) + \frac{K_{II}(t)}{\sqrt{2\pi r}}\sum\nolimits_{xy}^{II}(\Theta, v) + 0(1)
\end{aligned} \tag{11}$$

for mixed opening and inplane shear modes, and

$$\begin{aligned}
\sigma_{xz} &= \frac{K_{III}(t)}{\sqrt{2\pi r}}\sum\nolimits_{xz}^{III}(\Theta, v) + 0(1)\\
\sigma_{yz} &= \frac{K_{III}(t)}{\sqrt{2\pi r}}\sum\nolimits_{yz}^{III}(\Theta, v) + 0(1)
\end{aligned} \tag{12}$$

for antiplane shear mode.

In formulas (11) and (12), v is the velocity of propagation of the crack tip, r and Θ are moving polar coordinates with instantaneous origin at the crack tip (Fig. 1), while $K_I(t)$, $K_{II}(t)$ and $K_{III}(t)$ are dynamic stress intensity factors.

The displacement in the vicinity of a running crack tip for the case of mixed opening and inplane shear modes are:

$$u_x = \frac{2K_I(1+\delta_2^2)}{\sqrt{2\pi}\mu R_*(\delta_1,\delta_2)}\left(r_1^{1/2}\cos\frac{\Theta_1}{2} - \frac{2\delta_1\delta_2}{1+\delta_2^2}\, r_2^{1/2}\cos\frac{\Theta_2}{2}\right)$$

$$
\begin{aligned}
& +\frac{2K_{II}(1+\delta_2^2)\delta_2}{\sqrt{2\pi}\mu R_*(\delta_1,\delta_2)}\left(\frac{2}{1+\delta_2^2}\,r_1^{1/2}\sin\frac{\Theta_1}{2}-r_2^{1/2}\sin\frac{\Theta_2}{2}\right)\\
u_y \;=\; & \frac{2K_I(1+\delta_2^2)}{\sqrt{2\pi}\mu R_*(\delta_1,\delta_2)}\left(-\delta_1 r_1^{1/2}\sin\frac{\Theta_1}{2}+\frac{2\delta_1}{1+\delta_2^2}\,r_2^{1/2}\sin\frac{\Theta_2}{2}\right)\\
& +\frac{2K_{II}(1+\delta_2^2)}{\sqrt{2\pi}\mu R_*(\delta_1,\delta_2)}\left(\frac{2\delta_1\delta_2}{1+\delta_2^2}\,r_1^{1/2}\cos\frac{\Theta_1}{2}-r_2^{1/2}\cos\frac{\Theta_2}{2}\right)
\end{aligned}
\tag{13}
$$

For the case of antiplane shear, the following formula is used:

$$
u_z = w = \frac{2K_{III}}{\sqrt{2\pi}\mu}\,r_2^{1/2}\sin\frac{\Theta_2}{2} \tag{14}
$$

Here, $\delta_{1,2}^2 = 1-\epsilon_{1,2}^2$, $\epsilon_{1,2} = v/c_{1,2}$, $c_{1,2}$ are the velocities of elastic waves. The coordinate system (r_1,Θ_1) and (r_2,Θ_2) are shown in Fig. 2, and

$$
r_1e^{i\Theta_1} = x+\delta_1 y,\; r_2e^{i\Theta_2} = x+\delta_2 y
$$

where x, y is the system of moving Cartesian coordinates (see Fig. 1). $R_*(\delta_1,\delta_2)$ is the Rayleigh function and has the following form when expressed in terms of δ_1 and δ_2:

$$
R_*(\delta_1,\delta_2) = 4\delta_1\delta_2-(1+\delta_2^2)^2 \tag{15}
$$

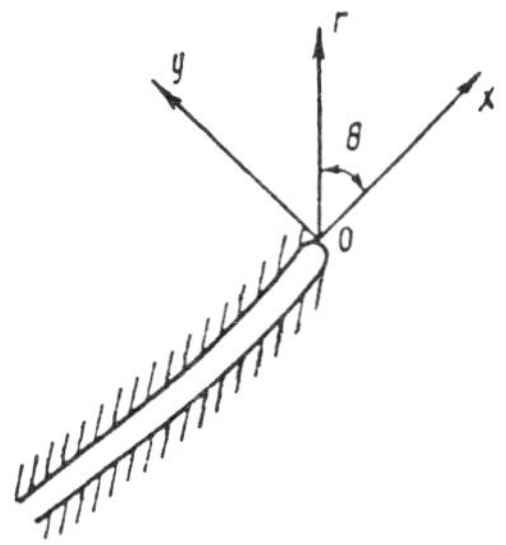

Figure 1: Local coordinate systems with the origin at the tip of a propagating crack

Formulae (13) and (14) have the following functions from (11) and (12) corresponding to them:

$$
\sum\nolimits_{xx}^{I},\ \sum\nolimits_{xy}^{I},\ \sum\nolimits_{yy}^{I},\ \sum\nolimits_{xx}^{II},\ \sum\nolimits_{xy}^{II},\ \sum\nolimits_{yy}^{II},\ \sum\nolimits_{xz}^{III}
$$

$$
\begin{aligned}
\sum\nolimits_{xx}^{I} &= \frac{(1+\delta_2^2)r^{1/2}}{R_*(\delta_1,\delta_2)}\left[(1+2\delta_1^2-\delta_2^2)\frac{\cos\Theta_1/2}{r_1^{1/2}}-\frac{4\delta_1\delta_2}{1+\delta_2^2}\,\frac{\cos\Theta_2/2}{r_2^{1/2}}\right]\\
\sum\nolimits_{xy}^{I} &= \frac{2(1+\delta_2^2)\delta_1 r^{1/2}}{R_*(\delta_1,\delta_2)}\left(\frac{\sin\Theta_1/2}{r_1^{1/2}}-\frac{\sin\Theta_2/2}{r_2^{1/2}}\right)
\end{aligned}
$$

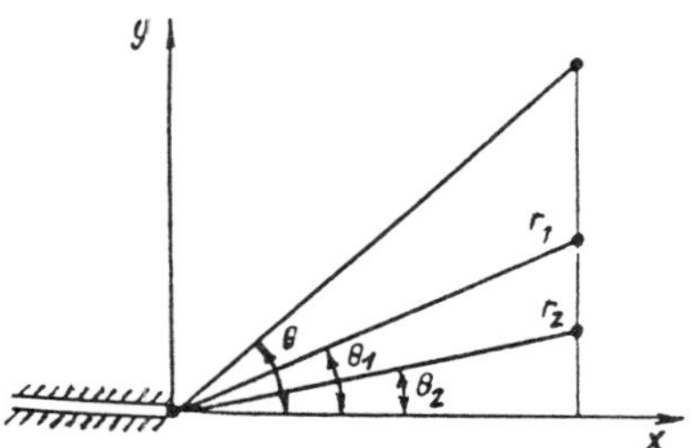

Figure 2: Coordinates (r_1, Θ_1) and (r_2, Θ_2) in the neighborhood of a crack tip

$$\sum\nolimits_{yy}^{I} = \frac{(1+\delta_2^2)r^{1/2}}{R_*(\delta_1,\delta_2)}\left[-(1+\delta_2^2)\,\frac{\cos\Theta_1/2}{r_1^{1/2}} + \frac{4\delta_1\delta_2}{1+\delta_2^2}\,\frac{\cos\Theta_2/2}{r_2^{1/2}}\right] \tag{16}$$

$$\begin{aligned}
\sum\nolimits_{xx}^{II} &= \frac{2(1+\delta_2^2)\delta_2 r^{1/2}}{R_*(\delta_1,\delta_2)}\left(-\frac{1+2\delta_1^2-\delta_2^2}{1+\delta_2^2}\,\frac{\sin\Theta_1/2}{r_1^{1/2}} + \frac{\sin\Theta_2/2}{r_2^{1/2}}\right)\\
\sum\nolimits_{xy}^{II} &= \frac{(1+\delta_2^2)r^{1/2}}{R_*(\delta_1,\delta_2)}\left[\frac{4\delta_1\delta_2}{1+\delta_2^2}\,\frac{\cos\Theta_1/2}{r_1^{1/2}} - (1+\delta_2^2)\,\frac{\cos\Theta_2/2}{r_2^{1/2}}\right]\\
\sum\nolimits_{yy}^{II} &= \frac{2(1+\delta_2^2)\delta_2 r^{1/2}}{R_*(\delta_1,\delta_2)}\left(\frac{\sin\Theta_1/2}{r_1^{1/2}} - \frac{\sin\Theta_2/2}{r_2^{1/2}}\right)
\end{aligned} \tag{17}$$

$$\begin{aligned}
\sum\nolimits_{xz}^{III} &= -\left(\frac{r}{r_2}\right)^{1/2} \sin\frac{\Theta_2}{2}\\
\sum\nolimits_{yz}^{III} &= \left(\frac{r}{r_2}\right)^{1/2} \cos\frac{\Theta_2}{2}
\end{aligned} \tag{18}$$

It should be observed that in the case of a limiting transition for $v \to 0$, we obtain the following familiar relations for a stationary crack:

$$\begin{aligned}
u_x &= \frac{K_I}{\mu}\sqrt{\frac{r}{2\pi}}\cos\frac{\Theta}{2}\left(1-2\nu+\sin^2\frac{\Theta}{2}\right) + \frac{K_{II}}{\mu}\sqrt{\frac{r}{2\pi}}\sin\frac{\Theta}{2}\left(2-2\nu+\cos^2\frac{\Theta}{2}\right)\\
u_y &= \frac{K_I}{\mu}\sqrt{\frac{r}{2\pi}}\sin\frac{\Theta}{2}\left(2-2\nu-\cos^2\frac{\Theta}{2}\right) + \frac{K_{II}}{\mu}\sqrt{\frac{r}{2\pi}}\cos\frac{\Theta}{2}\left(1-2\nu+\sin^2\frac{\Theta}{2}\right)
\end{aligned} \tag{19}$$

$$u_z = w = \frac{K_{III}}{\mu}\sqrt{\frac{2r}{\pi}}\sin\frac{\Theta}{2} \tag{20}$$

$$\sum\nolimits_{xx}^{I} = \cos\frac{\Theta}{2}\left(1 - \sin\frac{\theta}{2}\sin\frac{3\Theta}{2}\right)$$

$$\begin{aligned}\sum\nolimits_{yy}^{I} &= \cos\frac{\Theta}{2}\left(1+\sin\frac{\theta}{2}\sin\frac{3\Theta}{2}\right)\\ \sum\nolimits_{xy}^{I} &= \sin\frac{\Theta}{2}\cos\frac{\theta}{2}\cos\frac{3\Theta}{2}\end{aligned} \tag{21}$$

$$\begin{aligned}\sum\nolimits_{xx}^{II} &= -\sin\frac{\Theta}{2}\left(2+\cos\frac{\theta}{2}\cos\frac{3\Theta}{2}\right)\\ \sum\nolimits_{yy}^{II} &= \cos\frac{\Theta}{2}\sin\frac{\theta}{2}\cos\frac{3\Theta}{2})\\ \sum\nolimits_{xy}^{II} &= \cos\frac{\Theta}{2}\left(1-\sin\frac{\theta}{2}\sin\frac{3\Theta}{2}\right)\end{aligned} \tag{22}$$

$$\sum\nolimits_{xz}^{III} = -\sin\frac{\Theta}{2};\ \sum\nolimits_{yz}^{III} = \cos\frac{\Theta}{2} \tag{23}$$

K_I, K_{II} and K_{III} are called the stress intensity factors for opening modes, inplane shear, and antiplane shear, respectively. These factors play an exceptionally important role in fracture mechanics since they are the basic characteristics of the stress state of a body with a crack (in view of the universal nature of the asymptotic angular distributions of stresses and displacements in the vicinity of a crack tip). The stress intensity factors are determined by solving the problem of the theory of elasticity for a body with a crack by proceeding to the limit

$$K_I = \lim_{\substack{r\to 0\\ \Theta=0}} \sqrt{2\pi r}\sigma_{yy};\quad K_{II} = \lim_{\substack{r\to 0\\ \Theta=0}} \sqrt{2\pi r}\sigma_{xy};\quad K_{III} = \lim_{\substack{r\to 0\\ \Theta=0}} \sqrt{2\pi r}\sigma_{xz} \tag{24}$$

1.3 Energy release into the tip of a propagating crack

We shall now introduce the concept of the rate of release of mechanical energy or, in other words, the rate of energy flux into the tip of a crack, referred to a unit area of the crack:

$$G = \frac{\delta\Gamma_*}{\delta S} = 2\gamma \tag{25}$$

Here, $\delta\Gamma_*$ is the amount of energy required for the creation of a new fracture surface of area δS, and 2γ is the surface energy spent in causing the fracture. The latter quantity can be determined experimentally. The right-hand side of (25) is often replaced by G_c, the critical energy release rate. Dividing the numerator and denominator of (25) by δt, we get

$$G = \frac{\delta\Gamma_*/\delta t}{\delta S/\delta t} = \frac{\varepsilon}{v} = 2\gamma \tag{26}$$

where ε is the instantaneous energy flux into the tip of a crack propagating with a velocity v.

The quantity ε can be determined:

$$\varepsilon = \frac{1}{2\mu}\left\{\frac{v^3}{c_2^2 R_*(\delta_1, \delta_2)}[\delta_1 K_I^2(t) + \delta_2 K_{II}^2(t)] + \frac{v K_{III}^2(t)}{2}\right\} \tag{27}$$

The energy release rate G is obtained by referring the instantaneous energy flux ε to the rate of formation of the new surface:

$$G = \frac{1}{2\mu}\left\{\frac{v^2}{c_2^2 R_*(\delta_1, \delta_2)}[\delta_1 K_I^2(t) + \delta_2 K_{II}^2(t)] + \frac{K_{III}^2(t)}{2}\right\} \tag{28}$$

Thus, relation (28) connects the energy and force characteristics of the fracture process.

Since the energy release rate in the process of crack propagation is given by $G = G_c = 2\gamma$, the relation

$$2\gamma = \frac{1}{2\mu}\left\{\frac{v^2}{c_2^2 R_*(\delta_1, \delta_2)}[\delta_1 K_I^2(t) + \delta_2 K_{II}^2(t)] + \frac{K_{III}^2(t)}{2}\right\} \tag{29}$$

obtained from (25) and (28) can be used for determining the dependance of crack propagation on time. In this case, however, we must remember that the quantities K_I, K_{II} and K_{III} appearing in this relation are functionals of the velocity v (in general, this also applies to the quantity $\gamma = \gamma(v)$).

An analysis of expressions (28) and (29) leads to several useful conclusions. For mixed opening and inplane shear cracks, $G > 0$ in the velocity interval $0 < v < c_R$, and $G < 0$ in the interval $c_R < v < c_2$. Since the effective surface energy 2γ is positive, crack propagation at velocities exceeding the velocity of Rayleigh waves is not possible in the continuum model. If, on the other hand, $c_R < v < c_2$, the tensile stresses applied to the crack are superimposed. Obviously, the crack propagation rate in the case of antiplane shear cannot exceed c_2. It should be noted that, in practice, the velocity of crack propagation is restricted not by the velocity of Rayleigh waves, but by a smaller quantity varying for different materials between 0.2 and 0.5 times the velocity of shear waves. This is explained as being due to the effect of thermal expansion on the stress state and the subsequent formation of a considerable plastic region surrounding the crack tip or by using the concept of leading microfractures. Indeed, the process of microcrack formation hinders the growth of macroscopic cracks. If the fracture is not accompanied by the formation of microscopic cracks (for example, in the case of a pure cleavage of crystals), velocities close to the wave velocities are attained.

We shall now derive the law of crack propagation from (29) with the help of a simple example. Let us consider the propagation of a semi-infinite crack caused by antiplane shear in the field of a uniform shearing stress. In this case:[1]

$$K_{III} = \frac{2}{\sqrt{\pi}}\, q^{(3)}\sqrt{1 - \frac{v}{c_2}}\,\sqrt{2c_2 t} \tag{30}$$

Here and below, q^i $(i = 1, 2, 3)$ are the loads corresponding to three types of deformation.

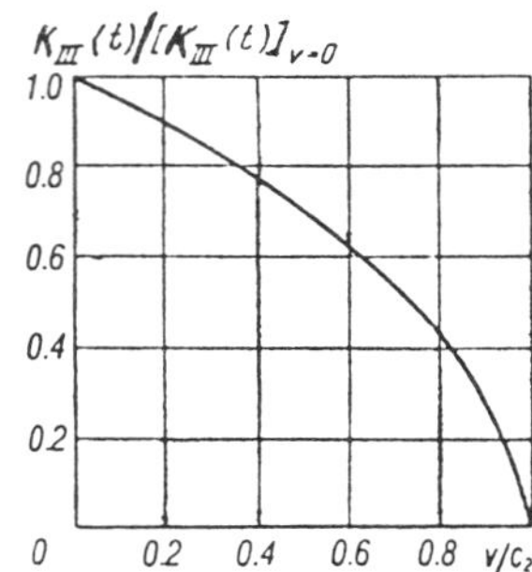

Figure 3: Dependence of stress intensity factor K_{III} on the rate of propagation of a semi-infinite crack ($\nu = 0.25$).

The plot showing the dependance of K_{III} on the crack propagation velocity is presented in Fig. 3. Substituting (30) into (29) for $K_I = K_{II} = 0$, we obtain

$$2\gamma = \frac{4q^{(3)^2}}{\pi\mu}\, c_2 t \sqrt{\frac{1-\frac{v}{c_2}}{1+\frac{v}{c_2}}} \tag{31}$$

Since the left-hand side of this equality is bounded for $t \to \infty$, the velocity of crack propagation asymptotically tends to the velocity of the shear waves as $t \to \infty$.

Let us consider the application of concentrated impact loads $q^{(3)}\delta(x + x_o)H(t)$ to an antiplane shear crack. In this case,

$$K_{III} = q^{(3)}\sqrt{\frac{2}{\pi(x+x_o)}\left(1 - \frac{v}{c_2}\right)}\ (c_2 t \geq x_o) \tag{32}$$

and we obtain from (29)

$$2\gamma = \frac{q^{(3)^2}}{\pi\mu}\sqrt{\frac{1-\frac{v}{c_2}}{1+\frac{v}{c_2}}}\,\frac{1}{x+x_o} \tag{33}$$

Obviously, if $q^{(3)}$ is less than a certain critical value, the crack does not start propagating at all since the right-hand side of (33) can only decrease as a result of crack propagation. If, however, the crack starts propagating, it is arrested as the tip reaches a point x_* where the right-hand side of (33) becomes smaller than 2γ.

The propagation of a semi-infinite crack in the field of tensile stress $q^{(1)}$ can be investigated in a similar manner.

It should be noted that proceeding to the limit as $v \to 0$ in relation (29), we arrive at the familiar static fracture criterion:

$$G_c = 2\gamma = \frac{1-\nu}{2\mu}\,(K_I^2 + K_{II}^2) + \frac{K_{III}^2}{2\mu} \tag{34}$$

If at a certain loading level the stress intensity factors obtained from the solution of the problem in the theory of elasticity attain such values that the right-hand side of (34) becomes equal to 2γ, the crack is able to propagate. Subsequent process of its propagation is described by (29).

In the particular case of an opening mode crack ($K_{II} = K_{III} = 0$), the energy criterion for fracture (34) contains one parameter, and can be written in the form

$$K_I = K_{Ic} \tag{35}$$

where K_{Ic} is the critical stress intensity factor (fracture toughness).

2 Steady-state vibrations. Analytical methods for determining stress intensity factors

Two types of problems are considered in the investigation of elastic stress fields in the vicinity of a crack tip being in a harmonic field. In the first type, harmonic loads are applied to crack faces, while in the second type a wave arrives from infinity and the singular stress field results from the diffraction of waves by the crack. The complete solution of the problem in the second case is presented as a sum of the incident wave (regular solution) and the scattered wave (singular solution). Obviously, the stress intensity factors are determined only by the solution for the scattered wave. In order to obtain this solution, say, for a stress-free opening mode crack in a symmetry plane, we most solve the boundary value problem for a half-plane with the following boundary conditions: on the crack, the tensile stress is equal to the stress due to the incident wave, but with the opposite sign; the normal displacement outside the crack is equal to zero; the shearing stress is also equal to zero everywhere on the boundary of the half-plane. It can be shown that, in the general case, the diffraction problems (in which the stress intensity factors are determined) can be reduced to the first type of problem. Consequently, we shall not accentuate the difference between diffraction problems and the problems of steady-state vibrations in the subsequent discussions.

It should be noted that a large number of problems have been solved to date for bodies with stationary cracks with the help of analytical methods. Solutions have been obtained for the problem of the steady state vibrations of a plane with a crack, a strip with a crack, a space with a crack 'tunnelled' by antiplane shear, a plane with a period system of cracks caused by normal tensile and shearing loads applied to their faces, a space with a semi-infinite crack produced by concentrated harmonic loads applied to their faces, etc. The most important solutions are given in Refs 1 and 2. A special feature of the results obtained for semi-infinite cracks is that the influence of inertia effects on the stress intensity factors cannot be estimated in this case, since these problems do not have a static analog (the solution for zero frequency does not exist). However, the solution of dynamic problems for finite cracks at zero frequency must coincide with the static solution for the corresponding load. This allows us to obtain qualitative and quantitative estimates for the effect of dynamic loading on the increased risk of brittle fracture. It is also interesting to study the influence of inertia effect in the three-dimensional case of a penny-shaped crack, and this problem will be solved and investigated below.

Let us study the vibrations of a medium containing a plane (circular) penny-shaped crack (Fig. 4). We shall first consider the axisymmetric vibrations corresponding to the waves of axial compression-dilatation or radial shear. Then, we shall take up the case of torsional vibratations. We introduce cylindrical polar coordinates in such a way that the crack lies in the plane $z = 0$ and the origin coincides with the center of the crack. In the axisymmetric case, we have two nonzero components of the displacement vector, viz., u_z and $u_r (u_\Theta = 0)$. Moreover, all derivatives with respect to Θ are equal to zero. The equations of motion for the potentials are written in the same way as eqn (8).

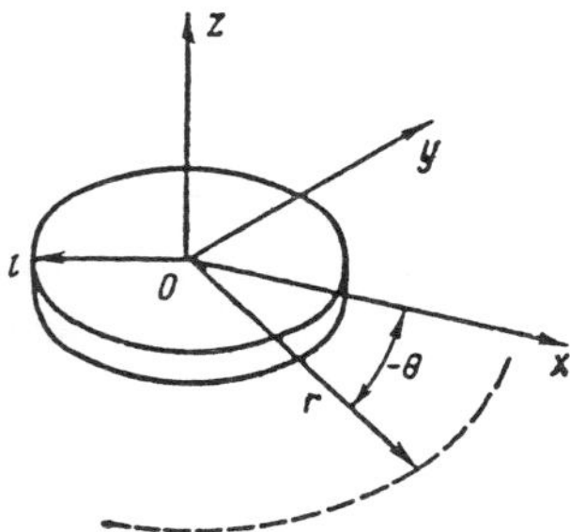

Figure 4: A penny-shaped crack in space

If a wave defined by the potentials $\varphi^{(i)}$ and $\psi^{(i)}$ is incident on the crack, the complete solution of the problem is equal to the superposition of the incident and scattered waves

$$\varphi(r,z,t) = \varphi^{(i)}(r,z,t) + \varphi^{(s)}(r,z,t) \tag{36}$$

$$\psi(r,z,t) = \psi^{(i)}(r,z,t) + \psi^{(s)}(r,z,t) \tag{37}$$

The singular nature of stresses is due to the scattered waves. Since the functions φ and ψ are assumed to vary harmonically in time, eqn (8) can be reduced to the form

$$(\nabla_1^2 + \alpha_1^2)\varphi = 0 \tag{38}$$

$$(\nabla_2^2 + \alpha_2^2)\psi = 0 \tag{39}$$

where

$$\nabla_j^2 = \frac{1}{r}\frac{\partial}{\partial r} r \frac{\partial}{\partial r} - \frac{(j-1)^2}{r^2} + \frac{\partial^2}{\partial z^2} \; (j = 1, 2) \tag{40}$$

Without any loss of generality, the potentials of incident waves can be presented in the form

$$\begin{aligned} \varphi^i &= w_1(r) \cos(\alpha_1 z) e^{-i\omega t} \\ \psi^i &= w_2(r) \cos(\alpha_2 z) e^{-i\omega t} \end{aligned} \tag{41}$$

where the functions w_1 and w_2 satisfy ordinary differetial equations

$$\frac{d^2 w_j}{dr^2} + \frac{1}{r}\frac{dw_j}{dr} - \frac{(j-1)^2}{r^2} w_j = 0 \;\; (j = 1, 2) \tag{42}$$

The solutions bounded at $r = 0$ are $w_1 = \varphi_o$ and $w_2 = \psi_o(r/l)$ where φ_o and ψ_o are constant amplitudes of the incident wave. Relations (41) then assume the form

$$\begin{aligned}\varphi^{(i)} &= \varphi_o \cos(\alpha_1 z)e^{-i\omega t} \\ \psi^{(i)} &= \psi_o(r/l)\cos(\alpha_2 z)e^{-i\omega t}\end{aligned} \tag{43}$$

In the investigation of axial dilatation-compression waves, we put $\psi^{(i)} = 0$. Then the stresses created by the incident wave having a potential $\varphi^{(i)}$ are equal to

$$\begin{aligned}\sigma_{rr}^{(i)} &= q^{(1)}(1 - 2n_*^4)\cos(\alpha_1 z)e^{-i\omega t},\ q^{(1)} = -\mu\alpha_2^2\varphi_o \\ \sigma_{\Theta\Theta}^{(i)} &= \sigma_{rr}^{(i)},\ \sigma_{zz}^{(i)} = q^{(1)}\cos(\alpha_1 z)e^{-i\omega t} \\ \sigma_{rz}^{(i)} &= \sigma_{r\Theta}^{(i)} = \sigma_{\Theta z}^{(i)} = 0,\ n_*^2 = \mathbf{c}_2/\mathbf{c}_1\end{aligned} \tag{44}$$

Since the crack is free from stresses and the symmetry conditions are satisfied for $z = 0$, we get

$$\begin{aligned}\sigma_{zz}^{(s)}(r,0,t) \ + \ \sigma_{zz}^{(i)}(r,0,t) &= 0,\ \ 0 \le r < l \\ u_z^{(s)}(r,0,t) &= 0,\ \ r > l \\ \sigma_{zz}^{(s)}(r,0,t) &= 0,\ \ r \ge 0\end{aligned} \tag{45}$$

Applying Hankel transforms to eqns (38) and (39), we obtain ($\beta_j^2 = s^2 - \alpha_j^2$)

$$\begin{aligned}\varphi^{(s)}(r,z,t) &= \int_0^\infty B_1^{(1)}(s)J_o(rs)\exp\big[-(\beta_1 z + i\omega t)\big]\mathrm{d}z,\ z \ge 0 \\ \psi^{(s)}(r,z,t) &= \int_0^\infty B_1^{(2)}(s)J_1(rs)\exp\big[-(\beta_2 z + i\omega t)\big]\mathrm{d}s,\ z \ge 0\end{aligned} \tag{46}$$

where J_o and J_1 are the zero- and first-order Bessel functions of the first kind, and $B_1^{(1)}$, $B_1^{(2)}$ are functions that have to be determined. Taking into account (45) and (46), we arrive at the dual integral equations

$$\begin{aligned}&\textstyle\int_0^\infty \ B_1(s)J_o(rs)\ \mathrm{d}s = 0,\ \ r \ge l \\ &\textstyle\int_0^\infty \ sf_1(s)B_1(s)J_o(rs)\ \mathrm{d}s = -\dfrac{q^{(1)}}{2\mu(1-n_*^4)},\ \ 0 \le r < l\end{aligned} \tag{47}$$

where

$$f_1(s) = \frac{(2s^2-\alpha_2^2)^2 - 4s^2\sqrt{s^2-\alpha_1^2}\sqrt{s^2-\alpha_2^2}}{2\alpha_2^2(1-n_*^4)s\sqrt{s^2-\alpha_1^2}} \tag{48}$$

and the unknown function B_1 is connected with $B_1^{(1)}$ and $B_1^{(2)}$ through the relations

$$\begin{aligned}B_1^{(1)}(s) &= \frac{2}{\alpha_2^2\sqrt{s^2-\alpha_1^2}}\left(s^2 - \frac{1}{2}\alpha_2^2\right)B_1(s) \\ B_1^{(2)}(s) &= \frac{2s}{\alpha_2^2}\ B_1(s)\end{aligned} \tag{49}$$

The system of integral equations obtained here can be solved by Copson's method.[2]

The solution of the system (47) can be written in the form

$$B_1(s) = \left(\frac{2}{\pi}\right)^{1/2} \int_0^l d_1(t) \sin(st) dt \tag{50}$$

where

$$d_1(\xi) = \frac{1}{(2\pi)^{1/2}} \frac{q^{(1)} l}{\mu(1-n_*^4)} D_1(\xi) \tag{51}$$

and $D_1(\xi)$ is the solution of the Fredholm integral equation

$$D_1(\xi) - \int_0^l K_1(\xi,\eta) D_1(\eta) d\eta = \xi \tag{52}$$

having a kernel

$$K_1(\xi,\eta) = \frac{2}{\pi} \int_0^\infty \left[f_1\left(\frac{s}{l}\right) + 1\right] \sin(s\xi) \sin(s\eta) ds \tag{53}$$

Since $D_1(\xi)$ and $K_1(\xi,\eta)$ are complex quantities, eqn (52) can be decomposed into two equations. Near a crack edge, the angular distribution of stresses in the plane of the crack is the same as in the plane problem on an opening mode crack. The stress intensity factor is equal to

$$K_1 = \frac{2}{\pi} q^{(1)} \sqrt{\pi l}\, D_1(1) \exp\left[-i\omega(t-\xi_1)\right] \tag{54}$$

where

$$\xi_1 = \frac{1}{\omega} \text{arc tan} \left[\frac{Im\ D_1(1)}{Re\ D_1(1)}\right] \tag{55}$$

The shear stress intensity factors are equal to zero and the solution for $\omega = 0$ coincides with the static solution $K_{Is} = (2/\pi) q^{(1)} \sqrt{\pi l}$.

The results of numerical computations are shown in Fig. 5. Let us now consider radial shearing vibrations. We put the potential $\varphi(i) = 0$ and describe the stresses of the incident wave in the form

$$\begin{aligned}
\sigma_{rr}^{(i)} &= -\frac{2q^{(2)}}{\alpha_2 l} \sin(\alpha_2 z) e^{-i\omega t}, \quad \left(q^{(2)} = -\mu\alpha_2^2 \psi_o\right) \\
\sigma_{\Theta\Theta}^{(i)} &= -\sigma_{rr}^{(2)}, \ \sigma_{zz}^{(i)} = \frac{4q^{(2)}}{\alpha_2 l} \sin(\alpha_2 z) e^{-i\omega t} \\
\sigma_{rz}^{(i)} &= -q^{(2)} \left(\frac{r}{l}\right) \cos(\alpha_2 z) e^{-i\omega t}, \quad \sigma_{r\Theta}^{(i)} = \sigma_{\Theta z}^{(i)} = 0
\end{aligned} \tag{56}$$

The boundary conditions and the symmetry conditions are

$$\sigma_{rz}^{(s)}(r,0,t) + \sigma_{rz}^{(i)}(r,0,t) = 0, \quad 0 \le r < l \tag{57}$$

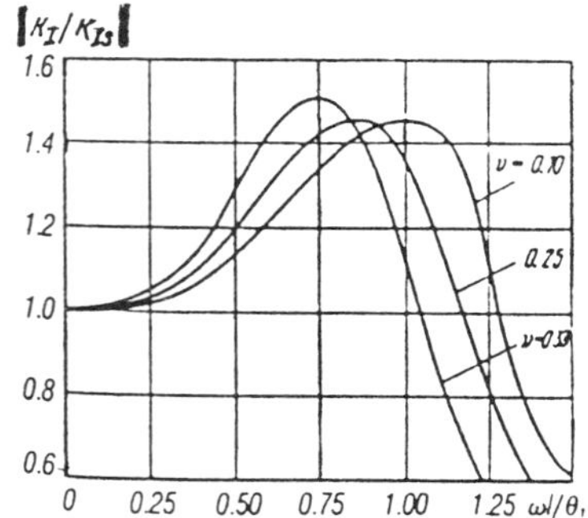

Figure 5: Opening stress intensity factor for the incidence of a dilatational wave on a penny-shaped crack.

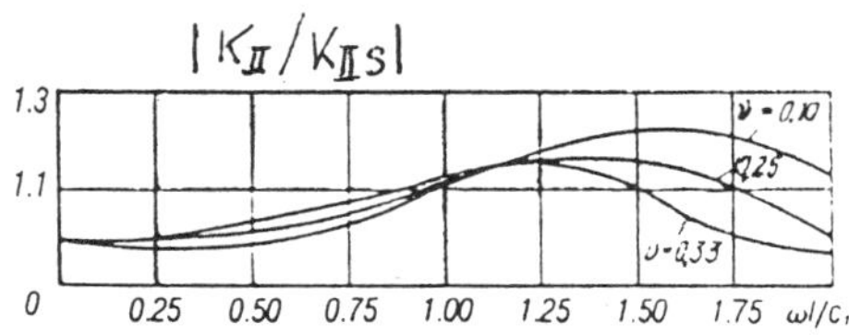

Figure 6: Inplane shear stress intensity factor for the case of incidence of a radial shear wave on a penny-shaped crack.

$$u_r^{(s)}(r, 0, t) = 0, \quad r \geq l$$

The resulting stress field is antisymmetric with respect to the plane $z = 0$; the angular distribution of stresses in this plane is the same as in plane problem on an inplane shear crack.

The results of numerical computations of the quantities K_{II} as functions of the wave number $\omega l / c_1$ are presented in Fig. 6.

Finally, let us consider a penny-shaped crack under the action of torsion waves. In this case, $u_r = u_z = 0$, while the remaining nonzero components of displacements and stresses are given by

$$u_\Theta = u_\Theta(r, z, t), \sigma_{r\Theta} = \mu\left(\frac{\partial u_r}{\partial r} - \frac{u_\Theta}{r}\right), \quad \sigma_{\Theta z} = \mu \frac{\partial u_\Theta}{\partial z} \tag{58}$$

The displacement u_Θ is determined from the equation

$$\frac{\partial^2 u_\Theta}{\partial r^2} + \frac{1}{r}\frac{\partial u_\Theta}{\partial r} - \frac{u_\Theta}{r^2} + \frac{\partial^2 u_\Theta}{\partial z^2} = \frac{1}{c_2^2}\frac{\partial^2 u_\Theta}{\partial t^2} \tag{59}$$

If u_Θ is harmonically dependent on time eqn (59) assumes the form

$$\frac{\partial^2 u_\Theta}{\partial r^2} + \frac{1}{r}\frac{\partial u_\Theta}{\partial r} - \frac{u_\Theta}{r^2} + \frac{\partial^2 u_\Theta}{\partial z^2} + \alpha_2^2 u_\Theta = 0 \tag{60}$$

Let the incident wave be defined by

$$u_\Theta(r, z, t) = w_3(r)\sin(\alpha_2 z)e^{-i\omega t}$$

where w_3 is the bounded solution of the following equation for $r = 0$:

$$\frac{d^2 w_3}{dr^2} + \frac{1}{r}\frac{dw_3}{dr} - \frac{1}{r^2}w_3 = 0 \tag{61}$$

and is equal to

$$w_3(r) = \frac{w_0 r}{l}$$

We then obtain

$$u_\Theta^{(i)}(r, z, t) = \frac{w_0 r}{l}\sin(\alpha_2 z)e^{-i\omega t} \tag{62}$$

$$\sigma_{\Theta z}^{(i)}(r, z, t) = \frac{q^{(3)} r}{l}\cos(\alpha_2 z)e^{-i\omega t}, \quad q^{(3)} = \mu\alpha_2 w_0 \tag{63}$$

The torsional moment of the incident wave is equal to

$$T^{(i)} = 2\pi\int_0^l r^2(\lim_{z\to 0}\sigma_{\Theta z}^{(i)})\mathrm{d}r = q^{(3)}\frac{J}{l}e^{-i\omega t} \tag{64}$$

where J is the polar moment of inertia of a circle of radius r. In the plane $z = 0$, we have

$$\sigma_{\Theta z}^{(i)} = \frac{T^{(i)} r}{l}$$

We shall seek the displacement $u_\Theta^{(s)}$ for the scattered wave in the form

$$u_\Theta^{(s)}(r, z, t) = \int_0^\infty B_3(s)J_1(rs)\exp\left[-(\beta_2 z + i\omega t)\right]\mathrm{d}s \tag{65}$$

where the unknown function $B_3(s)$ is determined from the boundary condition

$$\sigma_{\Theta z}^{(s)}(r, 0, t) + \sigma_{\Theta z}^{(i)}(r, 0, t) = 0, \quad 0 \le r < l \tag{66}$$

and from the condition for $z = 0$:

$$u_\Theta^{(s)}(r, 0, t) = 0, \quad r \ge l \tag{67}$$

This leads to the following system of dual integral equations:

$$\begin{aligned} &\textstyle\int_0^\infty \quad B_3(s)J_1(rs)\mathrm{d}s = 0, \quad r \ge l \\ &\textstyle\int_0^\infty \quad \beta_2 B_3(s)J_1(rs)\mathrm{d}s = \frac{\tau_2}{\mu}\frac{r}{l}, \quad 0 \le r < l \end{aligned} \tag{68}$$

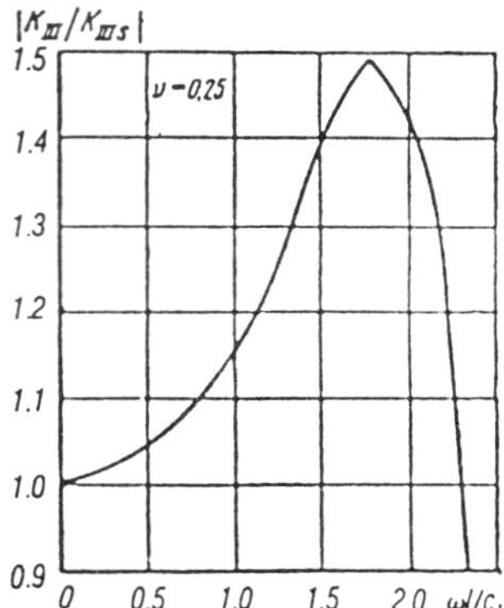

Figure 7: Antiplane shear stress intensity factor for the case of incidence of a torsion wave on a penny-shaped crack.

The required function $B_3(s)$ is expressed by

$$B_3(s) = \sqrt{s}\int_0^l \sqrt{t}\sqrt{\frac{2}{\pi}}\frac{2\tau_2 l}{3\mu}D_3\left(\frac{t}{l}\right)J_{\frac{3}{2}}(st)\mathrm{d}t \tag{69}$$

where $D_3(\xi)$ is the solution of the Fredholm integral equation of the second kind

$$D_3(\xi) + \int_0^l K_1(\xi,\eta)D_3(\eta)\mathrm{d}\eta = \xi^2 \tag{70}$$

having a symmetric kernel

$$K_3(\xi,\eta) = \sqrt{\xi\eta}\int_0^\infty \left[(s^2 - \alpha_2^2 l^2)^{1/2} - s\right]J_{\frac{3}{2}}(s\xi)J_{\frac{3}{2}}(s\eta)\mathrm{d}s \tag{71}$$

It can be shown that the angular distribution of the stress field in the plane $z = 0$ in the vicinity of the crack tip is the same as in the plane problem of a longitudinal shear crack. The results of the numerical computation of the stress intensity factor for the case of longitudinal shear, i.e.,

$$K_{III} = \frac{4}{3\pi}q^{(3)}\sqrt{\pi l}D_3(l)\exp\left[-i\omega(t-\xi_3)\right] \tag{72}$$

are shown in Fig. 7. Here,

$$\xi_3 = \frac{1}{\omega}\,\mathrm{arc\ tan}\left[\frac{\mathrm{Im}D_3(1)}{\mathrm{Re}D_3(1)}\right] \tag{73}$$

3 Steady-state vibrations. Application of the finite-element method for calculating stress intensity factors

We shall describe a method for calculating the dynamic stress intensity factors in plates with cracks subjected to steady-state vibrations.[1] This method is based on the presentation of these factors as a superposition of 'nominal' stress intensity factors corresponding to the normalized forms of free vibrations with certain weight factors. Another method for determining the stress intensity factors under harmonic loading is based on a direct step-by-step solution of the system of differential equations of motion.

However, the solution in this case is quite cumbersome and not suitable for analysis since the parameters of motion and the stress intensity factors in finite bodies depend on the radio $(\omega/\omega_i)^2$, where ω_i are the free-vibration frequencies and ω is the loading frequency.

The method of calculating stress intensity factors based on the superposition of free vibration modes is more convenient to use.

The importance of investigations of the role of cracks on the frequencies and shapes of free vibrations in plates and shells has been emphasized before. These investigations are necessary not only for determining the admissible levels of loads and frequencies, but also for diagnosing the size of defects in various structural elements.

It should be noted that the analytical frequency dependences of the stress intensity factors for infinite media cannot be directly applied to finite-sized bodies, as is often done in the case of static stress intensity factors and dynamic stress intensity factors for impact loading (for the initial interval of time up to the instant when a wave scattered at the boundary arrives at the crack tip).

Unlike analytic methods, the method of finite elements is applicable to finite-sized bodies. A special procedure was proposed for estimating the accuracy of the computational solution.

The finite-element equations of motion of an elastic body in the absence of attenuation under harmonic loading have the form

$$[M]\{x\}^{\cdot\cdot} + [K]\{x\} = \{f\}e^{i\omega t} \tag{74}$$

Here $[M]$ is the mass matrix, $[K]$ is the stiffness matrix, $\{x\}$ is the displacement vector, and $\{f\}$ is the load vector. For $\omega = 0$, we obtain the equilibrium equation

$$[K]\{x\} = \{f\} \tag{75}$$

Denoting by ω_i the eigenvalues in the ascending order, and by $\{x^{(i)}\}$ the eigenvectors from the generalized eigenvalue problem

$$[K]\{x\} = \omega^2[M]\{x\} \tag{76}$$

$$\{x^{(i)}\}^{\mathrm{T}}[M]\{x^{(j)}\} = \delta_{ij} \tag{77}$$

(the superscript 'T' denotes transposition), we can write the particular solution of (74) with the frequency of the perturbing force in the form

$$\{x(t)\} = \sum_i \{x^{(i)}\} \frac{(\{x^{(i)}\}^{\mathrm{T}}\{f\})}{\omega_i^2 - \omega^2} x^{(i)} \tag{78}$$

Consequently, the static solution can be presented as a superposition of free vibration forms:

$$\{x^{(s)}\} = \sum_i \frac{(\{x^{(i)}\}^{\mathrm{T}}\{f\})}{\omega_i^2} x^{(i)} \tag{79}$$

Let us denote by K_s the static stress intensity factor corresponding to $\{x^{(s)}\}$, by $K^{(i)}$ the intensity factors corresponding to $\{x^{(i)}\}$, and by $K(t)$ the dynamic stress intensity factor. The dimensionality of $K^{(i)}$ is determined by taking into account the normalization (77). Since the displacement vector together with a linear functional uniquely determines the stress intensity factor, we obtain

$$K(t) = \sum_i K^{(i)} \frac{(\{x^{(i)}\}^{\mathrm{T}}\{f\})}{\omega_i^2 - \omega^2} e^{i\omega t} \tag{80}$$

$$K_s = \sum_i K^{(i)} \frac{(\{x^{(i)}\}^{\mathrm{T}}\{f\})}{\omega_i^2} \tag{81}$$

Introducing the dimensionless coefficients

$$z_i = \frac{K^{(i)}(\{x^{(i)}\}^{\mathrm{T}}\{f\})}{K_s \omega^2} \tag{82}$$

we obtain from (79) and (80)

$$K(t) = K_s \sum_i z_i \frac{\omega_i^2}{\omega_i^2 - \omega^2} e^{i\omega t} \tag{83}$$

$$\sum z_i = 1 \tag{84}$$

The last equality serves as the criterion for the accuracy with which the dynamic stress intensity factor has been calculated, if the static stress intensity factor appearing in the expression for z_i is taken not in accordance with (81), but directly from the static system (75) of equilibrium equations. Moreover, the required number of vibrations modes in (83) is determined from eqn (84) in the frequency interval $0 \leq \omega < \omega_1$.

Thus, the error in the determination of the dynamic stress intensity factors is estimated from the difference $|\sum z_i - 1|$. In other words, the error in the determination of $K(t)$ is estimated by comparing the dynamic stress intensity factor at zero frequency with the static stress intensity factor determined from the equilibrium equation. The dimensionless coefficients z_i used for determining $K(t)$ do not depend in the two-dimensional case on the size of the plate, its thickness, or the Young's modulus, but they do depend on the ratio of the edges, configuration, and the relative crack-length, as well as on the Poisson's ratio.

The method of calculating the dynamic stress factors described above was applied to the case of a strip fixed at one end with an edge crack (Fig. 8), a free plate with a central horizontal crack, and a plate with an oblique central crack. The problem of calculating the dynamic stress intensity factors in a plate having an edge crack and fastened at one end may be encountered, for example in the investigations of the strength of turbine blades and wings of airplanes.

Suppose that a plate (see Fig. 8) is subjected to a harmonic extension-compression at the edge parallel to the line of fastening. The ratio of sides of the plate is $a/b = 1.17$. To determine the frequencies and the free longitudinal vibration modes for the plate, the latter is divided into a regular mesh of elements. The number of nodes was varied from 58 to 66 depending on the length of the crack. The special singular element was used. It was shown that any further increase in the number of degrees of freedom does not significantly affect the first eight frequencies and modes.

The results of computations of frequencies were identical to those obtained with the help of regular elements only in similar meshes. (In the case of regular elements, obviously, a considerable decrease in the element size in the vicinity of the crack tip is essential for calculating the stress intensity factors.) The accuracy of computation of eigenvectors was 10^{-4}.

The square of the ratio of vibration frequency ω_i^* of a plate without a crack is plotted in Fig. 9 as a function of the relative crack length for the first four frequencies. The dimensionless values of the square of the vibration frequency for a defect-free plate, computed for the ratio of sides equal to 1.17 and Poisson's ratio $v = 0.3$, are 0.33, 2.15, 2.81 and 9.34. The transition to dimensional frequencies is made using the relation

$$\omega_i^2 = \bar{\omega}_i^2 \frac{E}{\rho_o a B} \tag{85}$$

where $\bar{\omega}_i$ are dimensionless frequencies, E is Young's modulus, and ρ_o is the density.

It can be seen from Fig. 9 that the appearance and propagation of a crack considerably lowers the vibration frequencies. Consequently, even if the range of loading frequencies does not exceed the fundamental (lowest) frequency of free vibrations of a structure, the growth of the crack may lead to resonance. Resonance is most dangerous for high-frequency loading, since the free vibration frequencies in real structural elements are of the order $\sqrt{E/\rho_0}$, i.e., are rather high.

The obtained results show that the crack growth leads to a more rapid attenuation of the longitudinal vibration frequencies as compared to transverse vibration frequencies. For example, the first four longitudinal vibration frequencies in the case of a relative crack length of 0.33 decreased in comparison with the frequencies of a continuous plate by 19.4%, 40.0%, 7.8% and 14.0%.

The results of computations of vibration frequencies of plates with cracks can be used for calculating the admissible loading frequencies as well as for determining the defect size in structural elements from variations in the natural vibration spectrum. While calculating the maximum admissible loading frequencies by taking into account

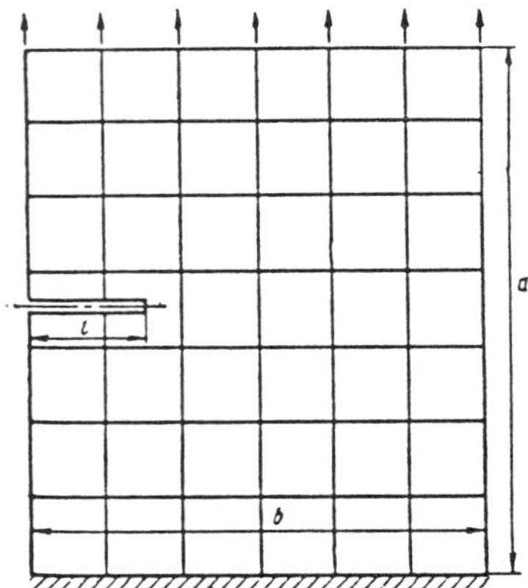

Figure 8: Division of a plate with an edge crack into finite elements

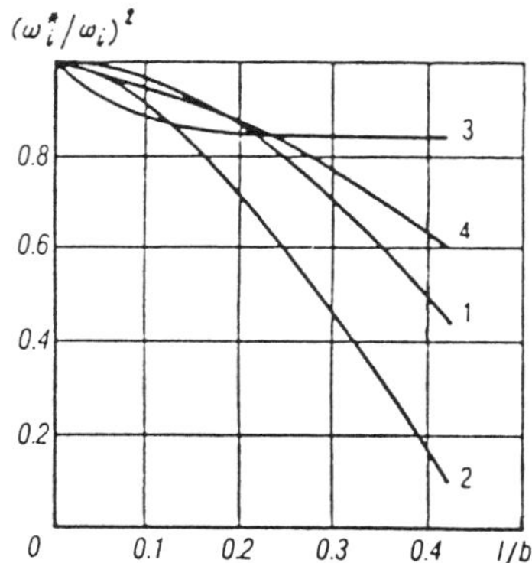

Figure 9: Dependence of the square of vibration frequency on crack length

the growth of defects, we should determine the free vibration frequencies corresponding to the largest crack size. Sometimes the corollaries of the Courant-Fisher theorem[3] are used to draw conclusions concerning the magnitudes of frequencies for small cracks. Suppose that a plate with a crack has vibration frequencies $\omega_1 \ll \omega_2 \ll \ldots$, while the frequencies of vibrations of a plate with a smaller crack, which can be treated upon finite-element discretization as the first plate with r constraints imposed on it, are given by $\omega_1' \leq \omega_2' \leq \ldots$. Then the following inequality holds:

$$\omega_i \leq \omega_i' \leq \omega_{i+r} \tag{86}$$

In general, it can be stated on the basis of the corollaries of the Courant-Fisher theorem that the frequencies of free vibrations of a structure can be increased not only by decreasing the crack length, but also by fastening parts of the boundary, decreasing the plastic zones, reducing the mass, and increasing the stiffness. On the other hand, these factors increase the risk of brittle fracture in some cases. Moreover, the inertia effects may cause the stress intensity factors to reach critical values before the onset of resonance. Hence, the calculations of admissible frequencies must also be carried out on the basis of an analysis of the dynamic stress intensity factors.

It should be observed that the use of stiffening ribs may serve as an effective means of reducing the risk of brittle fracture. This is so because, first, they increase the vibration frequency of plates and, second, they decrease the value of the stress intensity factors.

A comparison of the vibration modes for a cracked plate with the corresponding modes for a defect-free plate showed that the presence of a crack considerably changes the vibration modes. The fundamental vibration modes for a cracked plate are characterized by a considerable opening of the crack and, to a lesser extent, by a relative displacement of the crack faces. This is reflected in the stress intensity factors corresponding to normalized modes of vibrations: for the first four modes, all the ratios $K_{II}^{(i)}/K_I^{(i)}$ except the last one do not exceed unity and are equal to 0.24, 0.17, 0.77, and 1.35. Unlike transverse vibrations, the relative discontinuities in free vibration modes upon passing through the crack are quite large.

The effect of loading frequency on the stress intensity factors K_I for opening mode and K_{II} for inplane shear was investigated for the frequency interval $0 \leq \omega < \omega_1$, which is most important from the point of view of applications (the nonzero value of K_{II} during extension is due to pinching of the plate edge). These investigations showed that the opening mode stress intensity factor increases monotonically with loading frequency and ultimately exceeds the static value (for $\omega = 0$) on account of the structure of z_i in (83). For $l/b = 0.167$, these dimensionless coefficients are equal to 0.104, 1.102, -0.234, 0.060, 0, -0.043, 0.030, and 0.001. It can be seen that for $0 < \omega < \omega_1$, the amplitude ratio $K_1(t)/K_{1s}$ increases owing to quite large positive values of z_1 and z_2. The condition (84) is satisfied in this case with an accuracy of 98%. Similarly, the coefficients for $l/b = 0.25$ are 0.220, 0.845, -0.061, 0.019, -0.003, 0.020, 0, and 0. The sum of these coefficients is 1.040, which means that the error in this case is 4%. For long cracks, the error increases up to 6%. The dependences of $K_1(t)/K_{1s}$ on dimensionless frequency are shown in Fig. 10. The magnitude of the dynamic stress intensity factor for a given amplitude of the applied stress can be

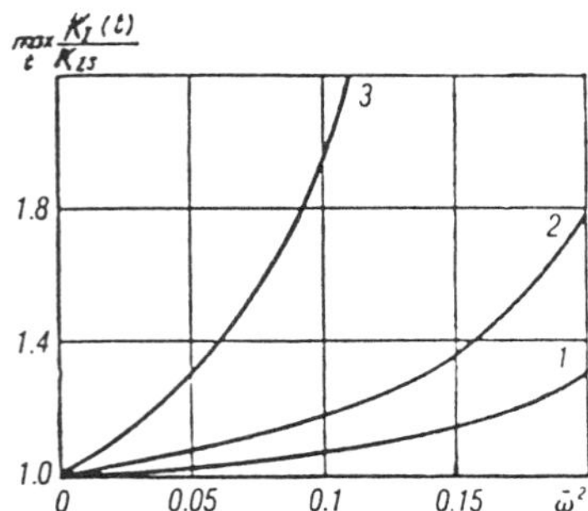

Figure 10: Dependence of the peak value of the ratio of dynamic and static opening stress intensity factors on the square of nondimensionalized vibration frequency (curves 1-3 correspond to relative crack lengths 0.167, 0.250, 0.417).

calculated with the help of these curves and the static intensity factor K_{1s} presented in the form $K_{1s} = q^{(1)}\sqrt{\pi l}F(l/b, a/b)$. If calculations are carried out by taking the ratio of the plate sides $a/b = 1.17$ and the values of relative crack length given above, the function $F(l/b, a/b)$ is found to be equal to 1.40, 1.65, and 2.53, respectively.

The accuracy with which condition (84) is satisfied for the stress intensity factors K_{II} is somewhat lower (the error in this case is between 6 and 14%). However, this degree of accuracy is sufficient for estimation and qualitative analysis of the stress intensity factor. It is found that the amplitude of the ratio $K_{II}(t)/K_{IIs}$ first increases with increasing frequency from 1 (for $\omega = 0$) to about 1.25, after which it decreases and reverses its sign (Fig. 11).

Calculations show that the ratio of the stress intensity factors of first and second kinds in the frequency range $0 \leq \omega \leq \omega_1$ is small. The modulus $\mid K_{II}/K_I \mid$ attains its highest value at $\omega = \omega_1$ and is equal to 26.6% and 19.0%, respectively, for $l/b = 0.25$ and 0.417. For $\omega^2 \leq 0.8\omega_1^2$, the modulus of this ratio does not exceed 7.5%, while its highest value for $\omega^2 \leq 0.9\omega_1^2$ is 13.7%. Consequently, the stress intensity factor for fracture is the most significant parameter in the range of operating frequencies. Both stress intensity factors increase indefinitely in absolute value as the loading frequency approaches the fundamental vibration frequency.

The obtained results confirm that as the vibration frequency increases, the magnitude of the fracture load decreases, i.e., the risk of brittle fracture grows.

It is well known that the interaction of turbine blades with a gas jet excites vibrations as a result of which the stresses in the turbine elements become unstationary. The frequencies of these vibrations may become indefinitely close to the resonance frequencies, since their upper limit is equal to the product of the rotational frequency and the number of blades,[4] while, according to the results obtained, the fundamental frequencies of free vibrations of plates are of the order of $\sqrt{E/\rho_0}$, which lies in the same range. This, as well as the above conclusions about the increased risk in brittle fracture upon an increase in the loading frequency, and about the decrease in the frequency of free vibrations due to crack growth, can be used to explain a number of cases described in literature on brittle fracture of turboengine elements resulting from operation at rotational frequencies exceeding the rated values[5] (including the fracture

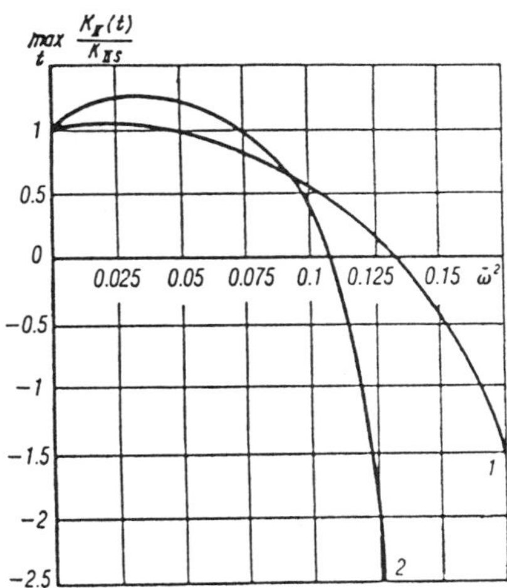

Figure 11: Dependence of the peak value of the ratio of dynamic (inplane shear) and static stress intensity factors on the square of nondimensionalized vibration frequency (curves 1 and 2 correspond to relative crack lengths 0.250 and 0.417, and the function $F(l/b, a/b)$ in the relation $K_{IIs} = q^1\sqrt{\pi l}F(l/b, a/b)$ is equal to 0.55 and 0.103, respectively, for these two cases).

during the first few months of operation when fatigue cracks cannot attain significant values).

Dynamic stress intensity factors in a square plate with an oblique central crack (Fig. 12) under harmonic extension-compression conditions were also investigated. The crack was inclined at an angle of 45° to the base of the plate and a load of unit intensity was applied to the horizontal edges.

Figure 13 shows the plots of the peak values of these stress intensity factors as functions of $\bar{\omega}^2$, the square of the dimensionless loading frequency in the interval $0 \leq \omega \leq \omega_1$. It can be seen that the stress intensity factors monotonically increase from their static values at $\omega = 0$ and tend to infinity as the fundamental vibration frequency is approached.

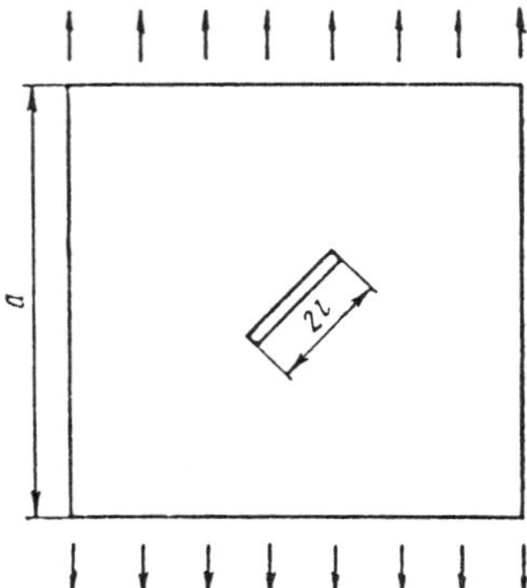

Figure 12: A square plate with an oblique crack ($2l = 4\sqrt{2} \times 10^3$mm, $a = 22 \times 10^3$mm, $E = 1Pa$, $\rho_0 = 0.1$kg/m^3).

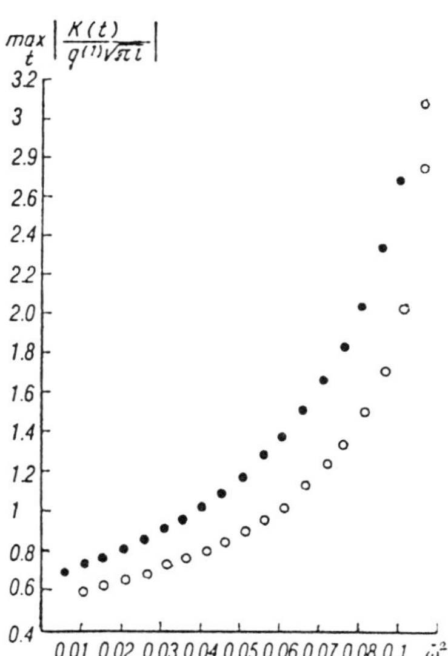

Figure 13: Dependence of stress intensity factors due to opening (light circles) and shear (dark circles) on the square of nondimensional loading frequency.

4 Impact loading. The methods of determining stress intensity factors

In this section we shall consider the problems of cracked bodies subjected to impact loading. As in the case of harmonic loading, the stress intensity factors increase in this case in comparison with their static values. This fact must be taken into consideration while designing machines and structures involving the application of the methods of fracture mechanics. Under the action of impact loading, the behavior of time-dependent dynamic stress intensity factors is more complicated than for harmonic loading. Thus, for example, the increase in the dynamic stress intensity factors for finite cracks takes place until the arrival at the crack tip of a wave scattered from the opposite tip of the crack.[6] In the case of semi-infinite cracks whose faces are subjected to a uniformly distributed tearing impact load, the stress intensity factor increases with time in proportion to $\sqrt{t}$ and becomes infinite as $t \to \infty$.[7] Another interesting fact worth mentioning here[8] is that the stress intensity factor for a plate with a semi-infinite crack whose faces are subjected to concentrated tearing impact forces assumes a constant (static) value after the passage of a certain time. As in the case of harmonic loading, the problem of impact loading of a body with a crack can be solved analytically to the end only for a few idealized formulations because of the complex nature of the mathematical analysis. Usually, such problems are considered for infinite media. The problem of the stress state at the tip of a semi-infinite crack whose faces are subjected to uniformly distributed tearing impact loads was first solved by Maue.[9] After this, Baker[10] obtained the solution of a two-dimensional problem in which a semi-infinite crack (appearing in a uniform tearing field at the instant $t = 0$) propagates with a constant velocity v; in this case, we obtain for $v = 0$ the solution for a stationary crack. In a certain sense, this is a 'calibration' problem, since the analytic and numerical solutions of the problem of impact opening of finite cracks for the initial interval of time (from the zero instant to the moment of arrival at the crack tip of the waves scattered from the boundary of the body or from the other tip of the crack) must coincide with the solution for a semi-infinite crack.

The problems of the behavior of finite cracks under impact loading were considered also in Refs 1, 2 and 11. The problem is reduced to the numerical solution of Fredholm integral equations for variables in the Laplace transform domain, while the inverse transform was carried out only for the main part of local stresses at the crack tip. A characteristic feature of this approach is that the solution for a finite crack remains finite as $t \to \infty$, and that after the attainment of the peak value at the instant when a wave emitted by the opposite tip of the crack arrives at the crack tip, the stress intensity factor oscillates about the static value with a decreasing amplitude. It should be emphasized once again that up to this instant, the solution for a finite crack coincides with that for a semi-infinite crack.

The mentioned approach does not lead to an exact determination of the points of discontinuity for the time derivative of the stress intensity factor, since the results are known to be smoothed out in the numerical inversion of the Laplace transform. Nevertheless, this method can be used to determine the maximum dynamic stress

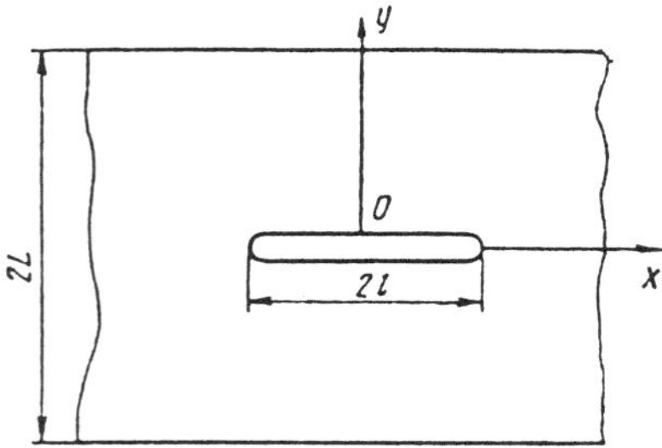

Figure 14: A strip with a crack

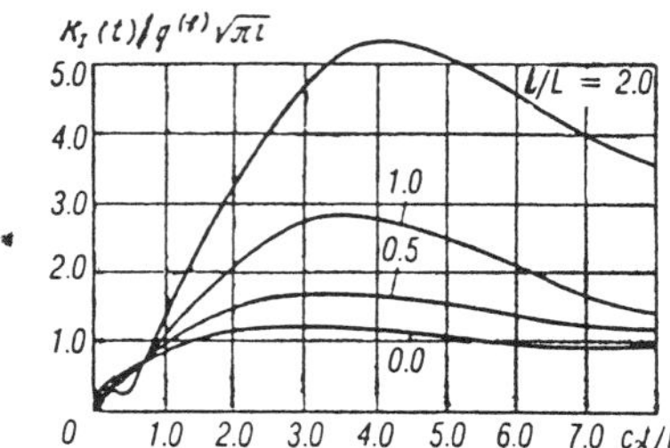

Figure 15: Stress intensity factors in a strip with a crack

intensity factor and other qualitative characteristics of its variation with time.

Let us consider the problem for an infinite strip with a crack (Fig. 14) and for a medium with a penny-shaped crack (Fig. 4). In the case of tearing mode, the boundary conditions have the following form:

$$\begin{aligned} \sigma_{yy}(x,0,t) &= -q^{(1)}H(t), \ \mid x \mid < l \\ u_y(x,0,t) &= 0, \ \mid x \mid \geq l \\ \sigma_{xy}(x,0,t) &= 0, \ \sigma_{yy}(x,\pm L,t) = 0, \ \sigma_{xy}(x,\pm L,t) = 0 \\ & \quad -\infty < x < \infty \end{aligned} \tag{87}$$

The variation of the stress intensity factor is shown in Fig. 15. It is clear that the oscillations observed for small values of t in the case of a narrow strip ($l/L = 2$) are due to the arrival of a wave scattered from the strip edge at the crack tip.

The results for an antiplane shear crack are shown in Fig. 16.

The solution of the problem of penny-shaped cracks due to inplane and antiplane shear is also presented. The results of calculations normalized with the help of the static values of the stress intensity factors are shown in Fig. 17.

The obtained results show that the stress intensity factors attain their peak values after a brief interval of time has elapsed following the application of the load. These values then oscillate with a decreasing amplitude in the vicinity of the static values. Unlike in the case of a plane crack, the initial wave has a toroidal form and the stress waves are generated at each point of the crack face. Consequently, the maximum

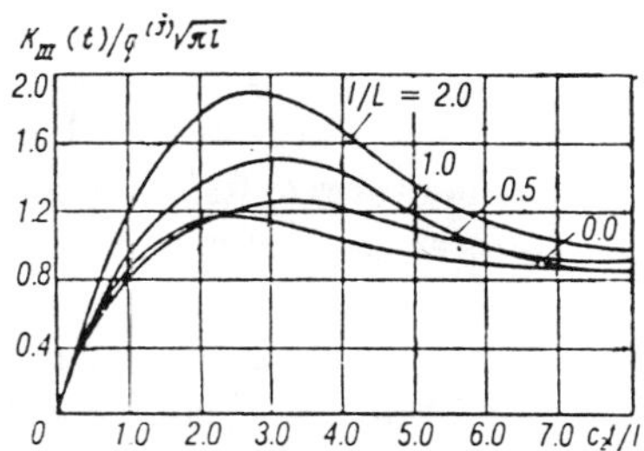

Figure 16: Stress intensity factors in a strip with a crack due to antiplane shear

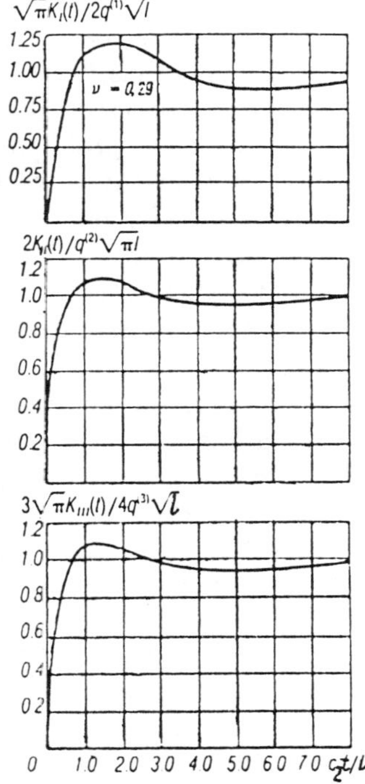

Figure 17: Time dependence of stress intensity factors (of three types) under impact loading of a penny-shaped crack.

values of the stress intensity factor are attained much more rapidly. For large intervals of time, when $c_2 l$ is considerably larger than the crack radius, the wave front gradually becomes spherical.

Here we also present the results of the numerical investigation of wave propagation in the cracked bodies obtained with the help of our program code which utilizes the finite element method together with singular elements and the implicit method of the integration on time.

The first group of calculations concerns to the modeling of wave interaction between the macrocrack and microcrack, and the tips of branched crack. Such calculations were stimulated by the experimental results[12,13] which have shown the necessity to account the complicated wave interaction between the cracks, specimen boundaries, other cracks and microcracks.

The second group of calculations has been made for the investigation of wave propagation in massive elements of metallurgical equipment undergoing impulse loads, namely in the rolls of rolling-mills and in the no-anvil hammers. Particularly the influence of the hammer shape on SIF dependencies was analysed.

Our analysis of the SIF determination was based on the finite element method. The simulation of stress singularities was provided by means of the singular finite element.[11] The finite element solution of wave problems is reduced to the solution of the system of the second-order differential equations. We have chosen for our purposes the unconditionally stable Θ-method which has been recommended in Ref. 15. For this method the step estimates based on the spectral analysis are known,[14] but our numerical experiments have shown that for the problems under consideration another estimates are useful. If one is interested in the wave propagation processes in details the step sometimes less than the time of wave propagation through the smallest element must be chosen. This step was usually less than the step recommended in Ref. 14. But if only the 'inertional effect' (in other words, the SIF amplification), is under investigation one may increase the step essentially.

We have investigated the plate with the central symmetrically branched crack (Fig. 18). The length of the branched part projection is equal to $0.2l$, where l is the half of the main crack length. The half of the branch angle is equal to $\pi/4$.

The calculated SIF K_I and K_{II} dependencies on time in the case of the tearing loads suddenly applied to the edges are presented in Fig. 19 ($2b$ is the height of the plate). Corresponding static values of SIF are: $K_{Is}/\sigma_o\sqrt{\pi l_*} = 1.47$, $K_{IIs}/\sigma_o\sqrt{\pi l_*} = 0.77$. Here l_* is equal to the main crack halflength plus the branch projection length. The total field in the plate is the superposition of the waves refracted from the boundaries and singular points, and all three types of waves (rarefaction, shear and Rayleigh) are essential. The oscillations of the SIF dependencies can be connected with the moments of waves arrival to the crack tip. The maximum ratio of dynamic SIF to static ones is about 2.

The distinguishing feature of the calculated dependencies is that they are identical to each other (when dynamic values are normalized by means of static ones, as it is shown in Fig. 20).

The next calculations refer to the case of the plate with macrocrack and two 'microcrack' behind its tips (Fig. 21). The 'microcrack' length is 1/20 of the macrocrack length, and its right tip (for the right half of the plate) coincides with the branch tip

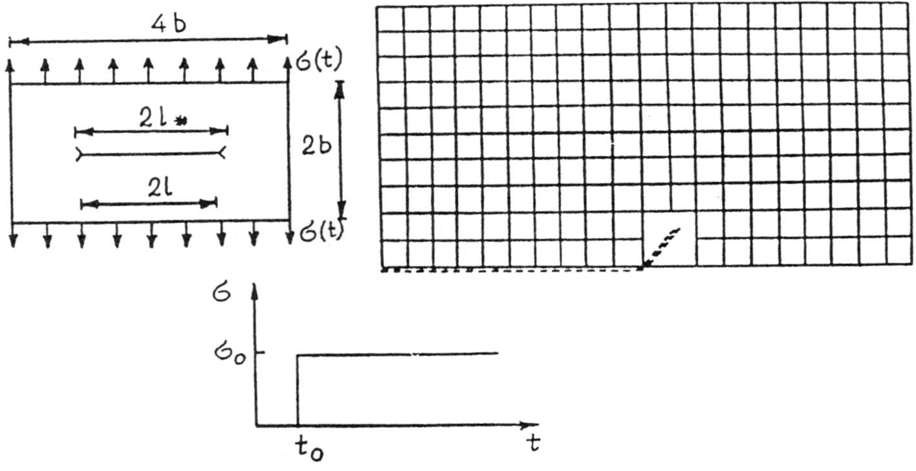

Figure 18: The branched crack in the plate

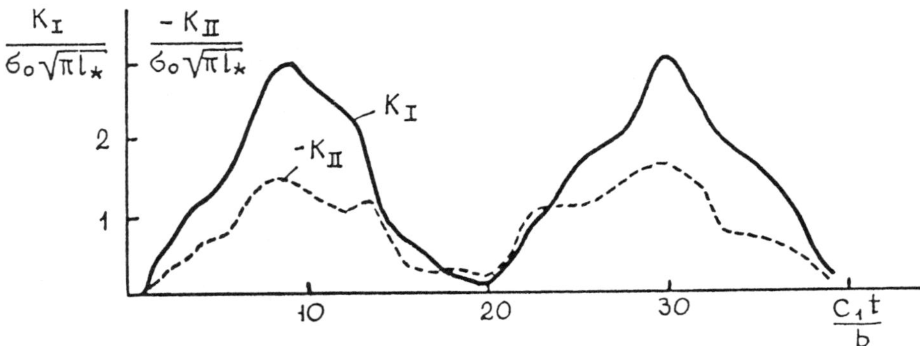

Figure 19: The dynamic SIF dependencies on time

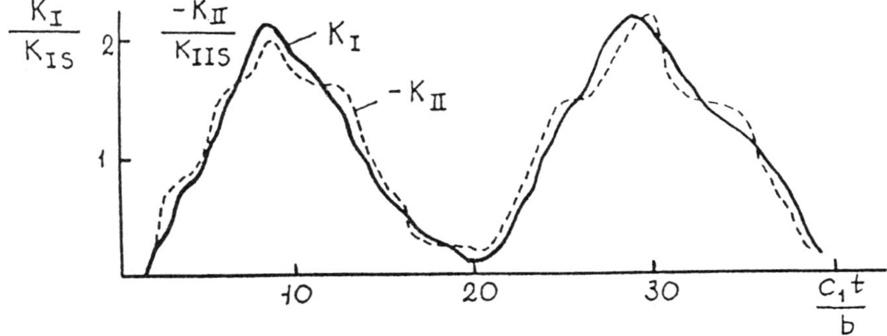

Figure 20: The ratio of the dynamic and static SIF

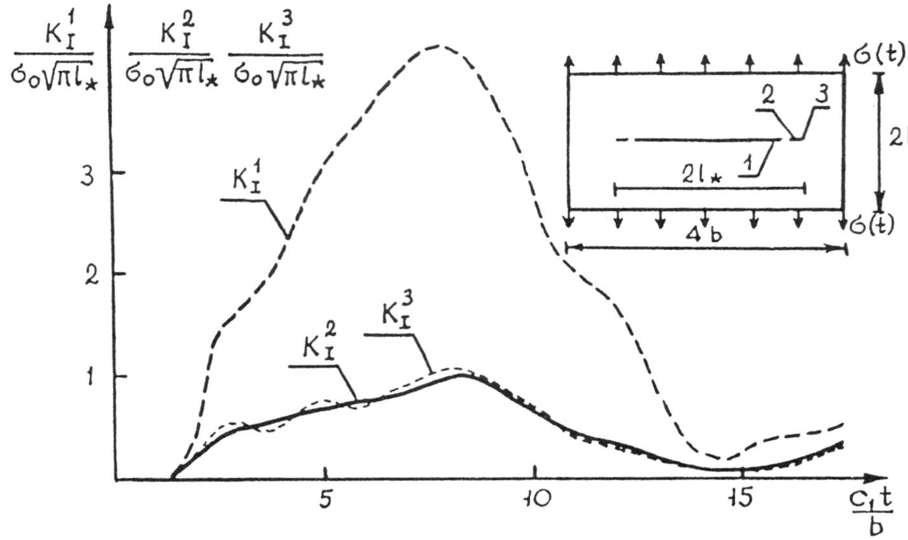

Figure 21: The plate with macrocrack and microcracks

projection from the previous problem. The distance between the cracks is approximately equal to the 'microcrack' length. The finite element grid contained of 119 elements and 139 nodes. The dependencies of the three SIF are presented in Fig. 21. The static values are $K_{Is}/\sigma_o\sqrt{\pi l_*} = 1.80;\ 0.51;\ 0.52$ for the tips 1, 2, 3 respectively.

These dependencies can be used for the detailed analysis of processes in the crack tip with the account of macroscopically observed microbranches and other microdefects. Particularly one can recommend to incorporate such defects behind the crack tip into the calculational models in practical cases and for the evaluation of the experimental data.

For the analysis of dynamic SIF in the rolls of rolling-mills we have considered the configuration shown in Fig. 22. The stress state of the rolls is essentially three-dimensional. Usually one can discover the penny-shaped cracks of the radii about 10-50 mm in it. Since the dynamic behavior of the crack is primarily due to the wave propagation in the diametrical cross-sections we can treat the problem as plane one in the first approach. The static analysis shows that the crack displayed in Fig. 22 is under the compression. But really the compression stresses are superimposed to the initial tensile stresses. The calculated roll was composed of two cylinders, and the ratio of the inner cylinder stiffness to the outer one stiffness was equal to 0.9. The crack length was equal to 0.005 of rolls radii.

The last calculations refer to the SIF in the no-anvil hammers striking the rigid foundation (Fig. 23). The ultrasonic examination usually displays that the cracks are localized near the weld and that they are penny-shaped. The cracks can have quite big size (up to 250 mm in diameter). The experimental results show that the stress state is characterized primarily with the wave propagation along the hammer in vertical direction. This fact allows us to reduce the problem to the plane one.

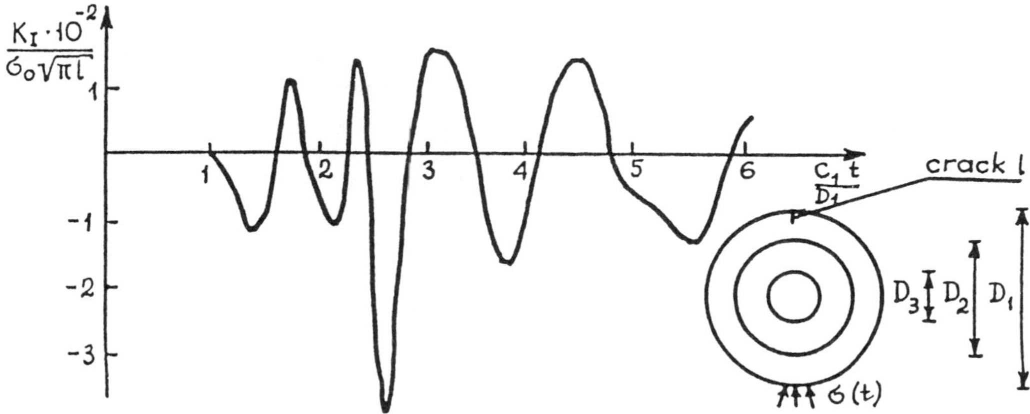

Figure 22: The crack in the roll and the SIF dependency

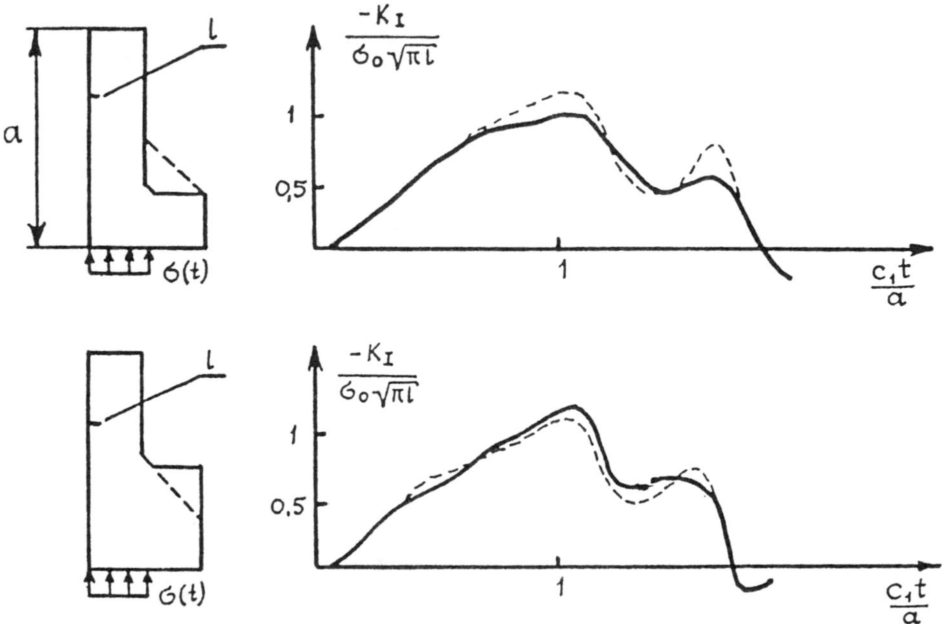

Figure 23: No-anvil hammers and the SIF dependencies

The detailed analysis should account for the solution of the contact problem, but we have applied to the bottom surface the stresses found from the experiment. Their dependence on time was described with the triangle impulse (duration up to 10-2s), so the SIF had the tendency to decrease.

The presented dependencies refer to the dynamic SIF normalized by means of the static (conditional) value. We can now obtain the approach to the dynamic SIF in three-dimensional case as the product of the displayed function to the static SIF in the three-dimensional case.

5 Electric fields in bodies with cuts

There is experimental evidence that a crack propagating in a real electroconducting body can be slowed down by an electric current pulse applied to the body. Finkel *et al.*[16,17] employed the rapid filming technique to monitor crack propagation in a conducting plate subjected to pulses ranging from 10^4 to 2×10^6 amperes/cm^2 in amplitude and from 1 to 100 microseconds in duration. It was found that the immediate consequence of such treatment was an intensive release of heat in the close vicinity of the crack tip (with a time rate of about 10^7 grad/s), which resulted in a local explosion and a subsequent increase of the curvature radius of the tip due to crater formation. For the longer pulses, the radius could increase by two or even three orders of magnitude within a few microseconds. Since this led, in turn, to stress release around the tip, the pulse treatment seems to suggest a useful means for both, suppressing the potential cracking centers and deceleration of cracks already present in the material.

It is therefore of interest to consider a magneto-thermoelastic problem for a crack-containing body, with special emphasis on crack tips as concentrators of electric and thermal fields. In the approach we shall follow[18,19] ponderomotive forces and temperature-deformation coupling effects are neglected and Maxwell's equation written for an electroconducting body are solved to determine the Joule heat density distribution. This latter is then inserted into a heat conduction equation, and after the (non-stationary) temperature field has been found, the quasistatic approximation can be used to determine the strained state of the material.

Suppose a direct current $\mathbf{J} = \{0, J_o, 0\}, J_o = \text{const}$, is being passed through a thin unbounded plate $-\infty < x_1,\ x_2 < \infty.\ \mid x_3 \mid < h,\ 2h$ being the thickness of the plate.[20] The electromagnetic field in the plate will clearly be defined by the vector $\mathbf{H}_0 = \{J_o x_2, 0, 0\}$ and $\mathbf{E}_o = 1/\sigma\{0, J_o, 0\}, \sigma$ denoting the conductivity of the material. We wish to calculate the perturbation produced in this field by a rectilinear finite-length crack $x_2 = 0,\ \mid x_1 \mid < l$ (Fig. 24) instantaneously appearing at $t = 0$. Magnetization and displacement currents will be neglected. If electric and magnetic fields of the form

$$\begin{aligned} \mathbf{E} &= \mathbf{E}_o + \mathbf{E}^*,\ \mathbf{H} = \mathbf{H}_o + \mathbf{H}^* \\ \mathbf{E}^* &= \{E_1^*(x_1, x_2, t),\ E_2^*(x_1, x_2, t), 0\} \\ \mathbf{H}^* &= \{0, 0, H_3(x_1, x_2, t)\} \end{aligned} \tag{88}$$

are assumed. Maxwell's equations may be written as

$$\frac{\partial E_1}{\partial x_1} + \frac{\partial E_2}{\partial x_2} = 0, \quad \frac{\partial E_2}{\partial x_1} - \frac{\partial E_1}{\partial x_2} = -\mu_a \frac{\partial H_3}{\partial t} \tag{89}$$

$$\frac{\partial H_3}{\partial x_2} = \sigma E_1 = j_1, \quad J_o - \frac{\partial H_3}{\partial x_1} = \sigma E_2 = j_2 \tag{90}$$

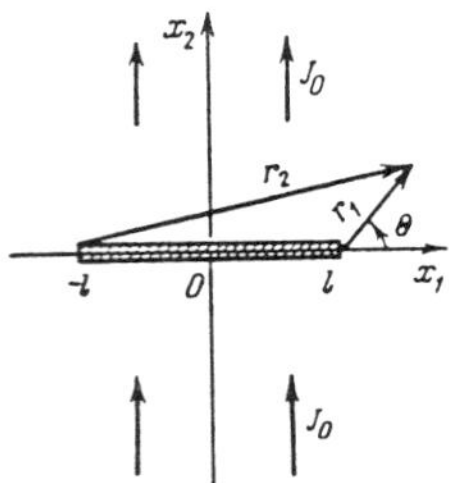

Figure 24: Plane with a rectilinear finite-length crack

where $j = \{j_1, j_2, 0\}$ is the electric current density. The first of eqns (89) is an identity because of (90), whereas the second yields

$$\frac{\partial^2 H_3}{\partial x_1^2} + \frac{\partial^2 H_3}{\partial x_2^2} = \sigma\mu_a \frac{\partial H_3}{\partial t} \tag{91}$$

where the function H_3 must satisfy the zero initial condition $H_3(x_1, x_2, 0) = 0$.

The boundary conditions for H_3 will be written under the assumption that the crack has no conduction in the direction x_2. We have

$$-J_o + \frac{\partial H_3}{\partial x_1}|_{x_2=0} = 0, \ | x_1 |< l \tag{92}$$

$$\frac{\partial H_3}{\partial x_2}|_{x_2=0} = 0, \ | x_1 |> l \tag{93}$$

Note that (93) follows from the symmetry of the electromagnetic field in the vicinity of the crack and corresponds to the vanishing of the j_2 component for the region $x_2 = 0, \ | x_1 |< l$.

The perturbed electromagnetic field is thus defined by the function $H_3(x_1, x_2, t)$ satisfying eqn (91) under the boundary conditions (92), (93) and the zero initial condition.

Application of the Laplace transform to the system (91)-(93) gives

$$\frac{\partial^2 \bar{H}_3}{\partial x_1^2} + \frac{\partial^2 \bar{H}_3}{\partial x_2^2} - k^2 \bar{H}_3 = 0 \tag{94}$$

$$\frac{\partial \bar{H}_3}{\partial x_1} \ |_{x_2=0} = \frac{J_o}{p}, \ | x_1 |< l; \ \frac{\partial \bar{H}_3}{\partial x_2} \ |_{x_2=0} = 0, \ | x_1 |> l \tag{95}$$

with

$$\bar{H}_3(x_1, x_2, p) = \int_0^\infty H_3(x_1, x_2, t) e^{-pt} dt \ (k^2 = p\sigma\mu_a)$$

Because of the symmetry of the magnetic field the solution is (94) may be taken in the form

$$\bar{H}_3(x_1,x_2,p)=\int_0^\infty A(\xi)e^{-x_2\sqrt{\xi^2+k^2}}\sin\xi x_1\mathrm{d}\xi\ (x_2>0,x_1>0) \tag{96}$$

and the conditions (95) lead to the dual integral equations

$$\int_0^\infty A(\xi)\sin\xi x_1\mathrm{d}\xi=\frac{J_0}{p}x_1,\ 0\le x_1<l \tag{97}$$

$$\int_0^\infty \sqrt{\xi^2+k^2}A(\xi)\ \sin\xi x_1\mathrm{d}\xi=0,\ l<x_1<\infty \tag{98}$$

for the function $A(\xi)$.

It proves possible to obtain a simple analytical expression for $H_3(x_1,x_2,t)$ for small times ($t\to 0$) which satisfies (91) and (92) exactly and approximates (93) rather accurately.

One finds that

$$\begin{aligned} A(\xi) \ \cong\ & \frac{2J_0\sqrt{k}}{\pi p\sqrt{2}}\Bigg[\left(\frac{\cos\xi l}{\xi}+l\sin\xi l\right)\frac{\sqrt{\sqrt{k^2+\xi^2}-k}}{\xi\sqrt{k^2+\xi^2}}+ \\ & +\left(\frac{\sin\xi l}{\xi}-l\sin\xi l\right)\frac{\sqrt{\sqrt{k^2+\xi^2}+k}}{\xi\sqrt{k^2+\xi^2}}\Bigg] \end{aligned} \tag{99}$$

which gives

$$\begin{aligned} H_3(x_1,x_2,p)\ \cong\ & \frac{J_0}{p\sqrt{\pi}}\Bigg\{x_1\sqrt{k}\int_{x_2}^\infty\frac{e^{-k\eta}\mathrm{d}\eta}{\sqrt{\eta+x_2}}+ \\ & +\frac{\sqrt{k}}{2}\left(x_1+\frac{1}{2k}\right)\int_{x_2}^{r_2}e^{-k\eta}\left(\frac{1}{\sqrt{\eta-x_2}}-\frac{1}{\sqrt{\eta+x_2}}\right)\mathrm{d}\eta\pm \\ & \pm\frac{\sqrt{k}}{2}\left(x_1-\frac{1}{2k}\right)\int_{x_2}^{r_1}e^{-k\eta}\left(\frac{1}{\sqrt{\eta-x_2}}-\frac{1}{\sqrt{\eta+x_2}}\right)\mathrm{d}\eta+ \\ & +\frac{1}{2\sqrt{k}}\left(\sqrt{r_2+x_2}-\sqrt{r_2-x_2}\right)e^{-kr_2}- \\ & -\frac{1}{2\sqrt{k}}\left(\sqrt{r_1+x_2}\mp\sqrt{r_1-x_2}\right)e^{-kr_1}\Bigg\} \end{aligned} \tag{100}$$

To revert we then use the relations[21]

$$\begin{aligned} \frac{1}{p^{3/4}}e^{-\alpha\sqrt{p}} \ &\rightleftharpoons\ \frac{2^{1/4}}{\sqrt{\pi}t^{1/4}}\,e^{-\frac{\alpha^2}{8t}}D_{-1/2}\left(\frac{\alpha}{\sqrt{2t}}\right) \\ \frac{1}{p^{5/1}}e^{-\alpha\sqrt{p}} \ &\rightleftharpoons\ \frac{2^{3/4}}{\sqrt{\pi}}\,t^{1/4}e^{-\frac{\alpha^2}{8t}}D_{-3/2}\left(\frac{\alpha}{\sqrt{2t}}\right) \end{aligned} \tag{101}$$

where $D_{-y}(z)$ is the parabolic cylinder function. For small t the result is

$$
\begin{aligned}
H_3(x_1,x_2,t) \cong\ & \frac{J_o l}{\pi}\Bigg\{ x_1' \frac{2^{\frac14}}{\tau^{\frac14}} \int_{x_2'}^{\infty} \frac{e^{-\eta^2/8\tau}}{\sqrt{\eta + x_2'}}\, D_{-\frac12}\left(\frac{\eta}{\sqrt{2\tau}}\right) \mathrm{d}\eta + \\
& + 2^{-\frac34}\tau^{-\frac14} \int_{x_2'}^{\rho_2} e^{-\eta^2/8\tau} \left(\frac{1}{\sqrt{\eta - x_2'}} - \frac{1}{\sqrt{\eta + x_2'}} \right) \times \\
& \times \left[x_1' D_{-\frac12}\left(\frac{\eta}{\sqrt{2\tau}}\right) + \frac{\sqrt{\tau}}{\sqrt{2}}\, D_{-\frac32}\left(\frac{\eta}{\sqrt{2\tau}}\right) \right] \mathrm{d}\eta \pm \\
& \pm 2^{-\frac34}\tau^{-\frac14} \int_{x_2'}^{\rho_1} e^{-\eta^2/8\tau} \left(\frac{1}{\sqrt{\eta - x_2'}} \mp \frac{1}{\sqrt{\eta + x_2'}} \right) \times \\
& \times \left[x_1' D_{-\frac12}\left(\frac{\eta}{\sqrt{2\tau}}\right) - \frac{\sqrt{\tau}}{\sqrt{2}}\, D_{-\frac32}\left(\frac{\eta}{\sqrt{2\tau}}\right) \right] \mathrm{d}\eta + \\
& + \tau^{\frac14} 2^{-\frac14} \left(\sqrt{\rho_2 + x_2'} - \sqrt{\rho_2 - x_2'} \right) e^{-\rho_2^2/8\tau}\, D_{-\frac32}\left(\frac{\rho_2}{\sqrt{2\tau}}\right) - \\
& - \tau^{\frac14} 2^{-\frac14} \left(\sqrt{\rho_1 + x_2'} \mp \sqrt{\rho_1 - x_2'} \right) e^{-\rho_1^2/8\tau}\, D_{-\frac32}\left(\frac{\rho_1}{\sqrt{2\tau}}\right) \qquad (102)
\end{aligned}
$$

where

$$x_1' = x_1/l,\ x_2' = x_2/l,\ \rho_1 = r_1/l,\ \rho_2 = r_2/l,\ \tau = t/l^2\sigma\mu_a$$

From (102) we see that H_3 on the line of the crack ($x_2 = 0$) is given by

$$
\begin{aligned}
H_3(x_1,0,t)\,|_{x_1<l} &= \frac{J_o l}{\pi} x_1' 2^{\frac12} \tau^{-\frac14} \int_0^{\infty} e^{-\eta^2/8\tau} D_{-\frac12}\left(\frac{\eta}{\sqrt{2\pi}}\right) \frac{\mathrm{d}\eta}{\sqrt{\eta}} = J_o l x_1' \qquad (103) \\
H_3(x_1,0,t)\,|_{x_1>l} &= \frac{J_o l}{J_1}\{\pi_1 x_1 - 2^{\frac14}\tau^{-\frac14} \int_0^{x_1'-1} e^{-\eta^2/8\tau} \times \\
& \times [x_1' D_{-\frac12}(\frac{\eta}{\sqrt{2\tau}}) - \frac{\sqrt{\tau}}{\sqrt{2}} D_{-\frac32}(\frac{\eta}{\sqrt{2\tau}})] \frac{\mathrm{d}\eta}{\sqrt{\eta}} - \\
& - 2^{\frac14}\tau^{\frac14} \sqrt{x_1' - 1}\, e^{-(x_1'-1)/8\tau} D_{-\frac32}(\frac{x_1'-1}{\sqrt{2\tau}})\} \qquad (104)
\end{aligned}
$$

Figure 25 shows $H_3(x_1,0,\tau)/J_o l$ versus $x_1' = x_1/l$ curves for $\tau = 0.001$ and 0.01 (curves 1,2) and $\tau \to \infty$, or stationary regime (curve 3). It is easily shown that in the latter case

$$
H_3(x_1,0,\infty)/J_o l = \begin{cases} x_1', 0 \le x_1' \le 1 \\ x_1' - \sqrt{(x_1')^2 - 1},\ 1 \le x_1' < \infty \end{cases}
$$

It follows from (103) and (104) that the magnetic field is finite at the tip of a nonconducting crack whereas the electric field components $j_1 = \sigma E_1$ and $j_2 = \sigma E_2$ undergo a singularity in this vicinity.

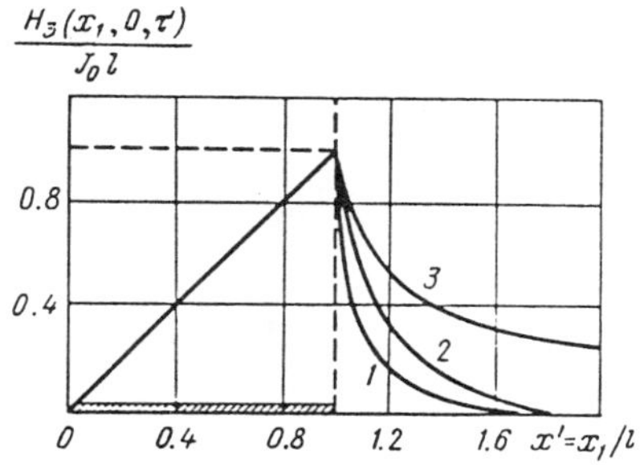

Figure 25: Change of the component of the magnetic field H_3 along the crack and on its prolongation (for a different time).

This readily yield the asymptotic behavior for $\rho_1 \to 0$ if we use polar coordinates ρ_1, Θ referred to the tip $x_1' = 1$, $x_2' = 0$ as origin. We have

$$j_1 \cong -\frac{J_0}{\pi}\frac{2^{\frac{1}{4}}}{\pi^{\frac{1}{4}}}\frac{\sin\frac{\Theta}{2}}{\sqrt{\rho_1}}e^{-\rho_1^2/8\tau}D_{-\frac{1}{2}}\left(\frac{\rho_1}{\sqrt{2\tau}}\right) \tag{105}$$

$$j_2 \cong -\frac{J_0}{\pi}\frac{2^{\frac{1}{4}}}{\pi^{\frac{1}{4}}}\frac{\cos\frac{\Theta}{2}}{\sqrt{\rho_1}}e^{-\rho_1^2/8\tau}D_{-\frac{1}{2}}\left(\frac{\rho_1}{\sqrt{2\tau}}\right) \tag{106}$$

$$0 \le \Theta \le \pi$$

Let us consider now an unbounded conducting plate occupying the region $-\infty < x_1,\ x_2 < \infty \ \mid x_3 \mid < h$, with a nonconducting thorough crack $\mid x_1 \mid < 1$, $\mid x_3 \mid < h$ in the plane $x_2 = 0$ as shown in Fig. 26. We neglect displacement currents in the Maxwell's equation of the problem and assume the magnetic field to be of the form

$$\mathbf{H} = \mathbf{H}^{(0)} + \mathbf{H}^* \tag{107}$$

where $\mathbf{H}^0 = \{H^{(0)}(x_3,t),0,0\}$ is the magnetic field that would be produced by the current $\mathbf{j}^{(0)} = \{0, j_2^{(0)}(x_3,t),0\}$ if the plate had no cracks; $\mathbf{H}^* = \{H_1^*, H_2^*, H_3^*\}$, $H_i^* = H_i^*(x_1,x_2,x_3,t)$ is the perturbation due to crack formation.

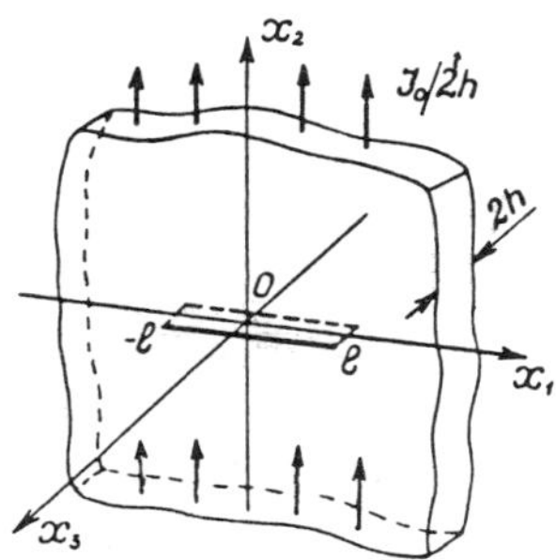

Figure 26: Unbounded conducting plate with a nonconducting thorough crack

The Maxwell's equations are

$$\nabla^2 \mathbf{H}^* - \sigma\mu_a \frac{\partial \mathbf{H}^*}{\partial t} = 0, \ \nabla^2 \mathbf{H}^{(0)} - \sigma\mu_a \frac{\partial \mathbf{H}^{(0)}}{\partial t} = 0 \tag{108}$$

$$\text{div } \mathbf{H}^* = 0, \ \text{div } \mathbf{H}^{(0)} = 0 \tag{109}$$

where

$$\nabla^2 = \frac{\partial^2}{\partial x_1^2} + \frac{\partial^2}{\partial x_2^2} + \frac{\partial^2}{\partial x_3^2}$$

In view of (107) the electric current components are

$$j_1 = \frac{\partial H_3^*}{\partial x_2} - \frac{\partial H_2^*}{\partial x_3}, \ j_2 = \frac{\partial H_1^{(0)}}{\partial x_3} + \left(\frac{\partial H_1^*}{\partial x_3} - \frac{\partial H_3^*}{\partial x_1} \right), \ j_3 = \frac{\partial H_2^*}{\partial x_1} - \frac{\partial H_1^*}{\partial x_2} \tag{110}$$

To simplify the analysis we present the solution of the static version of the present problem which is also of interest. If a direct current is being passed through an unbounded plate having a non-conducting cut, the magnetic field may be expressed as

$$\mathbf{H}_c = \mathbf{H}_c^{(0)} + \mathbf{H}_c^*,$$

$$\mathbf{H}_c^{(0)} = \left\{ \frac{J_o}{2h} x_3, 0, 0 \right\}, \ \mathbf{H}_c^* = \{0, 0, H_3(x_1, x_2)\} \tag{111}$$

where the function $H_3(x_1, x_2)$ is harmonic, and for the electric current density we obtain

$$\mathbf{j}_c = \{j_1(x_1, x_2), j_2(x_1, x_2), 0\},$$

$$j_1(x_1, x_2) = \frac{\partial H_3(x_1, x_2)}{\partial x_2}, \ j_2(x_1, x_2) = \frac{J_o}{2h} - \frac{\partial H_3(x_1, x_2)}{\partial x_1} \tag{112}$$

The function $H_3(x_1, x_2)$ obeys Laplace's equation in the half-plane $x_2 > 0$ under the conditions

$$j_2(x_1, 0) = -\frac{\partial H_3(x_1, 0)}{\partial x_1} + \frac{J_o}{2h} = 0, \ | x_1 | < l \tag{113}$$

$$j_1(x_1, 0) = \frac{\partial H_3(x_1, x_2)}{\partial x_2} |_{x_2=0} = 0, \ | x_1 | > l \tag{114}$$

at the face $x_2 = 0$, and must, of course, be finite at infinity. The solution is readily found to be

$$H_3(x_1, x_2) = \frac{J_o l}{2h} \int_0^\infty \frac{J_1(\xi l)}{\xi} e^{-\xi x_2} \sin\xi x_1 \mathrm{d}\xi \tag{115}$$

giving

$$\begin{aligned} j_1 &= -\frac{J_o l}{2h}\int_0^\infty J_1(\xi l)e^{-\xi x_2}\,\sin\xi x_1 \mathrm{d}\xi \\ j_2 &= \frac{J_o}{2h}-\frac{J_o l}{2h}\int_0^\infty J_1(\xi l)e^{-\xi x_2}\,\cos\xi_1 \mathrm{d}\xi \end{aligned} \tag{116}$$

After taking the integrals, it is convenient to change to the elliptic coordinates α and β defined by

$$x_1 = l\,\cos\,h\alpha\,\cos\beta,\ x_2 = l\,\cos\,h\alpha\,\sin\beta$$

$$(0 \le \alpha < \infty;\ -\pi < \beta < \pi)$$

which converts to (116) to

$$\begin{aligned} j_1 &= -\frac{J_o}{2h}\frac{\sin\beta\cos\beta}{(\cos h^2\alpha - \cos^2\beta)} \\ j_2 &= \frac{J_o}{2h}\frac{\sin h\alpha\,\cos h\alpha}{(\cos h^2\alpha - \cos^2\beta)} \end{aligned} \tag{117}$$

The Joule heat density will be given (in the stationary case) by

$$Q_c = \frac{1}{\sigma}(j_1^2 + j_2^2) = \frac{J_o^2}{4h^2\sigma}\left[1 + \frac{\cos 2\beta}{\cos h^2\alpha - \cos^2\beta}\right] \tag{118}$$

We note that at the crack tip $x_1 = l,\ x_2 = 0$ both j_1 and j_2 in (117) have a singularity which is easily separated out by changing to the polar coordinates r and $\Theta(x_1 - l = r\,\cos\Theta,\ x_2 = r\,\sin\Theta)$ and letting $r \to 0$. We obtain

$$j_1 \cong -\left(\frac{J_o}{2h}\right)\frac{\sin\frac{\Theta}{2}}{\sqrt{2\rho}},\ j_2 \cong \left(\frac{J_o}{2h}\right)\frac{\cos\frac{\Theta}{2}}{\sqrt{2\rho}} \quad (\rho \to 0) \tag{119}$$

where

$$\rho = r/l = \frac{1}{l}\sqrt{(x_1 - l)^2 + x_2^2}$$

6 Effects of a magnetic field on crack propagation in electroconducting bodies

The manner in which an electromagnetic field interacts with a crack-containing elastic body depends on a variety of factors, the material properties being of primary importance. For an electroconducting but nonmagnetic material, linear magnetoelastic equations are adequate for determining the strained state in the vicinity of the crack tip. If a body (usually assumed to be isotropic) is loaded mechanically and subjected to an external magnetic field, the problem involves Maxwell's equations, and the equations of motion of elastic medium in which Lorentz body forces (or ponderomotive forces) must be incorporated. These two types of equations must be complemented by Hooke's law

$$\sigma_{ij} = 2G\varepsilon_{ij} + \lambda\sigma_{ij}\varepsilon_{kk}, \varepsilon_{ij} = \frac{1}{2}\left(\frac{\partial u_i}{\partial x_j} + \frac{\partial u_j}{\partial x_i}\right) \tag{120}$$

the constitutive equations for moving isotropic medium[22]

$$\mathbf{D} = \varepsilon_a \mathbf{E} + \alpha \frac{\partial \mathbf{u}}{\partial t} \times \mathbf{H}, \ \mathbf{B} = \mu_a \mathbf{H} - \alpha \frac{\partial \mathbf{u}}{\partial t} \times \mathbf{E} \tag{121}$$

and the generalized Ohm's law

$$\mathbf{J} = \sigma\left(\mathbf{E} + \frac{\partial \mathbf{u}}{\partial t} \times \mathbf{B}\right) \tag{122}$$

where G and λ are the elastic constants, ε_a and μ_a are the primitivity and permeability, $\alpha = \varepsilon_a \mu_a - \varepsilon_o \mu_o$ and σ the conductivity of the material.

Maxwell's equations and equations of motion, together with relations (120)-(122), form a closed system of nonlinear magnetoelastic equations for a uniform isotropic medium. To actually solve the system, initial and boundary conditions must be applied. In the general case, the surface S enclosing an elastic body separates two media having different electromagnetic properties, and if no surface charges and no surface currents are present, the electromagnetic boundary conditions may be written as

$$[\mathbf{E}_\tau + \frac{\partial \mathbf{u}}{\partial t} \times \mathbf{B}_\tau] = 0, \ [\mathbf{H}_\tau - \frac{\partial \mathbf{u}}{\partial t} \times \mathbf{D}_\tau] = 0$$

$$[\mathbf{B}] \cdot \mathbf{n} = 0, \ [\mathbf{D}] \cdot \mathbf{n} = 0, \sigma\left(\mathbf{E} + \frac{\partial \mathbf{u}}{\partial t} \times \mathbf{B}\right)\mathbf{n} = 0 \tag{123}$$

Here $[A]$ denotes a jump of the quantity A on the surface S, $\mathbf{n}$ is the normal to the surface, and the subscript τ on a vector designates the tangential component with respect to S. The last of eqns (123) expresses a requirement that there be no electric current across S, and has the implication that the conductivity in the surrounding medium is zero.

The mechanical boundary conditions of the problem must take into account the Maxwell stresses

$$\tau_{ij} = E_i D_j + H_i B_j - \frac{1}{2}\delta_{ij}(E_k D_k + H_k B_k) \tag{124}$$

so that

$$[\sigma_{ij} + \tau_{ij}]n_j = 0 \tag{125}$$

on S.

In order to linearize the system of equations, consider a steady-state electromagnetic field acting on a stationary body. This field is found from the system of electro- and magnetostatic equations

$$\text{curl } \mathbf{E}_o = 0, \quad \text{div } \mathbf{D}_o = 0 \tag{126}$$

$$\textbf{curl } \mathbf{H}_o = \mathbf{J}_o, \quad \text{div } \mathbf{B}_o = 0, \quad \text{div } \mathbf{J}_o = 0$$

$$\mathbf{D}_o = \varepsilon_o \mathbf{E}_o, \ \mathbf{B}_o = \mu_a \mathbf{H}_o, \ \mathbf{J}_o = \sigma \mathbf{E}_o \tag{127}$$

subjected to the conditions

$$[\mathbf{E}_{0\tau}] = 0, \ [\mathbf{H}_{0\tau}] = 0, \ [\mathbf{B}_0]\mathbf{n} = 0, \ [\mathbf{D}_0]\mathbf{n} = 0 \tag{128}$$

and the corresponding stress-strain state of the body is governed by the equations

$$\frac{\partial \sigma_{ij}^{(o)}}{\partial x_j} + (\mathbf{J}_o \times \mathbf{B}_o)_i = 0 \tag{129}$$

of static elasticity theory.

The linear system of differential magnetoelastic equations is obtained by assuming solutions of the form

$$\mathbf{H} = \mathbf{H}_o + \mathbf{h}, \ \mathbf{E} = \mathbf{E}_o + \mathbf{e}, \ \mathbf{D} = \mathbf{D}_o + \mathbf{d}, \ \mathbf{B} = \mathbf{B}_o + \mathbf{b}, \ \mathbf{J} = \mathbf{J}_o + \mathbf{j} \tag{130}$$

where the lower-case letters denote perturbations, and by substituting into equations of motion, Maxwell's equations, and the constitutive relations (121) and (122). Keeping only first order terms we obtain

$$\begin{aligned} \text{curl } \mathbf{e} + \frac{\partial \mathbf{b}}{\partial t} = 0, \quad \text{div } \mathbf{d} &= 0 \\ \text{curl } \mathbf{h} - \frac{\partial \mathbf{d}}{\partial t} = \mathbf{j}, \quad \text{div } \mathbf{b} &= 0 \end{aligned} \tag{131}$$

$$\frac{\partial \sigma_{ij}}{\partial x_j} + (\mathbf{J}_o \times \mathbf{b} + \mathbf{j} \times \mathbf{B}_o)_i = \rho \frac{\partial^2 u_i}{\partial t^2} \tag{132}$$

$$\mathbf{j} = \sigma(\mathbf{e} + \mu_a \frac{\partial \mathbf{u}}{\partial t} \times \mathbf{H}_o), \ \mathbf{d} = \varepsilon_o \mathbf{e} + \alpha \frac{\partial \mathbf{u}}{\partial t} \times \mathbf{H}_o, \ \mathbf{b} = \mu_a \mathbf{h} - \alpha \frac{\partial \mathbf{u}}{\partial t} \times \mathbf{E}_o \qquad (133)$$

Substitution of (120) into (131) and (132) yields

$$\text{curl } \mathbf{e} + \mu_a \frac{\partial \mathbf{h}}{\partial t} - \alpha \frac{\partial^2 \mathbf{u}}{\partial t^2} \times \mathbf{E}_o = 0 \qquad (134)$$

$$\text{curl } \mathbf{h} - \varepsilon_a \frac{\partial \mathbf{e}}{\partial t} - \alpha \frac{\partial^2 u}{\partial t^2} \times \mathbf{H}_o = \sigma\left(\mathbf{e} + \mu_a \frac{\partial \mathbf{u}}{\partial t} \times \mathbf{H}_o\right)$$

$$\begin{aligned} \varepsilon_a \text{ div } \mathbf{e} + \alpha \text{ div } \left(\frac{\partial \mathbf{u}}{\partial t} \times \mathbf{H}_o\right) &= 0 \\ \mu_a \text{ div } \mathbf{h} - \alpha \text{ div } \left(\frac{\partial \mathbf{u}}{\partial t} \times \mathbf{E}_o\right) &= 0 \end{aligned} \qquad (135)$$

$$(\lambda + 2G)\text{grad div } \mathbf{u} - G \text{ rot rot } \mathbf{u} + \sigma\Big\{\mu_a \mathbf{E}_o \times \mathbf{h} + \mu_a \mathbf{e} \times \mathbf{H}_o -$$

$$- \alpha\left(\mathbf{E}_o \times \frac{\partial \mathbf{u}}{\partial t} \times \mathbf{E}_o\right) + \mu_a^2\left(\frac{\partial \mathbf{u}}{\partial t} \times \mathbf{H}_o \times \mathbf{H}_o\right)\Big\} = \rho \frac{\partial^2 \mathbf{u}}{\partial t^2} \qquad (136)$$

The boundary conditions (123) are linearized to give

$$\left[\mathbf{e}_\tau + \mu_a \frac{\partial \mathbf{u}}{\partial t} \times \mathbf{H}_{0\tau}\right] = 0, \qquad \left[\mathbf{h}_\tau + \epsilon_a \frac{\partial \mathbf{u}}{\partial t} \times \mathbf{E}_{0\tau}\right] = 0, \qquad (137)$$

$$\left[\mu_a \mathbf{h} - \alpha \frac{\partial \mathbf{u}}{\partial t} \times \mathbf{E}_0\right] \mathbf{n} = 0, \qquad \left[\epsilon_a \mathbf{e} + \alpha \frac{\partial \mathbf{u}}{\partial t} \times \mathbf{H}_0\right] \mathbf{n} = 0 \qquad (138)$$

For an electroconducting solid placed in a uniform magnetic field $\mathbf{H}_o$, the displacement currents in (131) may be neglected and, in some materials at least, α is very nearly zero. If further, $\mathbf{E}_o = 0$, and $\mathbf{J}_o = 0$, the linearized system of magnetoelastic equations reduces to

$$\text{curl } \mathbf{e} + \mu_o \frac{\partial \mathbf{h}}{\partial t} = 0, \ \ \text{div } \mathbf{e} = 0 \qquad (139)$$

$$\text{curl } \mathbf{h} = \mathbf{j}, \ \ \text{div } \mathbf{h} = 0$$

$$\frac{\partial \sigma_{ij}}{\partial x_j} + \mu_o(\mathbf{j} \times \mathbf{H}_o)_i = \rho \frac{\partial^2 u_i}{\partial t^2} \qquad (140)$$

$$\mathbf{j} = \sigma(\mathbf{e} + \mu_o \frac{\partial \mathbf{u}}{\partial t} \times \mathbf{H}_o) \qquad (141)$$

Eliminating $\mathbf{e}$ from (139) and (141) we obtain

$$\nabla^2 \mathbf{h} - \sigma\mu_o \frac{\partial \mathbf{h}}{\partial t} + \sigma\mu_o \operatorname{curl}\left(\frac{\partial \mathbf{u}}{\partial t} \times \mathbf{H}_o\right) = 0 \tag{142}$$

$$\frac{\partial \sigma_{ij}}{\partial x_j} + \mu_o(\operatorname{curl} \mathbf{h} \times \mathbf{H}_o)_i = \rho \frac{\partial^2 u}{\partial t^2} \tag{143}$$

where ∇^2 is the Laplacian.

The system of equations is greatly simplified if the material is a perfect electrical conductor. In this case

$$\mathbf{h} = \operatorname{curl}(\mathbf{u} \times \mathbf{H}_o) \tag{144}$$

because $\sigma \to \infty$, and (143) takes the simpler form

$$G\nabla^2 \mathbf{u} + (\lambda + G)\operatorname{grad}\operatorname{div}\mathbf{u} + \mu_o \operatorname{curl}\mathbf{h} \times \mathbf{H}_o = \rho \frac{\partial^2 \mathbf{u}}{\partial t^2} \tag{145}$$

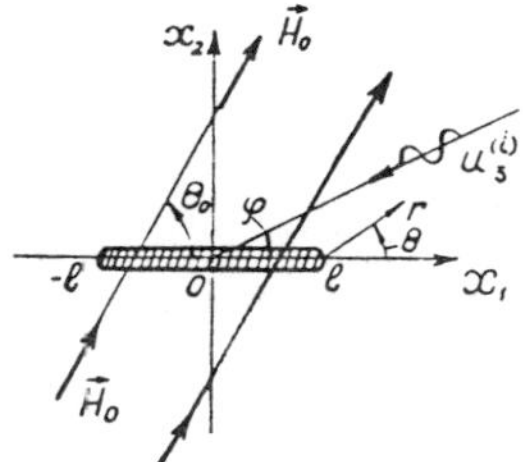

Figure 27: Ideally conducting media with a crack on which falls a plane harmonic shear wave.

The solution of a dynamic magnetoelastic problem in the framework of the above approach will be useful for studying the effects of a magnetic field on a conducting body with a crack. We reproduce here the principal results of the antiplane deformation problem for an unbounded perfectly conducting body containing a rectilinear crack. Suppose a uniform magnetic field

$$\mathbf{H}_o = (H_o \cos\Theta_o,\ \ H_o \sin\Theta_o, 0) \tag{146}$$

acts upon a medium in which a plane harmonic shear wave is propagating at an angle ϕ to the line of a crack (Fig. 27). In this case

$$\mathbf{u} = (0, 0, u_3(x_1, x_2)e^{-i\omega t}),\ \mathbf{h} = (0, 0, h_3(x_1, x_2)e^{-i\omega t})$$

and the system (144), (145) takes the form

$$h_3 = H_o \cos\Theta_o \frac{\partial u_3}{\partial x_1} + H_o \sin\Theta_o \frac{\partial u_3}{\partial x_2} \tag{147}$$

$$G\nabla^2 u_3 + \mu_o H_o \left(\frac{\partial h_2}{\partial x_2} \sin \Theta_o + \frac{\partial h_3}{\partial x_1} \cos \Theta_o \right) + \rho \omega^2 u_3 = 0 \tag{148}$$

where $\nabla^2 = \dfrac{\partial^2}{\partial x_1^2} + \dfrac{\partial^2}{\partial x_2^2}$ and ω is the frequency.

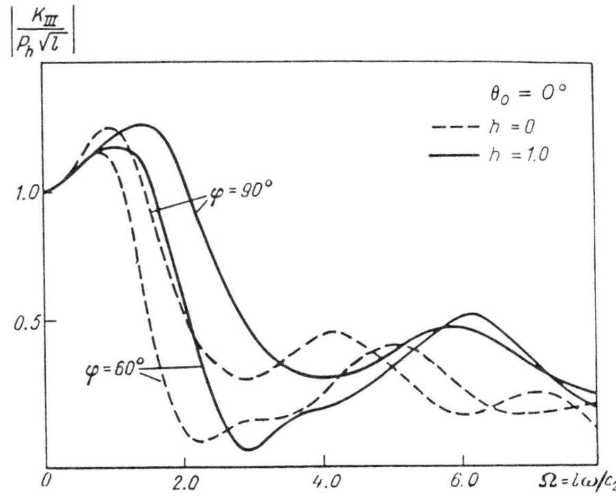

Figure 28: Dependence of the dynamic stress intensity factor on the frequency in the case of a magnetic field parallel to the crack.

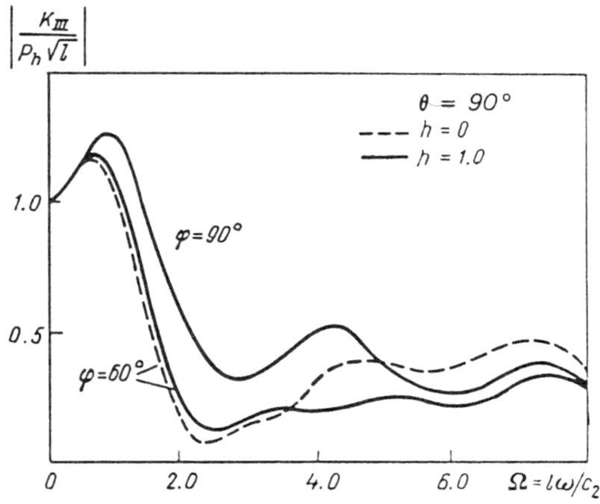

Figure 29: Dependence of the dynamic stress intensity factor on the frequency in the case of a magnetic field perpendicular to the crack.

For an incident harmonic magnetoelastic wave the solution is

$$u_3^{(i)} = W_o \exp\left[-i\omega \frac{x_1 \cos\phi + x_2 \sin\phi}{c_2\sqrt{1+h^2\cos^2(\phi-\Theta_o)}}\right] \tag{149}$$

$$h_3^{(i)} = -i\omega W_o \frac{H_o \cos(\phi-\Theta_o)}{c_2\sqrt{1+h^2\cos^2(\phi-\Theta_o)}} \times$$

$$\exp\left[-i\omega \frac{x_1 \cos\phi + x_2 \sin\phi}{c_2\sqrt{1+h^2\cos^2(\phi-\Theta_o)}}\right] \tag{150}$$

where W_o is the amplitude of the incident wave, $h^2 = \mu_o H_o^2/G$, and $c_2 = \sqrt{G/\rho}$ is the shear wave velocity.

Since the surface of the crack is mechanically free, the solution to (147) and (148) must obey the conditions

$$\sigma_{23}^{(S)}\big|_{x_2=0} = -P_h e^{-i\omega x_1 Z_h/c_2}, \ |\, x_1 \,| < l \tag{151}$$

$$u_3^{(S)}\big|_{x_2=0} = 0, \ |\, x_1 \,| > l \tag{152}$$

where

$$P_h = -\frac{iGW_o\, \omega \sin\phi}{c_2\sqrt{1+h^2\cos^2(\phi-\Theta_o)}}$$

$$Z_h = \frac{\cos\phi}{\sqrt{1+h^2\cos^2(\phi-\Theta_o)}}$$

For the reflected wave, the Fourier transformation with respect to x_1 divides the solution to (148) into two terms, one symmetric and the other antisymmetric in x_1,

$$u_3^{(S)} = \frac{2}{\pi}\int_0^\infty A_s(\alpha) e^{\beta x_2} \cos \alpha x_1 \mathrm{d}\alpha - \frac{2}{\pi} i \int_0^\infty A_a(\alpha) e^{\beta x_2} \sin \alpha x_1 \mathrm{d}\alpha$$

$$x_2 > 0 \tag{153}$$

where

$$\beta = -\sqrt{(1+h^2)(1+h^2\sin^2\Theta_0)^{-2}\alpha^2 - (\omega^2/c_2)^2(1+h^2\sin^2\Theta_0)}(1+h^2\sin^2\Theta_0)^{-1}$$

If we now write

$$\begin{aligned} A_s(\alpha) &= \frac{\pi}{2}\frac{(1+h^2\sin^2\Theta_0)P_h l}{G\sqrt{1+h^2}}\int_0^\infty \xi\psi_s(\xi)J_0(\alpha l\xi)\mathrm{d}\xi \\ A_a(\alpha) &= \frac{\pi}{2}\frac{(1+h^2\sin^2\Theta_0)P_h l}{G\sqrt{1+h^2}}\int_0^\infty \xi\psi_a(\xi)J_1(\alpha l\xi)\mathrm{d}\xi \end{aligned} \tag{154}$$

the boundary conditions (151) and (152) convert to a pair of integral Fredholm equations of the second kind for the functions $\psi_s(\xi)$ and $\psi_a(\xi)$. When these functions are found, it is possible to separate out the singularity ocurring in the tangential stresses in the vicinity of the crack tip ($x_1 = l,\ x_2 = 0$). Introducing polar coordinates r, Θ and taking the tip of the crack to be the origin we find

$$\sigma_{23}^{(S)} \cong K_{III} \frac{1}{\sqrt{2r}} H_c\left(h, \Theta_o, \Theta\right) \tag{155}$$

$$K_{III} = \lim_{x_1 \to l} \sqrt{2\pi(x_1 - l)}\sigma_{23}^{(S)}\Big|_{x_2=0} = \sqrt{\pi l}\, P_h(\Psi_s(1) - i\Psi_a(1)) \tag{156}$$

where

$$H_c(h, \Theta_0, \Theta) = \left\{ \frac{(1 + h^2\sin^2\Theta_0)[\sqrt{(}1 + h^2\sin\Theta_0)^4\cos^2\Theta_0 + (1 + h^2)^2\sin^2\Theta}{2[(1 + h^2\sin^2\Theta_0)\cos^2\Theta - (1 + h^2)\sin^2\Theta} \right.$$

$$\left. + \frac{(1 + h^2\sin^2\Theta_0)^2\cos\Theta]}{2[(1 + h^2\sin^2\Theta_0)\cos^2\Theta - (1 + h^2)\sin^2\Theta]} \right\}^{\frac{1}{2}}$$

In Figs 28 and 29 the dynamic stress intensity factor is shown as a function of frequency for two orientations of the vector $\mathbf{H}_o$ relative the crack line ($\Theta_o = 0,\ \Theta = 90^o$). From the diagram we see that for a magnetic field parallel to the crack the maximum of the coefficient shifts slightly toward the higher frequencies.

7 Comparison of theoretical and experimental results in dynamic fracture mechanics

In this section we shall describe the difference between dynamic fracture mechanics and quasi-static mechanics, and also discuss their principal aims and basic assumptions.

The following problems arise in the study of dynamic fracture mechanics. First, we must find the conditions under which a quasi-static or dynamic loading can cause catastrophic growth of a crack of a given size. Secondly, conditions of unloading under which crack growth can be arrested must be determined. Thirdly, the loading parameters and the properties of materials determining crack propagation must be specified. Finally, we must know the conditions under which a propagating crack branches out and the mechanism underlying this phenomenon. These problems are called the problems of starting, arrest, propagation, and branching of a crack, respectively. The task of dynamic fracture mechanics is to find a solution to these problems.

Thus, it is obvious that the range of topics covered by dynamic fracture mechanics is much wider than that of quasi-static fracture mechanics. While quasi-static fracture mechanics deals, as a rule, with formulating just the criterion for unstable crack propagation, in dynamic fracture mechanics a whole range of criteria, viz., the starting criterion, the criterion for crack arrest, bending and branching criteria, etc., have to be established. The phenomenological description of fracture dynamics, through the concept of a sharp arterial crack, results in the appearance of a large number of criterial stress intensity factors, like the stress intensity factor for crack start, which depends on the loading rate, the crack arrest stress intensity factor, and the intensity factor depending on the velocity of crack propagation. Some experimental results can be explained in this way, though serious disagreements are observed between theoretical and experimental results in some other cases. However, it must be noted that the experimental results themselves are conflicting. Quite often, one can find an experimental result in literature completely repudiating the result of another experiment.

7.1 Idealized model for quasi-brittle fracture and its inconsistencies

In the prevailing idealized model based on the ideas put forth by Griffith, Irwin, and others, the growth of a rectilinear crack in an elastic plane is usually considered. In this case, infinite stresses appear at the crack tip and the fracture process is assumed to take place at the crack tip itself. Moreover, it is assumed that the energy γ spent in creating a unit new surface area is a constant quantity characteristic of the material. On the basis of this assumption, the elastodynamic stress field at the crack tip is calculated and the energy balance equation is formulated. The stress at the crack tip has a $1/\sqrt{r}$–type singularity and the stress intensity factors depend on the crack velocity v. If this dependence is determined by solving the elastodynamic problem with a moving crack and is then substituted into the fracture criterion, the crack velocity can be determined, i.e., the crack behavior can be predicted. Depending on loading conditions, the crack may continue to propagate or it may get arrested. The starting criterion is also derived from the energy balance equation.

Thus, we can formulate the main features of the quasi-brittle dynamic fracture model as follows:

1. Stress fields at the crack tip are described with the help of stress intensity factors.
2. Criteria for start, propagation, and arrest are derived from the constancy of specific fracture energy.

The adequacy of this model can be judged by analyzing experimental data on the stress intensity factor and by comparing the conditions of crack start, propagation, and arrest with theoretical results.

It is found that, in the first place, the 'independence' of the criteria of crack start, propagation, and arrest is established beyond a shadow of doubt. This follows from the existence of a large number of critical stress intensity factors describing the start, propagation and arrest of cracks, which taken together do not satisfy the energy balance equation. Secondly, the relation between the instantaneous stress intensity factor and the instantaneous crack velocity is established from the energy balance equation under the assumption that the surface fracture energy is independent of the loading rate and the past history of the loading process. However, this relation, which establishes a one-to-one correspondence between K and v, is not confirmed by experiments. Besides, the idealized model does not provide a satisfactory explanation of the branching phenomenon.

A direct comparison of analytical and experimental results of investigations of the K vs. t dependence is carried out in Ref. [23] for a stationary crack which is set into motion by an impact load applied to its faces. The results obtained in Ref. [23] are discussed in Ref. [11].

Experiments were carried out on plates of Homalite-100 (perspex) using the caustics method. A very good agreement between theoretical and experimental values of $K_I(t)$ was obtained both before and after crack initiation (Fig. 30).

A similar experiment, carried out on the same material at a much faster loading rate and under higher stresses, showed that such an agreement exists only up to the instant when the crack starts to propagate (Fig. 31).

This discrepancy can be explained by taking into account the following circumstance.[13] Under an impact loading of the crack faces, the size of the region around the crack tip where the stress satisfies the theoretical criteria is equal to zero at the initial instant of time and increases with the velocity of propagation of elastic waves. Thus, in order to establish a zone size which could provide experimental information about the singular stressed state, a certain amount of time is required. This time is large in comparison with the time scale of the processes taking place during dynamic fracture, and increases with the crack velocity. Since the field is assumed to be steady during a theoretical decomposition of the stresses in the vicinity of the crack tip, the experimental methods based on this decomposition can be used only if the process is stabilized in a certain finite interval of time (before the moment of time under consideration). In other words, no stress waves should arrive at the crack tip during this period and there should be no abrupt variation in crack velocity.

Thus, in regions where the theoretical and experimental values of the stress fields coincide, the idealized model of dynamic fracture mechanics can be assumed to be

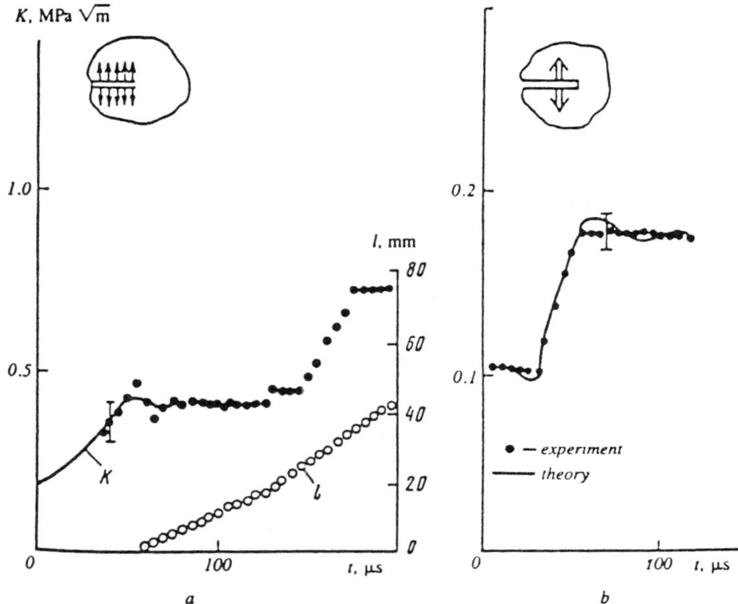

Figure 30: Time dependence of the stress intensity factor upon loading the crack by concentrated (a) and distributed (b) loads: comparison of theory and experiment (the vertical mark indicates start of the crack).

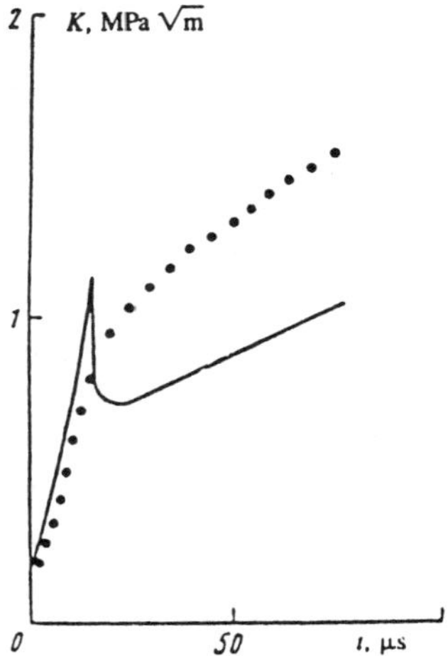

Figure 31: Time dependence of the stress intensity factor for a uniformly distributed load equal to double the load used in Fig. 30.

valid except during transient processes. For all other cases, the behavior of real materials under dynamic fracture differs from the idealized approach towards fracture at the crack tip.

7.2 Microscopic structural aspect of dynamic fracture

A number of inconsistencies between the idealized fracture model and the experiment can be explained by assuming that the fracture occurs in a certain region in front of the crack tip, as, for example, in the case of viscous bodies under quasi-static fracture. Usually, it is assumed that the main feature of deviation from idealized behavior is the plastic flow at the crack tip accompanied by the appearance of cavities which grow in the region of strong plastic deformation. This widespread point of view is undoubtedly valid for viscous materials. However, it is worth noting that even a brittle fracture does not lead to an ideal fracture for dynamic crack propagation. The existence of microfracture near the crack tip certainly influences the time dependence of the fracture process, i.e., affects the fracture rate, since a finite amount of time is required for establishing the region where the fracture process takes place.

In the experiments carried out on Homolite-100, which undergoes a brittle fracture upon static and dynamic loadings, a finite fracture region existed, while there was obviously no plastic deformation. At least three different observations lead to this result: the dependence of fracture surface roughness on the instantaneous stress intensity factor, highspeed photography of discrete fracture sources at the crack tip on the real time scale, and the stress waves generated by these discrete microscopic fractures.

Let us consider in detail the dependence of fracture surface roughness on loading level, investigated experimentally. It is convenient to divide the analysis into three stages. Observation of the fracture surface (after fracture) under low magnification reveals the main features of surface quality variation as a result of a variation of the macroscopic parameters of the process. The surface is then studied under an electron microscope, which shows its microscopic characteristics. The basic conclusions drawn after a visual inspection of the surface are confirmed at this stage. Finally, the same behavior is confirmed by photomicrography of the discrete fracture sources on real time scale, i.e., during the fracture process.

After the propagation of a crack at a high speed, three regions can be isolated on the fractured surface: 'mirror', 'mist' and 'hackle'. After passing the hackled region, the crack splits into several branches. The mirror zone is characterized by an absolutely smooth surface which completely reflects the light falling on it. In the mist region, the surface becomes rougher, while it becomes hackled in the last region. These results were obtained in for experiments on impact fracture of samples as a result of a step-by-step increase in the load.

A constant (time-independent) velocity was observed in all experiments. The stress intensity factor was the only parameter that varied in time. In the first case, the crack propagated along the symmetry line until the waves reflected at the crack faces began to interact with the crack tip. Since the waves were reflected from faces that were not fixed in the same way, the crack deviated from its rectilinear propagation path in such a way that the inplane shear stress intensity factor along its path was equal to zero. The fracture surface consisted mainly of the mirror region. Throughout the period of its propagation, the crack had the same velocity, equal to 363 m/s. At the instant of arrival of the scattered wave, i.e., in about 150 μs, the stress intensity factor began to increase sharply (Fig. 32) and the fractured surface became misty.

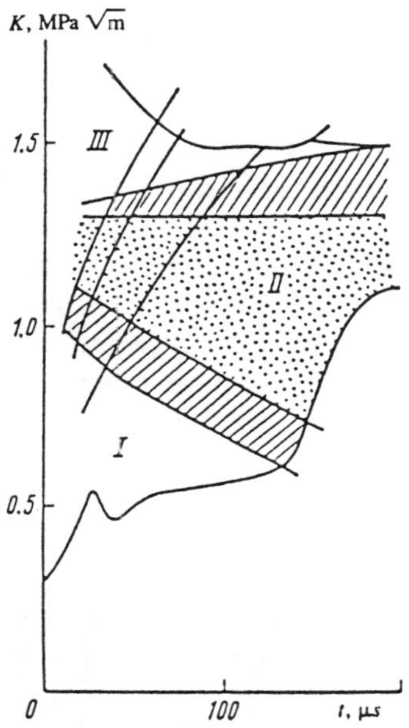

Figure 32: Dependence of K on t for four loading levels: I) mirror zone; II) mist zone; III) hackle zone.

In the second case (for a large value of the load), the crack propagated with a velocity of 410 m/s and the fracture surface consisted of three regions, including the hackled region. When the wave arrived at the crack tip ($\sim 150 \mu s$), branching sets in. A similar picture was observed in the last two experiments, but all processes occurred at a higher speed. The size of the mirror region at the crack faces was reduced, but the other two regions became larger in size.

The transition regions that exist between different regions are hatched in Fig. 32. It can be stated on the basis of the experimental results that the energy liberated at the crack tip is spent in the formation of a rough surface. It becomes clear that this circumstance may be the decisive contradiction in the idealized model of dynamic fracture, since according to this model the fracture energy is calculated by multiplying the energy density by the area of the fractured surface, which is equal to the product of the crack length and the sample thickness. In actual practice, if we take into account the roughness of the surface, the area of the fractured surface will have a different value. A direct attempt to correct the energy balance makes the fracture energy density a function of the past history of loading, and makes it extremely difficult to solve dynamic fracture mechanics problems.

The mechanism of rough surface formation becomes clear from a microscopic study of the fractured surface (Figs 33-35, the magnification is 7000). In the mirror region (Fig. 33), the crack encounters a large number of 10-25 μm cavities on its way. The interaction of the crack with these cavities is responsible for the start of a large number of cracks which, however, do not change the direction of arterial crack propagation. It can be stated that the microscopic cracks originating at the microcavities do not interact in the mirror region. In the mist region, the stress intensity factor increases and the stresses become sufficient for activating isolated cavities and triggering an interaction between them (Fig. 34). This is accompanied by the appearance of many parabolic figures, a characteristic feature of the interaction between cavities and cracks propagating at a constant velocity. These parabolas have different sizes and depths, thus indicating their propagation in three dimensions. Hence, even before the arrival of the arterial crack in the foggy region, a large number of microscopic defects appear with their orientations in different planes, endeavoring to change the direction of propagation of the crack. Finally, this process becomes even more intensive in the hackled region and covers an ever increasing region in front of the crack tip. Small 'rivulets' appear and grow in a direction perpendicular to the crack (Fig. 35).

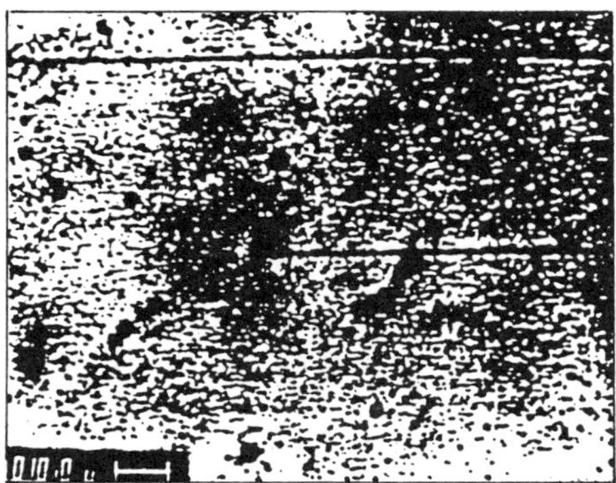

Figure 33: Microphotograph of the mirror zone

It can be concluded from here that a unit crack propagates in the mirror region in the beginning, and its behavior differs considerably from quasi-static growth. In the mist region, simultaneous uniform propagation of an ensemble of cracks takes place. In the hackled region, crack propagation follows the same physical process, but the size of the microfracture region increases. Thus, it can be stated that crack propagation for a high level of stresses is governed by the growth of microscopic cavities and cracks, their confrontation and interaction.

From this point of view it is possible to give an acceptable qualitative description of crack branching as a continuous process of evolution of leading microcracks. Indeed, let us take a look at the photomicrograph of crack branching (Fig. 36 for example). It is clearly seen that the branching process is a continuation of the intensive growth

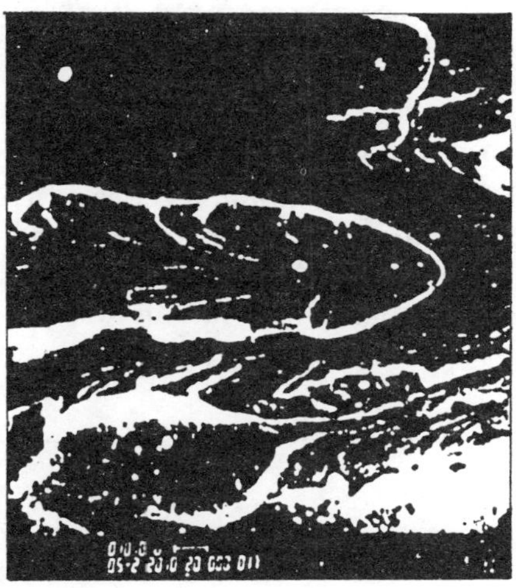

Figure 34: Microphotograph of the mist zone

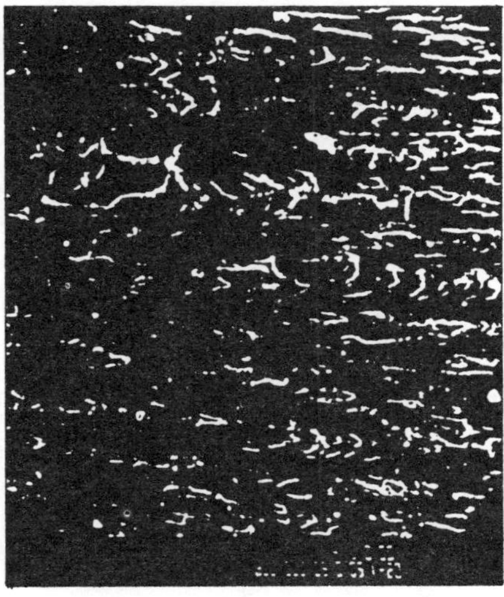

Figure 35: Microphotograph of the hackle zone

and interaction of microscopic defects taking place in the feather region. The onset of final branching is preceded by numerous attempts at branching; the microscopic cracks deviate from the direction of arterial crack propagation and are then arrested. A complex process of wave interaction occurs between these microscopic branches and the arterial crack. At a certain instant of time, the stresses acquire such a value that the crack finally branches out. Without doubt, this process is statistical and three-dimensional, but it also has deterministic features; the spread in the coordinate of the final branching point was just 1 mm in a series of five experiments. It must be emphasized that branching is an evolutionary process and does not involve just the conversion of a single mathematical cut into several cuts, as modelled in the problems of elastodynamics, but involves a gradual qualitative change in the fracture front according to the mechanism shown in Fig. 37.

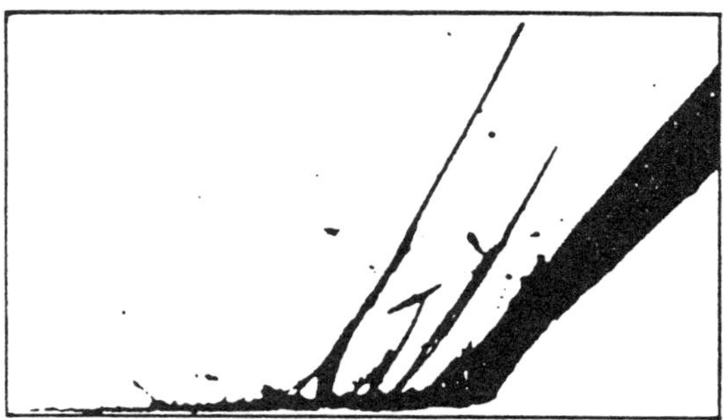

Figure 36: Magnified photograph of the crack branches

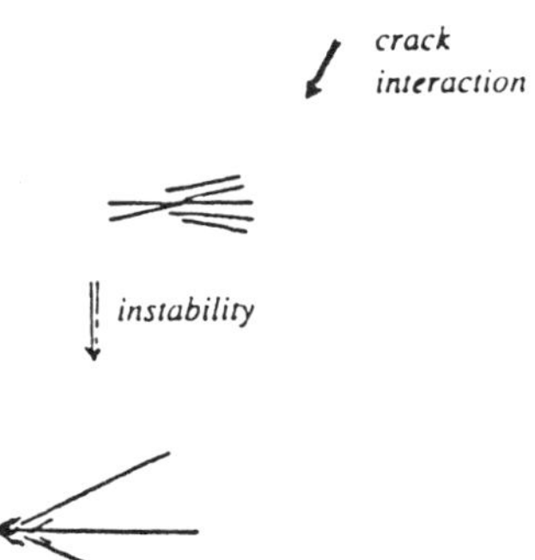

Figure 37: Branching mechanism

References

1. Parton, V.Z. & Boriskovsky, V.G. *Dynamic Fracture Mechanics, Vol. 1: Stationary cracks*, Hemisphere, New York, 1989.

2. Sih, G.C. (ed), *Elastodynamic Crack Problems*, Noordhoff, Leyden, 1977.

3. Lancaster, P. *Theory of Matrices*, Academic Press, New York, 1969.

4. Samoilovich, G.S. Excitation of vibrations in turbine blades, *Mashinostroyeniye*, Moscow, 1975 (in Russian).

5. Liebowitz, H. (ed), *Fracture*, Acad. Press, New York, London. Vol. I, 1968; Vol. II, 1968; Vol. III, 1971; Vol. IV, 1969; Vol. V, 1969; Vol. VI, 1969; Vol. VII, 1972.

6. Thau, S.A. & Lu, T.-H. Transient stress intensity factors for a finite crack in an elastic body caused by a dilatational wave, *Int. J. Solids and Struc.*, 1971, **7**(7), 731-750.

7. Cherepanov, G.P. *Mechanics of Brittle Fracture*, Nauka, Moscow, 1974 (in Russian).

8. Freund, L.B. The stress intensity factor due to normal impact loading of the faces of a crack, *Int. J. Eng. Sci.*, 1974, **12**(2), 179-189.

9. Maue, A.W. Die entspannungswelle bei plötzlichem einschnitt eines gespannten elastischen körperss, *Z. angew. Math. und Mech.*, 1954, **34**(1-2), 1-12.

10. Baker, B.R. Dynamic stresses created by moving crack, *Trans. ASME. J. Appl. Mech.*, 1962, **29**(3), 449-458.

11. Parton, V.Z. & Boriskovsky, V.G. *Dynamic Fracture Mechanics, Vol. 2: Propagating cracks*, Hemisphere, New York, 1990.

12. Knauss, W.G. & Ravi-Chandar, K. Some baric problems in stress wave dominated fracture, *Int. J. Fract.*, 1985, **27**, 127-144.

13. Knauss, W.G. & Ravi-Chandar, K. Fundamental considerations in dynamic fracture, *Eng. Fract. Mech.*, 1986, **23**, 9-20.

14. Laturelle, F.G. Finite element analysis of wave propagation in an elastic half-space under step loading, *Computers & Structures*, 1989, **32**, 721-735.

15. Bathe, K.J. *Finite Element Procedures in Engineering Analysis*, Prentice-Hall, Englewood Cliffs, NC, 1982.

16. Finkel, V.V., Golovnin Yu, I. & Slektov, A.A. On deceleration of rapid cracks by current pulses, *Dokl. Acad. Nauk SSSR*, 1976, **227**, 848-857.

17. Finkel, V.V., Golovnin Yu, I. & Slektov, A.A. Destruction of a crack tip by a strong electromagnetic field, *Dokl. Acad. Nauk SSSR*, 1977, **237**, 325-327.

18. Kudryavtsev, B.A., Parton, V.Z. & Rubinskii, B.D. Magnetothermoelastic field in a body with a half-infinite cut, *Prik. Mat. Mekh.*, 1980, **44**, 916-922.

19. Parton, V.Z., Kudryavtsev, B.A. & Rubinskii, B.D. Crack propagation in an electromagnetic field, *Dokl. Akad. Nauk SSSR*, 1980, **250**, 1096-1100.

20. Kudryavtsev, B.A., Parton, V.Z. & Rubinskii, B.D. Electromagnetic and thermoelastic fields in a conducting plate with a cut of finite length, *Izv. Akad. Nauk SSSR (Mekh. Tv. Tela)*, 1982, **1**, 110-118.

21. Bateman, H. & Erdelyi, A. *Tables of Integral Transforms*, McGraw-Hill, N.Y. *et al.*, 1954.

22. Parton, V.Z. & Kudryavtsev, B.A. *Electromagnetoelasticity*, Gordon and Breach, New York, 1988.

23. Ravi-Chandar, K. & Knauss, W.G. Dynamic crack-tip stresses under stress-wave loading: a comparison of theory and experiment, *Int. J. Fract.*, 1982, **20**, 209-222.

Chapter 5

Fundamentals of dynamics and BEM for dynamic fracture mechanics

J. Domínguez & R. Gallego

Escuela Superior de Ingenieros Industriales, Universidad de Sevilla, Av. Reina Mercedes s/n, 41012 Seville, Spain

Introduction

The main objetive of this chapter is to present the use of the Boundary Element Method (BEM) for the analysis of fracture mechanics problems for which the inertia effects are not negligible. To do so, the fundamentals of elastodynamics, its integral formulation for time and frequency domain problems and the application of this formulation to the solution of dynamic fracture problems through the BEM is presented. The first section summarized the basic equations of elastodynamics, the formulations in terms of different decompositions of the displacement field, the reciprocal theorem and the integral formulation of the problem, and finally, some simple ideas about elastic wave propagation problems. In the second section the ideas of the previous section are applied to the study of plane harmonic waves. Different cases of reflexion and refraction of waves on interfaces are considered. Finally, in the third section, the application of the BEM to the computation of dynamic Stress Intensity Factors (SIF) is presented. After some basic ideas of fracture mechanics are summarized, and the use of the singular Quarter Point Element (QPE) is reviewed, several SIF computations are presented. Three B.E. formulations are discussed: time- domain, frequency-domain and dual reciprocity.

1 Basic equations and fundamentals of wave propagation

The basic equations of linear elastodynamics are presented in this section. The formulations of the elastodynamic problem in terms of dilatation and rotation and in terms of displacement potentials are summarized. Reciprocal relations are derived to obtain the integral representation for the general elastodynamics. This representation is the starting point for the Boundary Element formulations. A brief discussion of the basic ideas of wave propagation for simple problems is also carried out. Only the basic equations and ideas needed to develop the Boundary Element solution of elastodynamics are presented. There is a large number of publications where the elastodynamics theory and the wave propagation phenomena are studied in depth. The interested reader may find a complete treatment of these subjects in Refs 66, 1, and 30.

1.2 Basic equations of linear elastodynamics

A system of fixed rectangular cartesian coordinates is used to present the theory. Light-faced letters stand for scalars while letters in boldface denote vectors and second-order tensors. The summation convention is used, whereby a repeated subindex implies a summation. Quantities with one or two subindexes denote components of a vector or of a second-order tensor, respectively.

The coordinate axes are denoted by x_j, where $j = 1, 2, 3$. The displacement vector at a point $\mathbf{x}$ and time t is $\mathbf{u}(\mathbf{x}, t)$. Commas indicate spatial differentiation and dots time derivatives.

Kinematic relations

The small strain tensor $\boldsymbol{\varepsilon}$, at a point $\mathbf{x}$ of a body Ω at time t, is defined by

$$\varepsilon_{ij} = \frac{1}{2}(u_{i,j} - u_{j,i}) \tag{1}$$

It is also useful to define the rotation tensor $\boldsymbol{\omega}$ by

$$\omega_{ij} = \frac{1}{2}(u_{i,j} - u_{j,i}) \tag{2}$$

It is obvious that $\boldsymbol{\varepsilon}$ is a symmetric tensor $\varepsilon_{ij} = \varepsilon_{ij}$ whereas $\boldsymbol{\omega}$ an antisymmetric one $(\omega_{ij} = -\omega_{ij})$.

Equilibrium equations

The Cauchy's first law of motion for the points of an elastic body is obtained from the balance of linear momentum.

$$\sigma_{ij,j} + \rho\, b_i = \rho\, \ddot{u}_i \tag{3}$$

where $\boldsymbol{\sigma}(\mathbf{x}, t)$ is the stress tensor, $\mathbf{b}(\mathbf{x}, t)$ is the body force per unit mass and $\rho(\mathbf{x})$ is the mass density. The balance of angular momentum implies that the stress tensor is symmetric $(\sigma_{ij} = \sigma_{ij})$.

Constitutive law

The linear relation between the components of the stress tensor and the components of the strain tensor for elastic isotropic solids, known as Hooke's law, is

$$\sigma_{ij} = \lambda\, \delta_{ij}\, \varepsilon_{kk} + 2\mu\, \varepsilon_{ij} \tag{4}$$

where λ and μ are two elastic constants known as Lamé's constants, and δ_{ij} is the Kronecker delta ($\equiv 1$ for $i = j$ and $\equiv 0$ for $i \neq j$).

The Lamé's constants can be written in terms of the more familiar Young's modulus E and Poisson's ratio ν

$$\mu = \frac{E}{2(1+\nu)}; \quad \lambda = \nu \frac{E}{(1+\nu)(1-2\nu)} \tag{5}$$

with which the constitutive law can be written as

$$\varepsilon_{ij} = -\frac{\nu}{E}\, \sigma_{kk}\, \delta_{ij} + \frac{1+\nu}{E}\, \sigma_{ij} \tag{6}$$

or

$$\sigma_{ij} = \frac{E}{(1+\nu)} \left[\frac{\nu}{(1-2\nu)}\, \delta_{ij}\, \varepsilon_{kk} + \varepsilon_{ij} \right] \tag{7}$$

For some particular problems (especially in soil mechanics) one may prefer to use the bulk modulus K. In these cases one defines the deviatoric stress and strain components

$$\sigma'_{ij} = \sigma_{ij} - \frac{1}{3}\, \sigma_{kk}\, \delta_{ij} \tag{8}$$

$$\varepsilon'_{ij} = \varepsilon_{ij} - \frac{1}{3}\, \varepsilon_{kk}\, \delta_{ij} \tag{9}$$

Thus the constitutive equations are expressed as

$$\sigma'_{ij} = 2\mu\, \varepsilon'_{ij}, \quad p = -K \varepsilon_{kk} \tag{10}$$

where p is the mean pressure.

$$p = -\frac{\sigma_{kk}}{3} \tag{11}$$

and

$$K = \lambda + \frac{2}{3}\mu = \frac{E}{3(1-2\nu)} \tag{12}$$

In general, for isotropic elastic materials, all material constants can be expressed in terms of two independent constants.

Field equations
The stress equilibrium eqns (3), kinematic relations (1) and Hooke's law (4) give a complete system of equations governing the motion of homogeneous, isotropic, linear elastic bodies. By substitution of (1) and (4) into (3) one obtains the displacement equations of motion (equations of Navier)

$$\mu\, u_{i,jj} + (\lambda + \mu)\, u_{j,ji} + \rho\, b_i = \rho\, \ddot{u}_i \tag{13}$$

In vector form:

$$\mu\, \nabla^2 \mathbf{u} + (\lambda + \mu) \boldsymbol{\nabla}\, \boldsymbol{\nabla} \cdot \mathbf{u} + \rho\, \mathbf{b} = \rho\, \ddot{\mathbf{u}} \tag{14}$$

or

$$(\lambda + 2\mu) \boldsymbol{\nabla}\, \boldsymbol{\nabla} \cdot \mathbf{u} - \mu \boldsymbol{\nabla} \times \boldsymbol{\nabla} \times \mathbf{u} + \rho\, \mathbf{b} = \rho\, \ddot{\mathbf{u}} \tag{15}$$

The equilibrium equations, the kinematic and constitutive relations and hence, the Navier's equation, must be satisfied at every interior point of the body Ω.

Boundary conditions
The stress components are projected at any point of the boundary Γ of the body Ω and produce surface tractions, denoted by p_i, such that

$$p_i = \sigma_{ij}\, n_j \quad \text{on } \Gamma \tag{16}$$

where n_j are the components of the unit outward normal to the boundary at the point.

The tractions are assumed to be given on the Γ_2 part of the boundary and they are the 'natural' boundary conditions

$$p_i = \bar{p}_i \quad \text{on } \Gamma_2 \tag{17}$$

and the displacements are assumed to be given on the Γ_1 part of the boundary and they are the 'essential' boundary conditions

$$u_i = \bar{u}_i \quad \text{on } \Gamma_1 \tag{18}$$

being $\Gamma_1 \cup \Gamma_2 = \Gamma$ and $\Gamma_1 \cap \Gamma_2 = \emptyset$.

Initial conditions
To complete the problem statement the initial conditions are defined in Ω at time $t = 0$

$$u_i(\mathbf{x}, 0) = u_{oi}(\mathbf{x}) \tag{19}$$

$$\dot{u}_i(\mathbf{x}, 0) = v_{oi}(\mathbf{x}) \tag{20}$$

for all the points $\mathbf{x}$ of the body Ω.

1.3 Formulation of the elastodynamic problem in terms of dilatation and rotation

The integration of the displacement equations of motion (13) for different bodies, boundary conditions and initial conditions constitute the essential target of the dynamic theory of linear elasticity. It was Poisson[69,70] who introduced what appears to be the earliest theorem concerning the general integration of the equations of motion. Since then, numerous researchers have presented different procedures to transform the Navier's equations into simpler mathematical formulations of the problem. In the following the integration of the equations is approached by a differentiation process, first presented by Stokes[78], which leads to simpler differential equations in terms of variables with a clear physical meaning: the dilatation θ and the rotation vector $\boldsymbol{\omega}$.

$$\theta = \varepsilon_{kk} = \boldsymbol{\nabla} \cdot \mathbf{u} \tag{21}$$

$$\boldsymbol{\omega} = \boldsymbol{\nabla} \times \mathbf{u} \tag{22}$$

Notice that the components of the rotation vector defined by (22) are the non-zero independent terms of the rotation tensor defined by (2) except for a factor of 2.

Taking the divergence of the equations of Navier (15) one obtains:

$$(\lambda + 2\mu)\boldsymbol{\nabla} \cdot \boldsymbol{\nabla}\boldsymbol{\nabla} \cdot \mathbf{u} + \rho\, \boldsymbol{\nabla} \cdot \mathbf{b} = \rho\, \boldsymbol{\nabla} \cdot \ddot{\mathbf{u}} \tag{23}$$

which can be written as

$$c_1^2\, \nabla^2\, \theta + \boldsymbol{\nabla} \cdot \mathbf{b} = \ddot{\theta} \tag{24}$$

where

$$c_1^2 = \frac{\lambda + 2\mu}{\rho} \tag{25}$$

Taking the curl of (15) using the identities for a vector field $\mathbf{v}$: $\nabla^2\mathbf{v} = \boldsymbol{\nabla}\boldsymbol{\nabla} \cdot \mathbf{v} - \boldsymbol{\nabla} \times \boldsymbol{\nabla} \times \mathbf{v}$ and $\boldsymbol{\nabla} \cdot (\boldsymbol{\nabla} \times \mathbf{v}) = 0$ the following equation is obtained

$$c_2^2\, \nabla^2\, \boldsymbol{\omega} + \boldsymbol{\nabla} \times \mathbf{b} = \ddot{\boldsymbol{\omega}} \tag{26}$$

where

$$c_2^2 = \frac{\mu}{\rho} \tag{27}$$

The two constants c_1 and c_2 have dimensions of velocity and depend on the density and elastic constants of the solid. For a given density c_1 and c_2 can be used to define the elastic characteristics of the solid. The ratio of these two constant is related to the Poisson's ratio by:

$$\frac{c_1^2}{c_2^2} = \frac{2(1-\nu)}{1-2\nu} \tag{28}$$

The equations of Navier have been transformed into eqns (24) and (26). These equations are both of the same type, the former being a scalar wave equation with propagation velocity c_1 and the latter a vector wave equation with propagation velocity c_2.

Since eqn (24) refers to the dilatation and eqn (26) refers to the rotation, c_1 and c_2 are called irrotational and equivolumial wave velocities, respectively. The type of solution of the wave equation and some of the characteristics of the wave propagation phenomena are discussed in subsection 1.7.

1.4 Formulation in terms of displacement potentials

The process of integration of the equations of Navier can be simplified somewhat by writing the field variable $\mathbf{u}$ in terms of potential functions. Several of these potentials have been proposed. Some of the most useful are those developed by Lamé,[48] Papkovich,[67] Neuber[61] and Iacovache.[39] A study of the different displacement potentials used for the integration of the equations of Navier, and the relations among them, may be found in the paper by Sternberg.[77] The Lamé potentials are among the simplest and most useful ones. They are based on the idea of the Helmholtz decomposition of a vector which states that any vector $\mathbf{b}$ piecewise differentiable in a finite open region, or in an infinite region, provided that it decreases to zero at large distances r from the origin at least as rapidly as r^{-2}, may be decomposed into irrotational and solenoidal parts, or, in other words, it admits the representation

$$\mathbf{b} = \nabla B + \nabla \times \mathbf{Q} \tag{29}$$

The field variables B and $\mathbf{Q}$ for a given vector $\mathbf{b}$ in a region Ω can be obtained from the following expressions:

$$B(\mathbf{x}, t) = -\frac{1}{4\pi} \int_\Omega \frac{1}{r} \nabla \cdot \mathbf{b}(\boldsymbol{\xi}, t) \, \mathrm{d}\Omega(\boldsymbol{\xi}) \tag{30}$$

$$\mathbf{Q}(\mathbf{x}, t) = \frac{1}{4\pi} \int_\Omega \frac{1}{r} \nabla \times \mathbf{b}(\boldsymbol{\xi}, t) \, \mathrm{d}\Omega(\boldsymbol{\xi}) \tag{31}$$

for 3-D, and

$$B(\mathbf{x}, t) = -\frac{1}{2\pi} \int_\Omega \ln\frac{1}{r} \nabla \cdot \mathbf{b}(\boldsymbol{\xi}, t) \, \mathrm{d}\Omega(\boldsymbol{\xi}) \tag{32}$$

$$\mathbf{Q}(\mathbf{x}, t) = \frac{1}{2\pi} \int_\Omega \ln\frac{1}{r} \nabla \times \mathbf{b}(\boldsymbol{\xi}, t) \, \mathrm{d}\Omega(\boldsymbol{\xi}) \tag{33}$$

for 2-D, where the function $\mathbf{Q}$ satisfies the condition $\nabla \cdot \mathbf{Q} = 0$ and $r \equiv \mid \mathbf{x} - \boldsymbol{\xi} \mid$. A proof of the above decomposition can be found in Ref. 1 among others.

Let us now assume a scalar function $\phi(\mathbf{x}, t)$ and a vector function $\boldsymbol{\Psi}(\mathbf{x}, t)$ and write

$$\mathbf{u} = \nabla \phi + \nabla \times \boldsymbol{\Psi} \tag{34}$$

Substitution of (34) into the field eqn (14), where the body force vector has been decomposed according to (29), yields

$$\mu\,\nabla^2[\boldsymbol{\nabla}\,\phi+\boldsymbol{\nabla}\times\boldsymbol{\Psi}]+(\lambda+\mu)\boldsymbol{\nabla}\,\boldsymbol{\nabla}\cdot[\boldsymbol{\nabla}\,\phi+\boldsymbol{\nabla}\times\boldsymbol{\Psi}]+\rho[\boldsymbol{\nabla}\,B+\boldsymbol{\nabla}\times\mathbf{Q}]=\rho[\boldsymbol{\nabla}\ddot{\phi}+\boldsymbol{\nabla}\times\ddot{\boldsymbol{\Psi}}] \tag{35}$$

and taking into account that $\boldsymbol{\nabla}\cdot\boldsymbol{\nabla}\times\boldsymbol{\Psi}=0$, one obtains

$$\boldsymbol{\nabla}[(\lambda+2\mu)\nabla^2\,\phi+\rho\,B]+\boldsymbol{\nabla}\times[\mu\,\nabla^2\boldsymbol{\Psi}+\rho\,\mathbf{Q}]=\rho\,\boldsymbol{\nabla}\,\ddot{\phi}+\rho\,\boldsymbol{\nabla}\times\ddot{\boldsymbol{\Psi}} \tag{36}$$

This equation is satisfied identically if ϕ and $\boldsymbol{\Psi}$ are solutions of

$$c_1^2\,\nabla^2\,\phi+B=\ddot{\phi} \tag{37}$$

$$c_2^2\,\nabla^2\,\boldsymbol{\Psi}+\mathbf{Q}=\ddot{\boldsymbol{\Psi}} \tag{38}$$

where $c_1^2=(\lambda+2\mu)/\rho$ and $c_2^2=\mu/\rho$.

Equations (37) and (38) are of the same kind of non-homogeneous wave equations with velocities c_1 and c_2 as those obtained with the representation in terms of dilatation and rotation. Equation (34) relates three components of the displacement vector to four other functions. This indicates that ϕ and the components of $\boldsymbol{\Psi}$ should be subjected to an additional constraint. This condition is normally taken as

$$\boldsymbol{\nabla}\cdot\boldsymbol{\Psi}=0 \tag{39}$$

which is consistent with the Helmholtz decomposition of a vector.

The question is now whether the solution of (37) and (38) is complete in the sense that every solution of the field equation can be obtained from those equations and the representation given by (34). The completeness has been proven by several authors since Somigliana. Most of them assume that the vector $\boldsymbol{\Psi}$ should satisfy $\boldsymbol{\nabla}\cdot\boldsymbol{\Psi}=0$. Compact proofs of the completeness may be found in Refs 1 and 30.

To end this subsection and to relate it with the previous one, it might be worthwhile to relate the dilation θ and the rotation $\boldsymbol{\omega}$ with the potentials ϕ and $\boldsymbol{\Psi}$. Assuming the usual constraint $\boldsymbol{\nabla}\cdot\boldsymbol{\Psi}=0$ it can be easily shown by substitution of (34) into (21) and (22) that

$$\theta=\boldsymbol{\nabla}\cdot\mathbf{u}=\nabla^2\,\phi \tag{40}$$

$$\boldsymbol{\omega}=\boldsymbol{\nabla}\times\mathbf{u}=-\nabla^2\,\boldsymbol{\Psi} \tag{41}$$

It can also be pointed out that each one of the two parts into which the displacement has been decomposed ($\mathbf{u}^1=\boldsymbol{\nabla}\phi$ and $\mathbf{u}^2=\boldsymbol{\nabla}\times\boldsymbol{\Psi}$) also satisfies the wave equation with body forces $\mathbf{b}^1=\boldsymbol{\nabla}B$ and $\mathbf{b}^2=\boldsymbol{\nabla}\times\mathbf{Q}$, respectively. This can be easily obtained by taking divergence of (37) and rotational of (38).

1.5 Reciprocal theorem in elastodynamics

The reciprocal theorem in elastodynamics is an extension of Betti's classical reciprocal theorem of elastostatics. The theorem, known as Graffi's Reciprocal Theorem, was stated by Graffi.[37] Wheeler and Sternberg[84] extended it to infinite regions. In the following the theorem will be proven starting from a weighted residual statement. In this way the process towards the final integral representation is consistent with that used frequently in the Boundary Element literature for many other different continuum mechanics static and dynamic problems. Nevertheless, the line of thought of the present proof is similar to that followed by Wheeler and Sternberg in the previously cited work where a rigorous proof for unbounded domains may be found.

Some properties of the Riemann convolution are reviewed first. Consider two scalar functions $g(\mathbf{x},t)$ and $h(\mathbf{x},t)$ continuous in the space $(\Omega \times T^+)$; where Ω is a region and t^+ the half open time interval $[0,\infty)$. The Riemann convolution is defined as

$$g * h = \int_0^t g(\mathbf{x}, t-\tau)\, h(\mathbf{x},\tau)\, \mathrm{d}\tau \quad \text{for all} \quad (\mathbf{x},t) \in \Omega \times T^+ \tag{42}$$

$$g * h = 0 \quad \text{for all} \quad (\mathbf{x},t) \in \Omega \times T^- \tag{43}$$

Some useful properties of the Riemann convolution are:

$$g * h = h * g \tag{44}$$

$$\frac{\partial}{\partial t}(g * h) = g * h + g(\mathbf{x},0)\, h(\mathbf{x},t) \tag{45}$$

$$\frac{\partial}{\partial x_k}(g * h) = g_{,k} * h + g * h_{,k} \tag{46}$$

The Riemann convolution of two vectors $\mathbf{u}(\mathbf{x},t)$ and $\mathbf{u}'(\mathbf{x},t)$ is defined as

$$\mathbf{u} * \mathbf{u}' = u_k * u'_k \tag{47}$$

and that of two tensors $\boldsymbol{\sigma}(\mathbf{x},t)$ and $\boldsymbol{\varepsilon}(\mathbf{x},t)$ as

$$\boldsymbol{\sigma} * \boldsymbol{\varepsilon} = \sigma_{kj} * \varepsilon_{kj} \tag{48}$$

The proof of the reciprocal theorem now starts from the equilibrium equation. For an elastodynamic state in a body Ω with boundary Γ, the following equation is satisfied at any time

$$\sigma_{kj,j} + \rho\, b_k - \rho\, \ddot{u}_k = 0 \tag{49}$$

Assuming a continuous vector function $\mathbf{u}^*(\mathbf{x},t)$ which is the displacement of another elastodynamic state with stresses $\boldsymbol{\sigma}^*$ and body forces $\mathbf{b}^*$ defined over Ω, one can write the following weighted residual statement:

$$\int_\Omega (\sigma_{kj,j} * u_k^*)\, d\Omega + \int_\Omega \rho(b_k * u_k^*)\, d\Omega - \int_\Omega \rho(\ddot{u}_k * u_k^*)\, d\Omega = 0 \tag{50}$$

which holds for any continuous function $u_k^*(\mathbf{x}, t)$. Using the rule of spatial derivation given by eqn (46) one may write

$$\frac{\partial}{\partial x_j}(\sigma_{kj} * u_k^*) = \sigma_{kj,j} * u_k^* + \sigma_{kj} * u_{k,j}^* \tag{51}$$

and using the divergence theorem

$$\int_\Omega \frac{\partial}{\partial x_j}(\sigma_{kj} * u_k^*)\, d\Omega = \int_\Gamma (\sigma_{kj} * u_k^*) n_j\, d\Gamma = \int_\Gamma p_k * u_k^*\, d\Gamma \tag{52}$$

A formula for the spatial integration by parts of the first convolution product of (50) may be obtained from (51) and (52)

$$\int_\Omega (\sigma_{kj,j} * u_k^*)\, d\Omega = -\int_\Omega (\sigma_{kj} * u_{k,j}^*)\, d\Omega + \int_\Gamma p_k * u_k^*\, d\Gamma \tag{53}$$

Taking into account symmetry of the strain tensor, Hooke's law and the distributivity and commutativity of convolutions, one can easily show

$$\sigma_{kj}^* * u_{kj} = \sigma_{kj}^* * \varepsilon_{kj} = \varepsilon_{kj} * \sigma_{kj}^* = u_{kj} * \sigma_{kj}^* \tag{54}$$

A second integration by parts of (53) yields:

$$\int_\Omega (\sigma_{kj,j} * u_k^*)\, d\Omega = \int_\Omega (u_k * \sigma_{kj,j}^*)\, d\Omega - \int_\Gamma (u_k * p_k^*)\, d\Gamma + \int_\Gamma (p_k * u_k^*)\, d\Gamma \tag{55}$$

To transform the acceleration term of eqn (50) the following identities can be obtained in accordance with (45).

$$\frac{\partial}{\partial t}(u_k * u_k^*) = \dot{u}_k * u_k^* + u_{ok}\, u_k^* \tag{56}$$

$$\frac{\partial^2}{\partial t^2}(u_k * u_k^*) = \frac{\partial}{\partial t}(\dot{u}_k * u_k^*) + u_{ok}\, \dot{u}_k^* \tag{57}$$

$$\frac{\partial^2}{\partial t^2}(u_k * u_k^*) = \ddot{u}_k * u_k^* + v_{ok}\, u_k^* + u_{ok}\, \dot{u}_k^* \tag{58}$$

where $u_{ok} = u_k(\mathbf{x}, 0)$ and $v_{ok} = u_k(\mathbf{x}, 0)$, and the reciprocal

$$\frac{\partial^2}{\partial t^2}(u_k^* * u_k) = \ddot{u}_k^* * u_k + v_{ok}^*\, u_k + u_{ok}^*\, \dot{u}_k \tag{59}$$

The first terms of (58) and (59) are identical. Hence

$$\ddot{u}_k * u_k^* = \ddot{u}_k^* * u_k + v_{ok}^*\, u_k + u_{ok}^* \dot{u}_k - v_{ok}\, u_k^* - u_{ok}\, \dot{u}_k^* \tag{60}$$

Substituting (55) and (60) into (50) yields

$$\int_\Omega (u_k * \sigma^*_{kj,j})\, \mathrm{d}\Omega - \int_\Gamma (u_k * p^*_k)\, \mathrm{d}\Gamma + \int_\Gamma (p_k * u^*_k)\, \mathrm{d}\Gamma + \int_\Omega \rho(b_k * u^*_k)\, \mathrm{d}\Omega -$$

$$\int_\Omega \rho(\ddot{u}^*_k * u_k + v^*_{ok}\, u_k + u^*_{ok}\, \dot{u}_k - v_{ok}\, u^*_k - u_{ok}\, \dot{u}^*_k)\, \mathrm{d}\Omega = 0 \qquad (61)$$

It was assumed that $\mathbf{u}^*(\mathbf{x}, t)$ was the displacement field of another elastodynamic state in Ω. Then

$$\sigma^*_{kj,j} - \rho\, \ddot{u}^*_k = -\rho\, b^*_k \qquad (62)$$

Substitution of (62) into (61) leads to the mathematical expression of the reciprocal theorem between two elastodynamic states.

$$\int_\Gamma (p_k * u^*_k)\, \mathrm{d}\Gamma + \int_\Omega \rho(b_k * u^*_k + u_{ok}\, \dot{u}^*_k + v_{ok}\, u^*_k)\, \mathrm{d}\Omega =$$

$$\int_\Gamma (p^*_k * u_k)\, \mathrm{d}\Gamma + \int_\Omega \rho(b^*_k * u_k + u^*_{ok}\dot{u}_k + v^*_{ok} u_k)\, \mathrm{d}\Omega \qquad (63)$$

1.6 Integral representation in elastodynamics

Let Ω be a regular elastic body with material properties ρ, c_1 and c_2 and with boundary Γ. Consider an elastodynamic state with displacements $\mathbf{u}(\mathbf{x}, t)$, prescribed tractions and displacements on the boundary, $\mathbf{p}(\mathbf{x}, t)$ on Γ_1 and $\mathbf{u}(\mathbf{x}, t)$ on Γ_2, respectively, body forces $\mathbf{b}(\mathbf{x}, t)$, and prescribed initial conditions $\mathbf{u}_0(\mathbf{x})$ and $\mathbf{v}_0(\mathbf{x})$. The so-called Love's[50] integral representation of the displacement field can be derived using the reciprocal theorem as done by Wheeler and Sternberg.[84] A second elastodynamic state corresponding to a unit impulse applied at $t = 0$ at point $\boldsymbol{\xi}$ in the direction of the x_l-axis in the infinite region is assumed. The state corresponds to a body force

$$\rho\, b^*_k = \delta(t)\, \delta(\mathbf{x} - \boldsymbol{\xi})\delta_{kl} \qquad (l \text{ fixed}) \qquad (64)$$

with zero initial conditions. The function δ is the Dirac delta function whereas δ_{kl} is the Kronecker delta.

The displacements in the infinite region corresponding to this second state considering each direction of load as independent, are denoted by

$$u^*_k = u^*_{lk}(\mathbf{x}, t; \boldsymbol{\xi})\, e_l \qquad (65)$$

and the stresses by

$$\sigma^*_{kj} = \sigma^*_{lkj}(\mathbf{x}, t; \boldsymbol{\xi})\, e_l \qquad (66)$$

where $\mathbf{x}$ denotes the point for which the displacement or the stress is written, t the instant of time, $\boldsymbol{\xi}$ the point where the loads is applied at $t = 0$ and e_l the l component of the unit vector in the direction of the load. The tractions at a point $\mathbf{x}$ on a surface with unit outward normal $\mathbf{n}$ are

$$p_k^* = \sigma_{lkj}^*(\mathbf{x}, t; \boldsymbol{\xi})\, n_j\, e_l = p_{lk}^*(\mathbf{x}, t; \boldsymbol{\xi})\, e_l \tag{67}$$

The reciprocal theorem is now applied between the actual state and that corresponding to the unit impulse load in the infinite region. Since both states should be defined on the same region Ω, one can assume that for the second state Ω is part of the infinite region. Taking into account the properties of the Dirac delta function when it is integrated in space and time (Jones[41]), the following expression is obtained from eqn (63)

$$c_{lk}(\boldsymbol{\xi}))\, u_k(\boldsymbol{\xi}, t) = \int_\Gamma (u_{lk}^* * p_k - p_{lk}^* * u_k)\, \mathrm{d}\Gamma + \int_\Omega \rho(u_{lk}^* * b_k)\, \mathrm{d}\Omega + \int_\Omega \rho(v_{ok} u_{lk}^* + u_{ok} \dot{u}_{lk}^*)\, \mathrm{d}\Omega \tag{68}$$

where $c_{lk}(\boldsymbol{\xi}) = \delta_{lk}$ when $\boldsymbol{\xi} \in \Omega$ and $c_{lk}(\boldsymbol{\xi}) = 0$ when $\boldsymbol{\xi} \notin \Omega$. In the cases of $\boldsymbol{\xi} \in \Gamma$ the boundary integrals include a singularity which can be taken out by a limiting process. From this process one may compute the value of $c_{lk}(\boldsymbol{\xi})$, which depends on the geometry of the boundary at point $\boldsymbol{\xi}$. The integrals are interpreted in the sense of Cauchy Principal Value.

The above integral representation has been obtained from the reciprocal theorem of elastodynamics. The point source solution, i.e. the fundamental solution, corresponding to the response of an infinite elastic region to a concentrated body force with time-dependent magnitude is originally attributed to Stokes.[78] The problem is known as Stokes's problem and the corresponding elastodynamic state, as the Stokes's state. The expressions for the displacements and tractions of the Stokes's problem can be found elsewhere.[30,27] A similar process can be followed for many other boundary value problems like those of elastostatics, Laplace's equation problems, scalar wave propagation, thermoelasticity, poroelasticity, etc. In all those cases one may obtain an integral representation of the problem using a reciprocal relation in combination with the solution of the infinite domain under the effect of a point source. Fundamental solutions for different dynamic problems, their corresponding integral representation, and B.E. implementation can be found in the book by Domínguez.[27]

1.7 Some simple elastic wave propagation problems

The field equations of linear elastodynamics become much simpler than the general Navier's equations for certain particular problems. In this subsection, some simple elastic wave propagation problems will be discussed to show the fundamental ideas of wave propagation.

The basic equation that governs wave propagation phenomena has been obtained above after some transformations of the equations of Navier. When zero body forces are assumed the wave equation can be written as

$$\nabla^2\, w = \frac{1}{c^2}\ddot{w} \tag{69}$$

where c is a constant with dimensions of velocity and w can be a scalar or a vector variable depending on the kind of wave. Equation (69) appears not only in elastodynamics but also in all basic wave propagation problems such as electromagnetism, pressure waves in fluids, etc.

1.7.1 Half-space under pressure

Assume an initially undisturbed linear elastic half-space ($x_1 \geq 0$) to which a uniform surface pressure $p(t)$ is applied (Fig. 1).

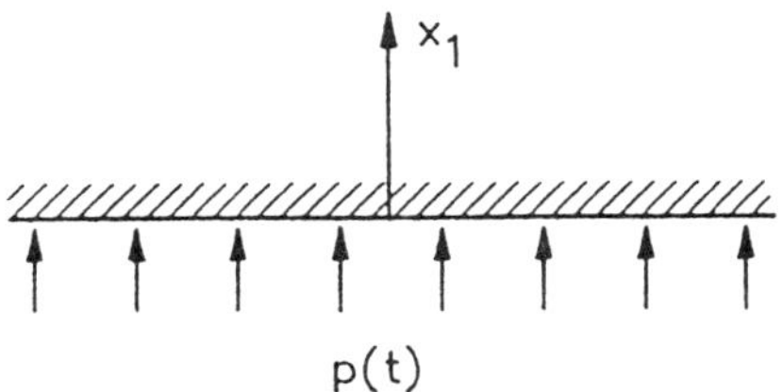

Figure 1: Half-space under normal pressure

Because of the symmetry no other displacement but u_1 exists and all the field variables are independent of x_2 and x_3. The only non-zero strain component is:

$$\varepsilon_{11} = \frac{\partial u_1}{\partial x_1} \tag{70}$$

Only normal stresses exist and all the stress components can be written in terms of

$$\sigma_{11} = (\lambda + 2\mu)\frac{\partial u_1}{\partial x_1} \tag{71}$$

From the first equilibrium equation one may write:

$$\frac{\partial^2 u_1}{\partial x_1^2} = \frac{1}{c_1^2}\frac{\partial^2 u_1}{\partial t^2} \tag{72}$$

where $c_1^2 = (\lambda + 2\mu)/\rho$.

The general solution of (72) can be easily obtained as follows: Calling $\alpha = t - x_1/c_1$ and $\beta = t + x_1/c_1$, eqn (72) becomes

$$\frac{\partial u_1}{\partial \alpha\, \partial \beta} = 0 \tag{73}$$

and consequently

$$u_1(x_1, t) = f(t - \frac{x_1}{c_1}) + g(t + \frac{x_1}{c_1}) \tag{74}$$

The term $f(t - x_1/c_1)$ represents a perturbation propagating in the positive x_1-direction whereas $g(t + x_1/c_1)$ represents a perturbation propagating in the negative x_1-direction. In order to visualize these meanings of the f and g functions one may assume a certain space distribution for a perturbation at $t = \bar{t}$; $f = f(\bar{t} - x_1/c_1)$.

The same space distribution will exist at time $\bar{t} + \Delta\, \bar{t}$ for points with another space coordinate x_1' if

$$f(\bar{t} - \frac{x_1}{c_1}) = f(\bar{t} + \Delta\bar{t} - \frac{x_1'}{c_1}) \tag{75}$$

which is satisfied as long as

$$\bar{t} - \frac{x_1}{c_1} = \bar{t} + \Delta\bar{t} - \frac{x_1'}{c_1} \tag{76}$$

or

$$x_1' = x_1 + c_1\, \Delta\bar{t} \tag{77}$$

In other words, whatever happens at a point x_1 at $t = \bar{t}$ will happen at the point $x_1 + c_1\Delta\bar{t}$ after $\Delta\bar{t}$. The perturbation propagates in the positive x_1-direction. In the case of $g(t + x_1/c_1)$, the function g reproduces itself after $\Delta\bar{t}$ at points that satisfy $\bar{t} + \Delta\bar{t} + x_1'/c_1 = \bar{t} + x_1/c_1$; i.e. $x_1' = x_1 - c_1\Delta\bar{t}$. The perturbation propagates in the negative x_1-direction.

To solve the problem completely one needs the initial conditions and the boundary conditions. The initial quiescence conditions are:

$$f(-x_1/c_1) + g(x_1/c_1) = 0 \quad \text{for } x_1 > 0 \tag{78}$$

$$f'(-x_1/c_1) + g'(x_1/c_1) = 0 \quad \text{for } x_1 > 0 \tag{79}$$

where primes stand for differentiation with respect to the argument. And the boundary condition at $x_1 = 0$.

$$-\frac{(\lambda + 2\mu)}{c_1}\,(f'(t) - g'(t)) = -p(t) \tag{80}$$

It can easily be shown[1] that the solution which satisfies the above conditions is:

$$u_1(x_1, t) = \frac{c_1}{\lambda + 2\mu}\int_0^{t-\frac{x_1}{c_1}} p(s)\; ds \quad \text{for } t > \frac{x_1}{c_1} \tag{81}$$

$$u_1(x_1, t) = 0 \quad \text{for } t < \frac{x_1}{c_1} \tag{82}$$

and the normal stress

$$\sigma_{11} = -p(t - \frac{x_1}{c_1}) \quad \text{for } t > \frac{x_1}{c_1} \tag{83}$$

$$\sigma_{11} = 0 \quad \text{for } t < \frac{x_1}{c_1} \tag{84}$$

It should be noticed that due to $p(t)$ applied on the surface, a perturbation propagates into the half-space along x_1 with a velocity c_1. A point at $x_1 = \bar{x}_1$ remains at rest until the perturbation reaches it; i.e. until $t = \bar{x}_1/c_1$.

1.7.2 Reflection and transmission

Assume now that the one dimensional perturbation strikes an infinite plane free surface normal to x_1 at $x_1 = a$. The perturbation is reflected into the body as shown below.

The stress of the incident wave is

$$(\sigma_{11})_I = f\left(t - \frac{x_1}{c_1}\right) \tag{85}$$

The reflected wave would be

$$(\sigma_{11})_R = g\left(t + \frac{x_1}{c_1}\right) \tag{86}$$

The boundary condition at $x_1 = a$, for $t \geq a/c_1$, is

$$(\sigma_{11})_I + (\sigma_{11})_R = 0 \tag{87}$$

then

$$g\left(t + \frac{a}{c_1}\right) = -f\left(t - \frac{a}{c_1}\right) \tag{88}$$

For any value of $t \geq a/c_1$ the function

$$g(s) = -f\left(s - \frac{2a}{c_1}\right) \tag{89}$$

being $s = t + a/c_1$. This means that the g function for any value of its argument is equal and opposite to the f function for an argument which is the same of g minus $2a/c_1$. Therefore

$$(\sigma_{11})_R = g\left(t + \frac{x_1}{c_1}\right) = -f\left(t - \frac{a}{c_1} + \frac{x_1 - a}{c_1}\right) \tag{90}$$

The reflected stress wave has the same shape of the incident wave with a change in sing after reflection. Tensile stresses become pressure stresses and viceversa. The wave travels in the negative x_1 direction with a time origin at $t = a/c_1$ and with a space origin at $x_1 = a$.

Assume now that at $x_1 = a$ instead of an infinite plane free surface normal to x_1 there is an infinite plane interface with another material with different properties (irrotational wave velocity $= c_1'$, density $= \rho'$) as shown in Fig. 2.

One can expect that part of the perturbation will be transmitted to the other side of the interface and part will be reflected back. The waves may be represented as

$$(\sigma_{11})_I = f\left(t - \frac{x_1}{c_1}\right) \tag{91}$$

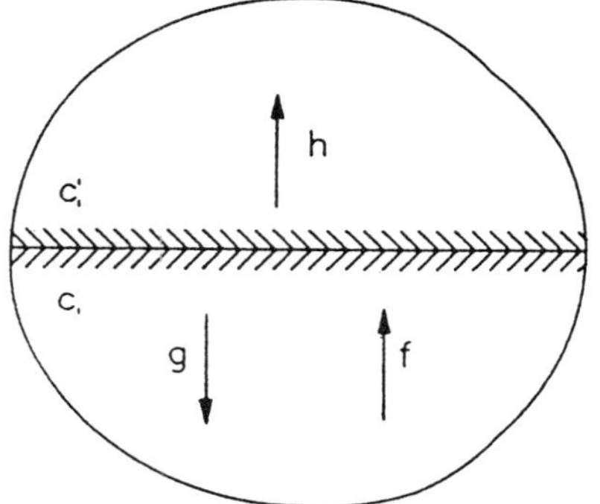

Figure 2: Wave reflection and transmission

$$(\sigma_{11})_R = g\left(t - \frac{a}{c_1} + \frac{x_1 - a}{c_1}\right) \tag{92}$$

$$(\sigma_{11})_T = h\left(t - \frac{a}{c_1} - \frac{x_1 - a}{c_1'}\right) \tag{93}$$

The equilibrium and compatibility boundary conditions at $x_1 = a$ are

$$(\sigma_{11})_I + (\sigma_{11})_R = (\sigma_{11})_T \tag{94}$$

$$\dot{u}_{1I} + \dot{u}_{1R} = \dot{u}_{1T} \tag{95}$$

The first boundary condition yields

$$f\left(t - \frac{a}{c_1}\right) + g\left(t - \frac{a}{c_1}\right) = h\left(t - \frac{a}{c_1}\right) \quad \text{for any } t \tag{96}$$

The second is established taking into account that

$$\frac{\partial u_{1I}}{\partial t} = -c_1 \frac{\partial u_{1I}}{\partial x_1} = -\frac{1}{c_1 \rho} (\sigma_{11})_I \tag{97}$$

and similarly

$$\frac{\partial u_{1R}}{\partial t} = \frac{1}{c_1 \rho}(\sigma_{11})_R \quad \text{and} \quad \frac{\partial u_{1T}}{\partial t} = -\frac{1}{c_1' \rho'} (\sigma_{11})_T \tag{98}$$

Hence, the second boundary condition can be written as

$$\frac{1}{c_1 \rho}\left[g\left(t - \frac{a}{c_1}\right) - f\left(t - \frac{a}{c_1}\right)\right] = -\frac{1}{c_1' \rho'} h\left(t - \frac{a}{c_1}\right) \tag{99}$$

From eqns (96) and (99) one obtains

$$g(s) = C_R f(s)$$

$$h(s) = C_T f(s) \tag{100}$$

where

$$C_R = (k-1)/(k+1), \quad C_T = 2k/(k+1) \text{ and } k = \rho' c_1'/\rho c_1$$

Then

$$(\sigma_{11})_R = C_R \cdot f\left(t - \frac{a}{c_1} + \frac{x_1 - a}{c_1}\right) \tag{101}$$

$$(\sigma_{11})_T = C_T \cdot f\left(t - \frac{a}{c_1} - \frac{x_1 - a}{c_1'}\right) \tag{102}$$

The coefficients c_R and c_T are known as the reflection and transmission coefficients, respectively. They indicate the amplitude of the reflected and transmitted waves, respectively in relation to the amplitude of the incident wave. The shape of the reflected and the transmitted waves are the same as that of the incident wave; i.e. the function f.

1.7.3 Half-space under shear

Assume now the same initially undisturbed linear elastic half-space ($x_1 \geq 0$), under the effects of a uniform shear applied on its surface, say in the x_2 direction. The displacement is now in a plane normal to the x_1-axis. The only non-zero stress and strain components are:

$$\sigma_{21} = \sigma_{12} = 2\mu\, \varepsilon_{12} \tag{103}$$

$$\varepsilon_{21} = \varepsilon_{12} = \frac{1}{2}\frac{\partial u_2}{\partial x_1} \tag{104}$$

For zero body forces, the second equilibrium equation in terms of the stresses becomes,

$$\sigma_{21,1} = \rho\, \ddot{u}_2 \tag{105}$$

Substituting (103) and (104) into (105) the homogeneous wave equation is obtained for u_2

$$\frac{\partial^2 u_2}{\partial x_1^2} = \frac{1}{c_2^2}\frac{\partial^2 u_2}{\partial t^2} \tag{106}$$

where $c_2^2 = \mu/\rho$. This equation is identical to (72) which was obtained for u_1 in the case of a uniform pressure excitation, with the only difference being that c_2 is used instead of c_1. The solution of (106) is the same as in the previous case.

$$u_2(x_1, t) = f\left(t - \frac{x_1}{c_2}\right) + g\left(t + \frac{x_1}{c_2}\right) \tag{107}$$

Waves propagate into the half-space along the x_1-direction. The transmission and reflection coefficients are the same except for the wave velocities which are now c_2 and c_2' instead of c_1 and c_1', respectively. The waves are in this case equivolumial with displacements perpendicular to the direction of propagation and a propagation velocity $c_2 = (\mu/\rho)^{1/2}$.

1.7.4 Two-dimensional problems

The two examples shown above are one-dimensional. Another situation which allows the simplification of the field equations is that of two-dimensional problems; i.e. problems in which the body forces, geometry and boundary conditions are independent of one coordinate, say x_3. All the field variables are independent of x_3 and the governing equations become uncoupled in two independent systems. The equilibrium equations can be written as:

$$\sigma_{3\beta,\beta} + \rho\, b_3 = \rho\, \ddot{u}_3 \tag{108}$$

$$\sigma_{\alpha\beta,\beta} + \rho\, b_\alpha = \rho\, \ddot{u}_\alpha \tag{109}$$

where the Greek indices take the values 1 and 2.

Taking into account that $u_3 = u_3(x_1, x_2, t)$

$$\sigma_{3\beta} = \mu\, u_{3,\beta} \tag{110}$$

and eqn (108) in terms of u_3 becomes

$$c_2^2\, u_{3,\beta\beta} + b_3 = \ddot{u}_3 \tag{111}$$

which is the wave equation governing the antiplane motion.

The other problem, uncoupled from the first, corresponds to the in-plane motion where displacements $u_1(x_1, x_2, t)$ and $u_2(x_1, x_2, t)$ exist, and $u_3 = 0$. The equilibrium eqn (109) and Hooke's law

$$\sigma_{\alpha\beta} = \lambda\, \varepsilon_{\gamma\gamma} + 2\mu\, \varepsilon_{\alpha\beta} \tag{112}$$

yield Navier's equation for two dimensions

$$\mu\, u_{\alpha,\beta\beta} + (\lambda + \mu)\, u_{\beta,\beta\alpha} + \rho\, b_\alpha = \rho\, \ddot{u}_\alpha \tag{113}$$

The in-plane problem defined above is a plane strain one. It is well known from basic elasticity that the so-called plane stress state can be defined for thin plane domains. The field eqn (113) remains the same for that case and only the stress and strain expressions in terms of u_1 and u_2 are different.

It should be noted that the two uncoupled problems appearing for two-dimensional situations are very different. In the first one, the displacements are governed directly by the wave equation and only equivolumial waves exist. In the second case, the Navier's field equations remain in their original form and will have to be transformed as in the three-dimensional case to obtain wave equations. Equivolumial and irrotational waves exist in this case.

1.7.5 Plane waves

A plane displacement wave propagating in a direction defined by a unit vector $\mathbf{q} = (l, m, n)$ is a perturbation for which all the points in any plane perpendicular to $\mathbf{q}$ have the same displacement at the same time. The displacement vector in this case is of the form

$$\mathbf{u} = f(t - (\mathbf{q} \cdot \mathbf{x})/c)\, \mathbf{d} \tag{114}$$

where $\mathbf{d}$ is the unit vector defining the direction of motion and $\mathbf{x}$ is the position vector. Notice that the product $\mathbf{q} \cdot \mathbf{x}$ is a constant for all the points on a plane perpendicular to the vector $\mathbf{q}$.

The characteristics of the motion of the points under the effects of plane waves in elastic media can be analyzed using the field equations in terms of dilatations θ and rotations ω as presented in subsection 1.3. These variables have to satisfy eqns (24) and (26) respectively, which for zero body forces become

$$\nabla^2 \theta = \frac{1}{c_1^2}\, \ddot{\theta} \tag{115}$$

$$\nabla^2 \boldsymbol{\omega} = \frac{1}{c_2^2}\, \ddot{\boldsymbol{\omega}} \tag{116}$$

The general solution of eqn (115) with the form of a plane wave is

$$\theta = f(t - \mathbf{q} \cdot \mathbf{x}/c_1) = f(t - [l x_1 + m x_2 + n x_3]/c_1) \tag{117}$$

where $l^2 + m^2 + n^2 = 1$ to satisfy eqn (115).

The displacements associated to these waves can be written as

$$u_1^\theta = c_1\; l\; g(t - \mathbf{q} \cdot \mathbf{x}/c_1) \tag{118}$$

$$u_2^\theta = c_1\; m\; g(t - \mathbf{q} \cdot \mathbf{x}/c_1) \tag{119}$$

$$u_3^\theta = c_1\; n\; g(t - \mathbf{q} \cdot \mathbf{x}/c_1) \tag{120}$$

where the function $g(s)$ is such that

$$g'(s) = -f(s) \tag{121}$$

The motion takes place along the direction of propagation $\mathbf{q} = (l, m, n)$ with a velocity of propagation c_1 (Fig. 3). The vector $\mathbf{d}$ of eqn (114) is in this case $\mathbf{d} = \mathbf{q}$. These waves are called longitudinal waves, irrotational waves (due to $\boldsymbol{\nabla} \times \mathbf{u}^\theta = 0$, or P-waves ('Primae waves'). The latter name was given in seismology because in the event of an earthquake they are the first waves to arrive at a certain location.

The general solution of eqn (116) with the form of a plane wave is

$$\boldsymbol{\omega} = \mathbf{A}\; h\; (t - \mathbf{q}' \cdot \mathbf{x}/c_2) \tag{122}$$

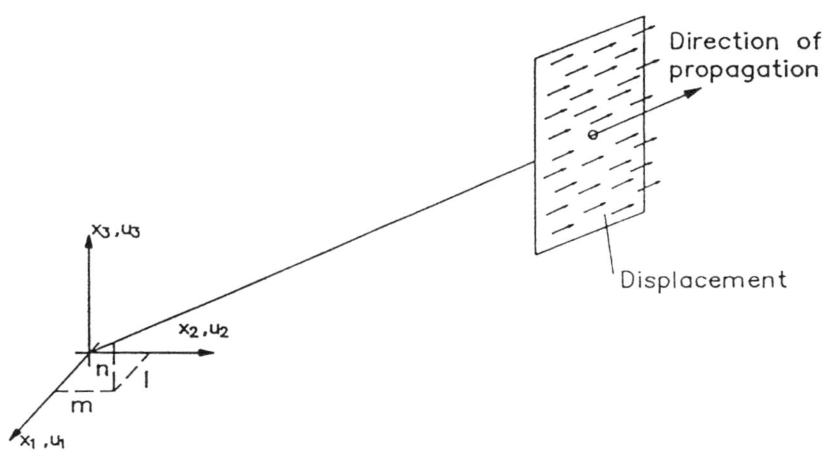

Figure 3: Displacements corresponding to longitudinal waves

where $\mathbf{q}' = (l', m', n')$ indicates the direction of propagation with $l'^2 + m'^2 + n'^2 = 1$, and $\mathbf{A}$ is a vector of amplitudes which determines the amplitude $\mid \mathbf{A} \mid$ and direction of the vector $\boldsymbol{\omega}$. From the definition of $\boldsymbol{\omega}$ one can write $\boldsymbol{\nabla} \cdot \boldsymbol{\omega} = 0$ and hence

$$A_1\, l' + A_2\, m' + A_3\, n' = 0 \tag{123}$$

The rotation vector is perpendicular to the direction of propagation. The displacements associated to these waves can be written as

$$u_1^{\omega} = c_2(n' A_2 - m' A_3)\, r(t - \mathbf{q}' \cdot \mathbf{x}/c_2) \tag{124}$$

$$u_2^{\omega} = c_2(l' A_3 - n' A_1)\, r(t - \mathbf{q}' \cdot \mathbf{x}/c_2) \tag{125}$$

$$u_3^{\omega} = c_2(m' A_1 - l' A_2)\, r(t - \mathbf{q}' \cdot \mathbf{x}/c_2) \tag{126}$$

where the function $r(s)$ is such that $r'(s) = -h(s)$. Equations (124) to (126) indicate that the motion $\mathbf{u}^{\omega}$ has no component along the direction of propagation; i.e. the vector of displacements is perpendicular to the direction of propagation (Fig. 4). Thus, the direction of propagation, the rotation and the displacement are three orthogonal vectors. This type of waves, which propagate with velocity c_2, are called shear waves or equivolumial waves as mentioned above. They are also called S-waves ('Secundae waves') because in the event of an earthquake they arrive after the P-waves.

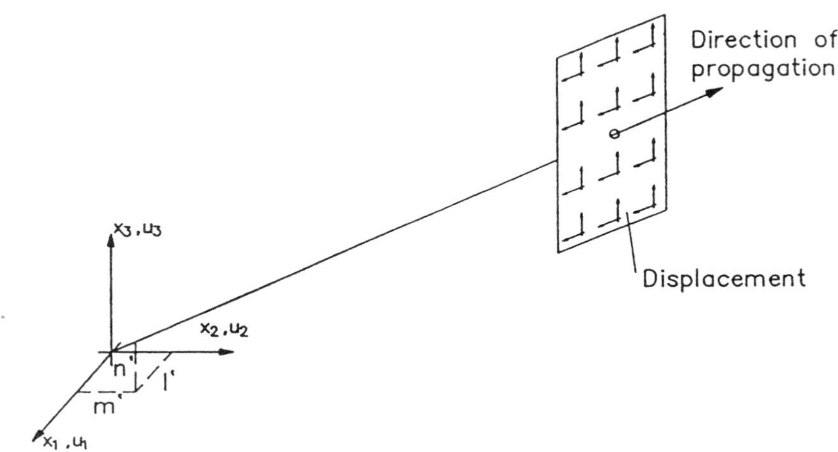

Figure 4: Displacements corresponding to shear waves

If the cartesian axes are chosen in a way such that the direction of propagation of the plane P-waves is contained in the $(x_1\, x_2)$-plane, the motion is in the plane and one has a two-dimensional in-plane problem with $u_3 \equiv 0$ and all the $\partial/\partial x_3 \equiv 0$. If the cartesian axes are chosen in a way such that the direction of propagation of the plane S-waves is contained in the $x_1\, x_2$-plane, the motion due to the S-waves can be decomposed in two parts, one contained in the $x_1\, x_2$-plane and the other along the x_3-axis. The two parts are uncoupled. One is a two-dimensional in-plane problem and the other an antiplane problem.

1.7.6 Time-harmonic waves

Time-harmonic waves are those whose time dependence is of the type $\cos(\omega t)$, $\sin(\omega t)$; where ω is the circular frequency. Using complex notation the time dependence can be written as $\exp(i\omega t)$ or $\exp(-i\omega t)$. In this case only the real or the imaginary part of the function has a physical meaning. The study of time-harmonic waves in linear media is very interesting because by virtue of the superposition principle, other perturbations can be written in terms of harmonic waves. Using the Fourier series one can represent periodic waves as a series of harmonic waves. Non-periodic perturbations can also be written in terms of harmonic waves using the Fourier transform.

Going back to the one-dimensional problem of the half-space under uniform pressure, assume that the pressure has a harmonic time dependence. In general

$$p(t) = P\, e^{i\omega t} \tag{127}$$

The displacement solution of eqn (72) is

$$u_1(x_1,t) = A\, e^{i\frac{\omega}{c}(ct-x_1)} + A'\, e^{i\frac{\omega}{c}(ct+x_1)} \tag{128}$$

where A and A' are amplitudes independent of x_1 and t. The functions $\exp[i(\omega/c)(ct-x_1)]$ and $\exp[i(\omega/c)(ct+x_1)]$ clearly belong to the general forms $f(t-x_1/c)$ and $g(t+x_1/c)$, and represent travelling waves in the positive and negative x_1-direction, respectively. All the reasoning already done for the general functions $f(t-x_1/c)$ and $g(t+x_1/c)$ is valid for this particular and simpler case.

The constant c takes the values c_1 and c_2 for irrotational and equivolumial waves, respectively. The argument $(\omega/c)(ct-x_1)$ is called the phase of the wave and c the phase velocity. The displacement u_1 is a time-harmonic function with a time period T, where $T = 2\pi/\omega$, and a wave length $\lambda = c\,(2\pi/\omega) = c\,T$. The quantity $k = \omega/c$ is known as wavenumber. It indicates the number of wavelengths within a length 2π $(k = 2\pi/\lambda)$. When the phase velocity, i.e. the propagation velocity, does not depend on the frequency, the system is said to be non-dispersive. Elastic materials are non-dispersive while non-elastic materials are, in general, dispersive. A certain perturbation can always be decomposed into time-harmonic components. If all the components have the same propagation velocity independent of their frequency, the perturbation maintains its shape (does not disperse) as it propagates through the body.

Equation (128) represents a one-dimensional time-harmonic wave. Time-harmonic plane waves have the form

$$\mathbf{u} = A\, e^{\frac{i\omega}{c}(ct-\mathbf{q}\cdot\mathbf{x})}\mathbf{d} \tag{129}$$

where A is the real value or complex amplitude, $\mathbf{d}$ defines the direction of the motion and $\mathbf{q}$ the direction of propagation. Plane time-harmonic waves are a particular case of the plane waves discussed above. The results of that discussion are applicable. In particular, it may be remarked that there are two types of time-harmonic waves in the infinite domain, the P-waves and the S-waves with phase velocities c_1 and c_2, respectively. The displacements follow the propagation direction for the P-waves and are perpendicular to the propagation direction for the S-waves. A more detailed study of plane harmonic waves is included in the next section which is dedicated to harmonic problems.

2 Plane harmonic problems

2.1 Plane harmonic waves in elastic solids

In the previous section a plane elastic wave propagating in a direction determined by a unit vector $\mathbf{q} = (l, m, n)$ was defined as a perturbation for which all the points in any plane perpendicular to $\mathbf{q}$ have the same displacement at the same time. The displacement vector was written as

$$\mathbf{u} = f(t - (\mathbf{q} \cdot \mathbf{x})/c)\, \mathbf{d} \tag{130}$$

where $\mathbf{d}$ is the unit vector defining the direction of motion.

For time-harmonic body forces and boundary conditions, the resulting displacement vector is also time-harmonic and in the case of plane waves the displacements can be written as

$$\mathbf{u} = A\, e^{i\frac{\omega}{c}(ct - \mathbf{q}\cdot\mathbf{x})}\mathbf{d} \tag{131}$$

where only the real or the imaginary parts of $\mathbf{u}$ have a physical meaning and A is the amplitude which, in general, can be real-valued or complex, but is independent of $\mathbf{x}$ or t.

It was shown in the previous section, using the representation in terms of dilatation and rotation, that there are two types of plane waves: longitudinal waves and shear waves, propagating with velocities $c_1 = [(\lambda + 2\mu)/\rho]^{(1/2)}$ and $c_2 = (\mu/\rho)^{(1/2)}$, respectively. The existence of these two types of plane waves in an infinite domain can also be demonstrated using a simple mathematical reasoning as done below in accordance with Achenbach[1] and Eringen and Suhubi.[30] The following discussion is done for time harmonic waves. However, it can be easily checked that it is valid for general plane disturbances.

Using the wave velocities as basic elastic constants of the body, Navier's equation (see eqn 14) for zero body forces can be written as

$$c_2^2\nabla^2\mathbf{u} + (c_1^2 - c_2^2)\boldsymbol{\nabla}\, \boldsymbol{\nabla}\cdot\mathbf{u} = \ddot{\mathbf{u}} \tag{132}$$

Substituting the displacement expression (131) into (132) one obtains

$$(c_2^2 - c^2)\mathbf{d} + (c_1^2 - c_2^2)(\mathbf{q}\cdot\mathbf{d})\mathbf{q} = 0 \tag{133}$$

Since $\mathbf{d}$ and $\mathbf{q}$ are independent vectors, eqn (133) can only be satisfied if:

$$\mathbf{q} = \pm\mathbf{d} \quad \text{and then} \quad c^2 = c_1^2 \tag{134}$$

or

$$\mathbf{q}\cdot\mathbf{d} = 0 \quad \text{and then} \quad c^2 = c_2^2 \tag{135}$$

which means that the displacement follows a direction $\mathbf{d}$ which is either the same as the direction of propagation ($\mathbf{q} = \pm\mathbf{d}$), in which case the waves propagate with a velocity $c = c_1$, or it follows a direction **d** perpendicular to the direction of propagation ($\mathbf{q} \cdot \mathbf{d} = 0$) in which case the waves propagate with a velocity $c = c_2$.

As was already mentioned in the previous section, the first type of waves is known as dilatational, irrotational, longitudinal or P-waves and the second as equivolumial, shear, transversal or S-waves. In the following subsections, c_P and c_S are used to denote the P and S-wave velocity respectively, when the symbols c_1 and c_2 may become ambiguous.

If the cartesian axes are selected in a way such that the direction of propagation **q** is in the x_1x_2-plane ($n = 0$), then all the displacements and derived variables are independent of x_3

$$\mathbf{u} = A\ e^{i\frac{\omega}{c}(ct - lx_1 - mx_2)}\mathbf{d} \tag{136}$$

Displacements corresponding to P-waves follow the direction **q** and hence, only have u_1 and u_2 components. The problem is a two-dimensional in-plane one. In the case of S-waves, the displacement vector **u** is perpendicular to the direction of propagation **q**. This vector can be decomposed into two normal components: one in the x_3 direction and the other following the intersection between the x_1x_2-plane and the plane perpendicular to **q** (Fig. 5). The displacement along x_3 : $u_3(x_1, x_2, t)$, and its derived variables constitute an independent antiplane problem.

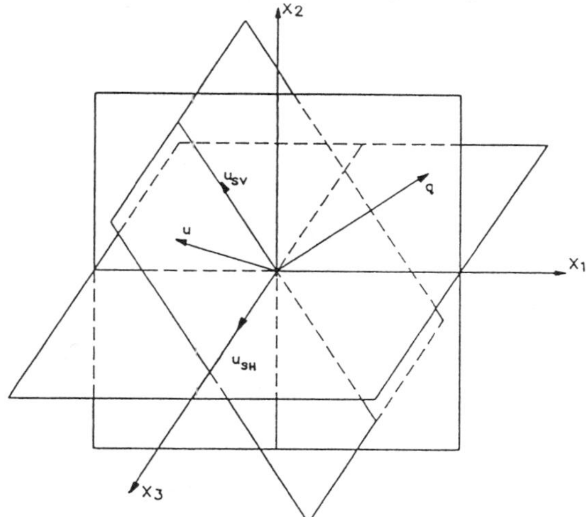

Figure 5: Plane S-waves

The displacement in the x_1x_2-plane, also independent of x_3, defines a two-dimensional in-plane problem. Since the antiplane and the in-plane elastic problems are uncoupled, one can talk about two different kinds of S-waves, the SH-waves (horizontally polarized S-waves) which produce the u_3 displacement and the SV-waves (vertically polarized S-waves) which produce the displacement in the x_1x_2-plane. When P and S-waves propagate in a plane, the P and SV-waves are parts of an in-plane field whereas the SH-waves produce an anti-plane motion. Due to these facts, when SH-waves propagate in a plane and find a change of material or an external boundary, only antiplane motion and hence, SH-waves, will appear after

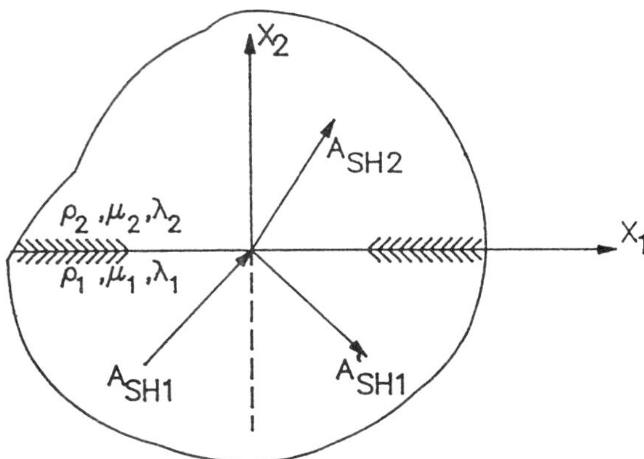

Figure 6: SH-waves in a plane with an infinite interface along the x_1-axis

the refraction or reflection process. On the other hand, when one of the in-plane motion waves, SV or P, propagates in a plane and finds a change of material or a free boundary, both types of waves SV and P can appear after the refraction or reflection process regardless of which one of these two was the incident wave. In the next subsection, the reflection and refraction of plane harmonic waves will be briefly discussed to allow the reader a better understanding of the forthcoming discussions and examples. A more complete treatment of this matter can be found in the general books on elastic wave propagation already mentioned (Refs 1, 30 and 57).

2.2 Reflection and refraction of plane harmonic elastic waves

To make the mathematical treatment simpler, plane problems with one dimensional geometry and boundary conditions are assumed; i.e. problems where all the material properties and boundary conditions depend only on one of the cartesian directions, say x_2.

2.2.1 *SH*-waves

These waves are associated to the antiplane motion. Assume a uniform elastic half-plane ($x_2 \leq 0$) with material properties ρ_1, μ_1, λ_1, under the effects of steady time-harmonic plane SH-waves coming from the far field and propagating with a direction $\mathbf{q}_1 = (l_1, m_1)$. The half-plane $x_2 \geq 0$ will be considered first to be free of any material and second, to be another elastic region with properties ρ_2, μ_2, λ_2. In both cases, when the incident waves find the boundary $x_2 = 0$, waves will be reflected into the region $x_2 \leq 0$ (Fig. 6).

The displacements in the region with properties ρ_1, μ_1, λ_1 will be of the form

$$u_3 = A_{SH1}\, e^{i\frac{\omega}{c_{S1}}(c_{S1}t - l_1x_1 - m_1x_2)} + A'_{SH1}\, e^{i\frac{\omega}{c_{S1}}(c_{S1}t - l'_1x_1 + m'_1x_2)} \tag{137}$$

where a positive sign has been adopted for m'_1 because the reflected waves travel following a direction with negative x_2-component. A_{SH1} and A'_{SH1} are the amplitudes of the incident and reflected waves, respectively and $c_{S1} = (\mu_1/\rho_1)^{1/2}$ the shear wave velocity.

When the motion of the boundary points is excited by incident waves with a certain x_1 variation, the reflected waves have the same x_1 variation; otherwise, no boundary condition uniform along the x_1-axis could be satisfied (notice that the functions are harmonic in space). Then $l'_1 = l_1$ and $m'_1 = (1 - l'^2_1)^{1/2} = m_1$. The displacement u_3 can be written as

$$u_3 = \left(A_{SH1}\, e^{-i\frac{\omega}{c_{S1}}m_1 x_2} + A'_{SH1}\, e^{i\frac{\omega}{c_{S1}}m_1 x_2} \right) f(x_1, t) \tag{138}$$

where

$$f(x_1, t) = e^{(i\omega/c_{S1})(c_{S1}t - l_1 x_1)}$$

The incident and the reflected waves form the same angle with respect to the normal to the boundary.

The shear stress is

$$\sigma_{32} = \sigma_{23} = \mu\, u_{3,2} = \mu\, i\frac{\omega}{c_{S1}} m_1 \left(-A_{SH1} e^{-i\frac{\omega}{c_{S1}}m_1 x_2} + A'_{SH1} e^{i\frac{\omega}{c_{S1}}m_1 x_2} \right) f(x_1, t) \tag{139}$$

The displacements and stresses along the $x_2 = 0$ boundary in terms of the wave amplitudes can be written in matrix form as

$$\begin{bmatrix} u_3 \\ \sigma_{32} \end{bmatrix}_1 = \begin{bmatrix} 1 & 1 \\ -i\frac{\omega}{c_{S1}} m_1 \mu_1 & i\frac{\omega}{c_{S1}} m_1 \mu_1 \end{bmatrix} \begin{bmatrix} A_{SH1} \\ A'_{SH1} \end{bmatrix} \tag{140}$$

where the term $f(x_1, t)$ has been dropped. The reflected wave amplitude A'_{SH1} for a free boundary can be obtained from the second equation by doing $\sigma_{32} = 0$. Then $A'_{SH1} = A_{SH1}$. If the boundary $x_2 = 0$ is clamped, the amplitude A'_{SH1} is obtained from the first equation by doing $u_3 = 0$. Then $A'_{SH1} = -A_{SH1}$.

When the region $x_2 \geq 0$ is another elastic solid with properties ρ_2, μ_2, λ_2 and $c^2_{S2} = \mu_2/\rho_2$, in addition to the incident and reflected waves there will be plane waves refracted into the $x_2 > 0$ part of the body.

The u_3 displacement in this region is

$$u_3 = A_{SH2}\, e^{i\frac{\omega}{c_{S2}}(c_{S2}t - l_2 x_1 - m_2 x_2)} \tag{141}$$

where A_{SH2} is the amplitude of the refracted wave and $q_2 = (l_2, m_2)$ its direction of propagation. To satisfy the equilibrium and compatibility conditions over the boundary $x_2 = 0$, the variation of these displacements along the x_1-axis must be the same as those of the incident and reflected waves. Then

$$\frac{l_1}{c_{S1}} = \frac{l'_1}{c_{S1}} = \frac{l_2}{c_{S2}} \tag{142}$$

and the displacements in the $x_2 \geq 0$ region can be written as

$$u_3 = A_{SH2}\, e^{-i\frac{\omega}{c_{S2}}m_2 x_2}\, f(x_1, t) \tag{143}$$

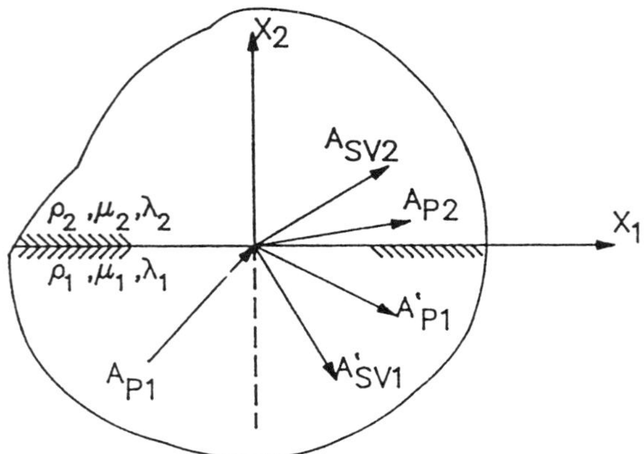

Figure 7: Reflection and refraction of P-waves

The shear stresses are

$$\sigma_{32} = \sigma_{23} = -\mu\, i\frac{\omega}{c_{S2}} m_2\, A_{SH2}\, e^{-i\frac{\omega}{c_{S2}} m_2 x_2} f(x_1, t) \tag{144}$$

The displacements and stresses along the boundary $x_2 = 0$ as part of the $x_2 \geq 0$ region can be written as

$$\begin{bmatrix} u_3 \\ \sigma_{23} \end{bmatrix}_2 = \begin{bmatrix} 1 \\ -i\frac{\omega}{c_{S2}} m_2 \mu_2 \end{bmatrix} A_{SH2} \tag{145}$$

Using (140) and (145) and the conditions

$$\begin{bmatrix} u_3 \\ \sigma_{23} \end{bmatrix}_1 = \begin{bmatrix} u_3 \\ \sigma_{23} \end{bmatrix}_2 \quad \text{along } x_2 = 0 \tag{146}$$

one may determine the amplitudes of the reflected and the refracted waves (A'_{SH1} and A_{SH2}, respectively) in terms of the incident wave amplitude (A_{SH1}). The direction of the reflected and the refracted waves are determined by eqn (142).

2.2.2 *SV* and *P*-waves

Assume an elastic half-plane ($x_2 \leq 0$) with material properties ρ_1, μ_1, λ_1 under the effects of steady time-harmonic plane *P*-waves coming from the far field and propagating with a direction $\mathbf{q}_1 = (l_1, m_1)$. The half-plane $x_2 \geq 0$ will again be considered first, to be free of any material and second, to be another elastic region with properties ρ_2, μ_2, λ_2. When incident *P*-waves find the boundary $x_2 = 0$, both types of waves *P* and *SV* will, in general, be reflected into the region $x_2 \leq 0$ (Fig. 7).

The u_1 displacement in the region $x_2 \leq 0$ can be written as

$$\begin{aligned} u_1 \;=\;& l_1 A_{P1}\, e^{i\frac{\omega}{c_{P1}}(c_{P1}t - l_1 x_1 - m_1 x_2)} + l'_1 A'_{P1}\, e^{i\frac{\omega}{c_{P1}}(c_{P1}t - l'_1 x_1 + m'_1 x_2)} - \\ & - m''_1 A'_{SV1}\, e^{i\frac{\omega}{c_{S1}}(c_{S1}t - l''_1 x_1 + m''_1 x_2)} \end{aligned} \tag{147}$$

where $\mathbf{q}_1' = (l_1', -m_1')$ and $\mathbf{q}_1'' = (l_1'', -m_1'')$ are the wave propagation directions of the reflected P and SV-waves, respectively, A'_{P1} and A'_{SV1} the corresponding wave amplitudes and c_{P1} and c_{S1} the P and S-wave velocities, respectively. To satisfy any uniform boundary condition along the x_1-direction, the variation along this direction of all the waves must be the same.

Then

$$\frac{l_1}{c_{P1}} = \frac{l_1'}{c_{P1}} = \frac{l_1''}{c_{S1}} \tag{148}$$

and consequently, $m_1' = m_1$ and $m_1'' = (1 - (c_{S1}/c_{P1})^2\, l_1^2)^{1/2}$

Calling

$$f(x_1, t) = e^{i\frac{\omega}{c_{P1}}(c_{P1}t - l_1 x_1)} \tag{149}$$

eqn (147) can be written as

$$u_1 = \left(l_1 A_{P1}\, e^{-i\frac{\omega}{c_{P1}} m_1 x_2} + l_1 A'_{P1}\, e^{i\frac{\omega}{c_{P1}} m_1 x_2} - m_1'' A'_{SV1}\, e^{i\frac{\omega}{c_{S1}} m_1'' x_2} \right) f(x_1, t) \tag{150}$$

Similarly the u_2 displacement can be written as

$$u_2 = \left(m_1 A_{P1}\, e^{-i\frac{\omega}{c_{P1}} m_1 x_2} - m_1 A'_{P1}\, e^{i\frac{\omega}{c_{P1}} m_1 x_2} - l_1'' A'_{SV1}\, e^{i\frac{\omega}{c_{S1}} m_1'' x_2} \right) f(x_1, t) \tag{151}$$

The stress components can be easily derived from (150) and (151)

$$\begin{aligned} \sigma_{22} &= \lambda_1(u_{1,1} + u_{2,2}) + 2\mu\, u_{2,2} = -i\,\frac{\omega}{c_{P1}} \Big[(\lambda_1 + 2\mu_1\, m_1^2)\, A_{P1}\, e^{-i\frac{\omega}{c_{P1}} m_1 x_2} + \\ &\quad + (\lambda_1 + 2\mu_1\, m_1^2)\, A'_{P1}\, e^{i\frac{\omega}{c_{P1}} m_1 x_2} + 2\mu_1\, l_1\, m_1''\, A'_{SV1}\, e^{i\frac{\omega}{c_{S1}} m_1'' x_2} \Big] f(x_1, t) \end{aligned}$$

$$\begin{aligned} \sigma_{21} &= \mu_1(u_{1,2} + u_{2,1}) = i\,\frac{\omega}{c_{S1}} \Big[-2\mu_1\, l_1''\, m_1\, A_{P1}\, e^{-i\frac{\omega}{c_{P1}} m_1 x_2} + \\ &\quad + 2\mu_1\, l_1''\, A'_{P1}\, m_1\, e^{i\frac{\omega}{c_{P1}} m_1 x_2} + \mu_1 (l_1''^2 - m_1''^2)\, A'_{SV1}\, e^{i\frac{\omega}{c_{S1}} m_1'' x_2} \Big] f(x_1, t) \end{aligned} \tag{152}$$

The displacements and stresses along the $x_2 = 0$ boundary can be written in matrix form as

$$\begin{bmatrix} u_1 \\ u_2 \\ \sigma_{22} \\ \sigma_{21} \end{bmatrix}_1 = \mathbf{T} \begin{bmatrix} A_{P1} \\ A'_{P1} \\ A'_{SV1} \end{bmatrix} \tag{153}$$

where

$$\mathbf{T} = \begin{bmatrix} l_1 & l_1 & -m_1'' \\ m_1 & -m_1 & -l_1'' \\ -i\frac{\omega}{c_{P1}}(\lambda_1 + 2\mu_1\, m_1^2) & -i\frac{\omega}{c_{P1}}(\lambda_1 + 2\mu_1\, m_1^2) & -i\frac{\omega}{c_{P1}} 2\mu_1\, l_1\, m_1'' \\ -i\frac{\omega}{c_{S1}} 2\mu_1\, l_1''\, m_1 & i\frac{\omega}{c_{S1}} 2\mu_1\, l_1''\, m_1 & i\frac{\omega}{c_{S1}}\mu_1(l_1''^2 - m_1''^2) \end{bmatrix}$$

and the term $f(x_1, t)$ has been dropped.

The reflected wave amplitudes A'_{P1} and A'_{SV1} for a free boundary condition, can be obtained in terms of the incident wave amplitude A_{P1} from the last two equations by making $\sigma_{22} = \sigma_{21} = 0$. Other boundary conditions on $x_2 = 0$ also permit obtaining A'_{P1} and A'_{SV1} in terms of A_{P1} using eqn (153).

When the region $x_2 \geq 0$ is another elastic solid with properties ρ_2, μ_2, λ_2, in addition to incident and reflected waves there will be SV and P-waves refracted into the $x_2 > 0$ part of the plane. In general, there will be a refracted P-wave with amplitude A_{P2} and wave propagation direction $\mathbf{q}_2 = (l_2, m_2)$, and a refracted SV-wave with amplitude A_{SV2} and wave propagation direction $\mathbf{q}'_2 = (l'_2, m'_2)$.

To satisfy the equilibrium and compatibility conditions along the boundary $x_2 = 0$, all the displacements must have the same variation along the x_1-axis.

Then

$$\frac{l_1}{c_{P1}} = \frac{l_1'}{c_{P1}} = \frac{l_1''}{c_{S1}} = \frac{l_2}{c_{P2}} = \frac{l_2'}{c_{S2}} \tag{154}$$

The displacements in the region $x_2 \geq 0$ can be written as

$$u_1 = \left[l_2\, A_{P2}\, e^{-i\frac{\omega}{c_{P2}} m_2 x_2} + m_2'\, A_{SV2}\, e^{-i\frac{\omega}{c_{S2}} m_2' x_2} \right] f(x_1, t)$$

$$u_2 = \left[m_2\, A_{P2}\, e^{-i\frac{\omega}{c_{P2}} m_2 x_2} - l_2'\, A_{SV2}\, e^{-i\frac{\omega}{c_{S2}} m_2' x_2} \right] f(x_1, t) \tag{155}$$

and the stresses

$$\sigma_{22} = \frac{-i\omega}{c_{P2}} \left[\left(\lambda_2 + 2\mu_2\, m_2^2 \right) A_{P2}\, e^{-i\frac{\omega}{c_{P2}} m_2 x_2} - \right.$$

$$\left. -2\mu_2\, l_2\, m_2'\, A_{SV2}\, e^{-i\frac{\omega}{c_{S2}} m_2' x_2} \right] f(x_1, t)$$

$$\sigma_{21} = \frac{i\omega}{c_{S2}} \left[-\, 2\mu_2 l_2' m_2\, A_{P2}\, e^{-i\frac{\omega}{c_{P2}} m_2 x_2} + \right.$$

$$\left. +\, \mu_2 (l_2'^2 - m_2'^2)\, A_{SV2}\, e^{-i\frac{\omega}{c_{S2}} m_2' x_2} \right] f(x_1, t) \tag{156}$$

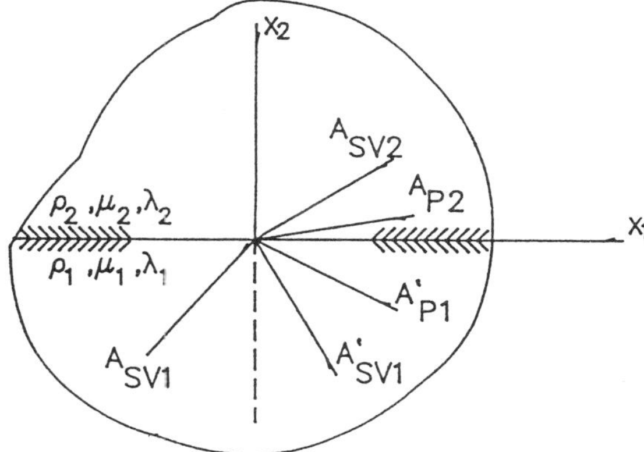

Figure 8: Reflection and refraction of SV-waves

The displacements and stresses along the $x_2 = 0$ boundary are

$$\begin{bmatrix} u_1 \\ u_2 \\ \sigma_{22} \\ \sigma_{12} \end{bmatrix}_2 = \begin{bmatrix} l_2 & m'_2 \\ m_2 & -l'_2 \\ -i\frac{\omega}{c_{P2}}(\lambda_2 + 2\mu_2 m_2^2) & i\frac{\omega}{c_{P2}}2\mu_2 l_2 m'_2 \\ -i\frac{\omega}{c_{S2}}2\mu_2 l'_2 m_2 & i\frac{\omega}{c_{S2}}\mu_2(l'^2_2 - m'^2_2) \end{bmatrix} \begin{bmatrix} A_{P2} \\ A_{SV2} \end{bmatrix} \tag{157}$$

Using eqns (153) and (157) and the continuity condition along the boundary $x_2 = 0$

$$\begin{bmatrix} u_1 \\ u_2 \\ \sigma_{22} \\ \sigma_{12} \end{bmatrix}_1 = \begin{bmatrix} u_1 \\ u_2 \\ \sigma_{22} \\ \sigma_{12} \end{bmatrix}_2 \tag{158}$$

one has four equations to determine the reflected waves amplitudes A'_{P1}, A'_{SV1} and the refracted waves amplitudes A_{P2}, A_{SV2} in terms of the incident wave amplitude A_{P1}.

Assume now that the incident wave coming from the far field is an SV-wave with an amplitude A_{SV1} and direction of propagation $\mathbf{q}_1 = (l_1, m_1)$.

When the incident SV-waves find the boundary $x_2 = 0$, both types of waves P and SV will be reflected into the region $x_2 \leq 0$ in this case (Fig. 8).

The reflected P and SV-waves have amplitudes A'_{P1} and A'_{SV1}, respectively. The propagation directions are $\mathbf{q}'_1 = (l'_1, -m'_1)$ for the P-waves and $\mathbf{q}''_1 = (l''_1, -m''_1)$ for the SV-waves. In order to satisfy the boundary conditions, all the waves must have the same x_1-variation again.

$$\frac{l_1}{c_{S1}} = \frac{l'_1}{c_{P1}} = \frac{l''_1}{c_{S1}} \tag{159}$$

and then: $m'_1 = (1 - (c_{P1}/c_{S1})^2\, l_1^2)^{1/2}$ and $m''_1 = m_1$.

Calling

$$f(x_1,t) = e^{i\frac{\omega}{c_{S1}}(c_{S1}t - l_1 x_1)} \tag{160}$$

the following expressions can be written for displacements and stresses in the $x_2 \leq 0$ plane

$$u_1 = \left(m_1 A_{SV1}\, e^{-i\frac{\omega}{c_{S1}}m_1 x_2} - m_1\, A'_{SV1}\, e^{i\frac{\omega}{c_{S1}}m_1 x_2} + l'_1\, A'_{P1}\, e^{i\frac{\omega}{c_{P1}}m'_1 x_2} \right) f(x_1,t)$$

$$u_2 = \left(- l_1 A_{SV1}\, e^{-i\frac{\omega}{c_{S1}}m_1 x_2} - l_1\, A'_{SV1}\, e^{i\frac{\omega}{c_{S1}}m_1 x_2} - m'_1\, A'_{P1}\, e^{i\frac{\omega}{c_{P1}}m'_1 x_2} \right) f(x_1,t) \tag{161}$$

and

$$\begin{aligned} \sigma_{22} \;=\; & \frac{-i\omega}{c_{P1}} \Big[- 2\mu\, l'_1\, m_1\, A_{SV1}\, e^{-i\frac{\omega}{c_{S1}}m_1 x_2} + 2\mu_1\, l'_1\, m_1\, A'_{SV1}\, e^{i\frac{\omega}{c_{S1}}m_1 x_2} + \\ & +(\lambda_1 + 2\mu_1\, m_1'^2)\, A'_{P1}\, e^{i\frac{\omega}{c_{P1}}m'_1 x_2} \Big] f(x_1,t) \end{aligned}$$

$$\begin{aligned} \sigma_{12} \;=\; & \frac{i\omega}{c_{S1}} \Big[\mu_1\, (l_1^2 - m_1^2)\, A_{SV1}\, e^{-i\frac{\omega}{c_{S1}}m_1 x_2} + \mu_1\, (l_1^2 - m_1^2)\, A'_{SV1}\, e^{i\frac{\omega}{c_{S1}}m_1 x_2} + \\ & +2\mu_1\, l_1\, m'_1\, A'_{P1}\, e^{i\frac{\omega}{c_{P1}}m'_1 x_2} \Big] f(x_1,t) \end{aligned} \tag{162}$$

The displacements and stresses along the $x_2 = 0$ boundary are

$$\begin{bmatrix} u_1 \\ u_2 \\ \sigma_{22} \\ \sigma_{21} \end{bmatrix}_1 = \mathbf{T} \begin{bmatrix} A_{SV1} \\ A'_{SV1} \\ A'_{P1} \end{bmatrix} \tag{163}$$

where

$$\mathbf{T} = \begin{bmatrix} m_1 & -m_1 & l'_1 \\ -l_1 & -l_1 & -m'_1 \\ i\frac{\omega}{c_{P1}}2\mu_1\, l'_1\, m_1 & -i\frac{\omega}{c_{P1}}2\mu_1\, l'_1\, m_1 & -i\frac{\omega}{c_{P1}}(\lambda_1 + 2\mu_1\, m_1'^2) \\ i\frac{\omega}{c_{S1}}\mu_1(l_1^2 - m_1^2) & i\frac{\omega}{c_{S1}}\mu_1(l_1^2 - m_1^2) & i\frac{\omega}{c_{S1}}2\mu_1\, l_1\, m'_1 \end{bmatrix}$$

and the term $f(x_1,t)$ has been dropped.

The reflected wave amplitudes can be obtained in terms of the incident wave amplitude using eqn (163) and the conditions on the $x_2 = 0$ boundary.

When the region $x_2 \geq 0$ is another elastic solid with properties ρ_2, μ_2 and λ_2 the incident SV-waves over the boundary $x_2 = 0$ produce, in addition to the reflected

P and SV-waves, P and SV-waves refracted into the region $x_2 \geq 0$ occupied by the second material. Again the amplitude and the propagation direction of the refracted P-waves will be called A_{P2} and $\mathbf{q}_2 = (l_2, m_2)$, respectively. The amplitude and the propagation direction of the SV-waves are A_{SV2} and $\mathbf{q}_2' = (l_2', m_2')$, respectively. The condition of having the same variation along the x_1-direction for all the waves leads, in this case, to

$$\frac{l_1}{c_{S1}} = \frac{l_1'}{c_{P1}} = \frac{l_1''}{c_{S1}} = \frac{l_2}{c_{P2}} = \frac{l_2'}{c_{S2}} \tag{164}$$

The displacements and stresses in the region $x_2 \geq 0$ are given by the general eqns (155) and (156) as in the case of incident P-waves, since in both cases there are refracted P and SV-waves.

The amplitude of the reflected and refracted waves can be determined, as usual, by enforcing the continuity conditions along $x_2 = 0$ (eqn 185) and by using eqn (163) for the first material displacements and tractions and eqn (157) for the second material displacements and tractions.

The detailed analysis of the different situations that can take place in a reflection-refraction process depending on the type of incident wave, angle of incidence and properties of the materials at both sides of the boundary, is very illustrative; however, it is out of the scope of this chapter. The interested reader may find it in Ref. 1. In this section only a brief analysis of the situations which give rise to the so-called surface waves will be carried out in the next subsection.

2.3 Surface waves

The equations presented in the above section for plane harmonic waves are valid no matter whether the values of l, m, n, l', m', n' and l'', m'', n'' are real or complex. The physical interpretation of a direction of propagation must be, however, reconsidered if they are not real. Assume for example that $l = (1+\alpha^2)^{1/2}$, $m = i\alpha$ and $n = 0$. The motion corresponds to a wave propagating in the x_1 direction with a velocity of propagation $c/(1+\alpha^2)^{1/2}$. The amplitude of the motion decreases or increases exponentially with x_2 depending on the sign of α. When all the coefficients are real, one talks properly of body waves. When some of them are complex, the equations represent generalized surface waves.

Given a certain incident body wave, surface waves appear when the angles of incidence, type of wave and material properties are such that the boundary continuity conditions yield to a director cosine greater than one for any reflected or refracted wave.

One of the simplest situations giving rise to surface waves may appear when SH-waves are reflected and refracted between two domains. The ratio between the angles of the different waves is given by eqn (142)

$$\frac{l_1}{c_{S1}} = \frac{l_1'}{c_{S1}} = \frac{l_2}{c_{S2}}$$

where l_1, l_1' and l_2 correspond to the incident, reflected and refracted waves, respectively. Then $l_1' = l_1$ and $l_2 = l_1\, c_{S2}/c_{S1}$. When $c_{S2} > c_{S1}$, it may happen that

$l_2 = l_1\, c_{S2}/c_{S1} > 1$; then $m_2 = \pm i(l_1^2\, c_{S2}^2/c_{S1}^2 - 1)^{1/2}$ and the transmitted wave is of the form

$$u_3 = A_{SH2}\, e^{-\frac{\omega}{c_{S2}}(l_1^2 \frac{c_{S2}^2}{c_{S1}^2}-1)^{1/2}\, x_2}\; e^{i\frac{\omega}{c'_{S2}}(c'_{S2}t-x_1)} \tag{165}$$

where $c'_{S2} = c_{S2}/l_2$. This equation represents a movement in the $x_2 \geq 0$ region whose amplitude diminishes exponentially with the distance to the interface. The plus sign for the first exponential does not have any physical meaning because the motion of the $x_2 \geq 0$ region cannot increase beyond bounds when x_2 tends to infinity. In the particular case where $l_2 = l_1\, c_{S2}/c_{S1} = 1$, $m_2 = 0$ and the reflected wave propagates along the x_1 direction.

There can also be surface waves at the interface of two solids in the case of in-plane motion. These surface waves are called Stoneley waves after the scientist who first investigated the propagation of such non-dispersive waves.[80] A study of the conditions for which Stoneley waves exist may be found in Cagniard.[15]

Consider now an elastic half-space and the possibility of plane SH, P or SV-waves traveling with a certain angle. It can be said, in accordance with the previous section, that SH-waves are always reflected as SH-waves by the free boundary and P-waves are reflected as SV and P-waves. SV-waves, will in general be reflected as P and SV-waves; however, for values of l such that $l' = l\, c_P/c_S > 1$ the reflected P-wave is a surface wave propagating with a velocity $c = c_P/l'$ and whose amplitude decays with the depth into the material.

The observation of seismograms showed that in the event of an earthquake, in addition to P and SV-waves there are other waves, with a smaller phase velocity, which can be predominant at certain distances from the source. It was Lord Rayleigh[71] who first studied the existence of waves traveling along the surface of a half-space with displacements decaying exponentially with depth. Assume a plane perturbation in a half-plane with the following displacement components

$$\begin{aligned} u_1 &= A_1\, e^{-bx_2}\, e^{i\frac{\omega}{c}(ct-x_1)} \\ u_2 &= A_2\, e^{-bx_2}\, e^{i\frac{\omega}{c}(ct-x_1)} \\ u_3 &= 0 \end{aligned} \tag{166}$$

where b has a positive real part.

By substitution of u_1 and u_2 into the Navier's equation for zero body forces, one has a system of two homogeneous equations which only has solution if

$$\left[c_P^2\, b^2 - (c_P^2 - c_S^2)\frac{\omega^2}{c^2}\right]\left[c_S^2\, b^2 - (c_S^2 - c^2)\frac{\omega^2}{c^2}\right] = 0 \tag{167}$$

which leads to $b_1 = (\omega/c)(1 - c^2/c_P^2)^{1/2}$ and to $b_2 = (\omega/c)(1 - c^2/c_S^2)^{1/2}$. Substitution of b_1 and b_2 into (166) gives expressions for u_1 and u_2 in terms of the amplitudes A_1 and A_2 and the phase velocity c. Using the definition of the stresses one obtains σ_{22} and σ_{12} also in terms of A_1, A_2 and c. The boundary conditions $\sigma_{22} = \sigma_{12} = 0$ on the free surface yield a new homogeneous system which has non-trivial solutions when

$$\left(2-\frac{c^2}{c_S^2}\right)^2-4\left(1-\frac{c^2}{c_P^2}\right)^{1/2}\left(1-\frac{c^2}{c_S^2}\right)^{1/2}=0 \tag{168}$$

which is known as the Rayleigh equation for the phase velocity. It can be shown (see Ref. 1) that this equation has two real roots with opposite signs. Obviously, the positive is the only one with a physical meaning. Since the ratio c_P^2/c_S^2 only depends on the Poisson's ratio, the Rayleigh wave velocity computed from (168) will also depend on this parameter. Its value goes from $0.862c_S$ to $0.955c_S$ when the Poisson's ratio varies from 0 to 0.5.

There was still evidence of another kind of waves in the recorded seismograms. They were observed to be dispersive and to have a large transverse displacement component. It was Love[51] who first explained the existence of antiplane shear waves which travel in a superficial layer on top of an elastic half-plane with a shear wave velocity greater than that of the layer. These waves are dispersive and are known as Love waves. Detailed analyses of Love and Rayleigh waves can be found in the abovementioned general books on elastodynamics and wave propagation. Any further study is beyond the scope of this chapter.

2.4 Reciprocal theorem for time-harmonic elasticity

The Reciprocal Theorem between two general elastodynamic states, presented in subsection 1.5, is now recalled. Let Ω be an elastic region with boundary Γ and density ρ. Let two elastodynamic states be defined over Ω. The displacements, tractions, body forces, initial displacements and initial velocities of the first state are denoted by $\mathbf{u}$, $\mathbf{p}$, $\mathbf{b}$, $\mathbf{u}_0$ and $\mathbf{v}_0$, respectively. Those of the second state are represented by $\mathbf{u}^*$, $\mathbf{p}^*$, $\mathbf{b}^*$, $\mathbf{u}_0^*$ and $\mathbf{v}_0^*$. The following reciprocal relation can be written.

$$\int_\Gamma (p_i * u_i^*)\mathrm{d}\Gamma + \int_\Omega \rho\,(b_i * u_i^* + u_{oi}\,\dot{u}_i^* + v_{oi}\,u_i^*)\,\mathrm{d}\Omega =$$

$$\int_\Gamma (p_i^* * u_i)\mathrm{d}\Gamma + \int_\Omega \rho\,(b_i^* * u_i + u_{oi}^*\,\dot{u}_i + v_{oi}^*\,u_i)\,\mathrm{d}\Omega \tag{169}$$

Assume that the body forces and boundary conditions for both states are harmonic in time with the same angular frequency ω. For the first state, for instance, the body forces can be written as

$$\mathbf{b}(\mathbf{x},t)=\mathbf{b}(\mathbf{x},\omega)\,e^{i\omega t} \tag{170}$$

where the complex notation has been adopted with the convention that only the real or the imaginary part of the variables have a physical meaning. The amplitude $\mathbf{b}(\mathbf{x},\omega)$ is, in general, a function of the frequency ω and may be real or complex-valued. The displacement field for harmonic body forces and boundary conditions can be assumed to be

$$\mathbf{u}(\mathbf{x},t)=\mathbf{u}^T(\mathbf{x},t)+\mathbf{u}(\mathbf{x},\omega)\,e^{i\omega t} \tag{171}$$

where $\mathbf{u}^T$ is the transient part of the solution and $\mathbf{u}(\mathbf{x},\omega)\exp(i\omega t)$ is the steady-state displacement. Due to the existence of damping in all the physical systems it is reasonable to assume that after some time the transient part of the solution disappears. Given certain initial conditions $\mathbf{u}_0(\mathbf{x})$ and $\mathbf{v}_0(\mathbf{x})$ the displacement field solution of the problem can be written as the superposition of two as indicated by eqn (171). The transient part can be obtained from the homogeneous field equations with homogeneous boundary conditions and initial conditions

$$\mathbf{u}^T(\mathbf{x}, 0) = \mathbf{u}_0 - \mathbf{u}(\mathbf{x}, \omega) \tag{172}$$

$$\dot{\mathbf{u}}^T(\mathbf{x}, 0) = \mathbf{v}_0 - i\omega\mathbf{u}(\mathbf{x}, \omega) \tag{173}$$

thus, the steady-state problem has as initial conditions the steady-state values of the displacement $\mathbf{u}(\mathbf{x}, \omega)$ and velocity $i\omega\mathbf{u}(\mathbf{x}, \omega)$.

If the two elastodynamic states are time harmonic with the same frequency ω, the convolution products of eqn (169) transform into dot products and, if the initial condition values are the same as the steady-state values, the initial condition terms on both sides of eqn (169) become identical. Thus, the reciprocal relation for time-harmonic elastodynamic states of the same frequency can be written as

$$\int_\Gamma p_i\, u_i^*\, \mathrm{d}\Gamma + \int_\Omega \rho\, b_i\, u_i^*\, \mathrm{d}\Omega = \int_\Gamma p_i^*\, u_i\, \mathrm{d}\Gamma + \int_\Omega \rho\, b_i^*\, u_i\, \mathrm{d}\Omega \tag{174}$$

where the field variables u_i, p_i, b_i, u_i^*, p_i^*, and b_i^* are real or complex-valued and depend on $\mathbf{x}$ and ω. Notice that for a given frequency the above expression of the Reciprocal Theorem coincides with that of elastostatics.

3 Boundary elements for dynamic fracture mechanics

3.1 Some basic fracture mechanics ideas

An ideal crack can be geometrically defined as two coplanar surfaces, inside or at the edge of a continuous region, sharing a line connection (the crack front). The cracks are usually flat, but may have curved surfaces. In three-dimensional problems, the crack front forms a curve which, in the case of flat cracks, is in a plane. Some crack problems can be assumed to have a two-dimensional geometry; i.e. a geometry independent of one of the coordinates (say x_3). In those cases, the two coplanar surfaces are represented by two coincident lines in the x_1x_2-plane (Fig. 9). The crack front is a line perpendicular to the x_1x_2-plane, represented by the point (crack tip) where the two lines representing the crack surfaces are connected. Crack problems with a two-dimensional geometry in the x_1x_2-plane and loading conditions independent of the x_3-coordinate can be, as usual in planar elasticity, under plane stress, plane strain or antiplane conditions.

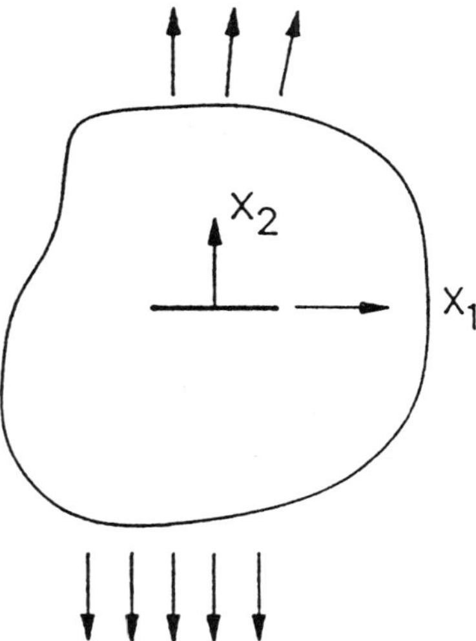

Figure 9: Plane crack geometry

The part of continuum mechanics dedicated to the analysis of cracks is known as fracture mechanics. An extensive treatment of its fundamentals is out of the scope of this chapter. The interested reader may find a comprehensive study of fracture mechanics in Refs 13 and 43.

The displacement field at any point of a crack front can be decomposed into three kinds of displacements as shown in Fig. 10. The first one, known as Mode-I response, corresponds to a symmetric opening displacement (Fig. 10a). The second one, known as Mode-II response, corresponds to a sliding displacement (Fig. 10b) and the third one (Fig. 10c), known as Mode-III response, corresponds to antiplane shear. Plane stress and plane strain problems may include Mode-I and Mode-II deformation, antiplane problems only have Mode-III deformation, while three-dimensional problems may include all three modes.

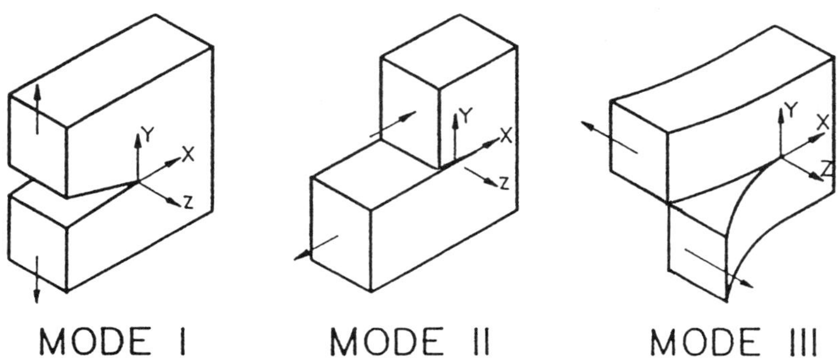

Figure 10: Modes of crack front displacement

The stresses near the tip of a crack under load assuming linear elastic behavior were given by Williams[85] as a series of the form

$$\sigma = a_0\, r^{-1/2} + a_1 + a_2\, r^{1/2} + \;\dots\dots \tag{175}$$

and the displacements

$$u = b_0 + b_1\, r^{1/2} + b_2\, r + \;\dots\dots \tag{176}$$

According to Irwin,[40] the state of stress near a crack tip is characterized by a parameter known as the stress intensity factor (SIF). Taking into account that for small values of r (Fig. 11) only the first term of σ and the second term of u (the first one is a rigid body motion) are significant, the leading term for the stresses near the crack tip can be written, for plane strain, as

$$\begin{aligned}
\sigma_{11} &= \frac{K_I}{\sqrt{2\pi r}} \cos\frac{\theta}{2}\left(1 - \sin\frac{\theta}{2}\sin\frac{3}{2}\theta\right) - \frac{K_{II}}{\sqrt{2\pi r}} \sin\frac{\theta}{2}\left(2 + \cos\frac{\theta}{2}\cos\frac{3}{2}\theta\right) \\
\sigma_{22} &= \frac{K_I}{\sqrt{2\pi r}} \cos\frac{\theta}{2}\left(1 + \sin\frac{\theta}{2}\sin\frac{3}{2}\theta\right) + \frac{K_{II}}{\sqrt{2\pi r}} \cos\frac{\theta}{2}\sin\frac{\theta}{2}\cos\frac{3}{2}\theta \\
\sigma_{12} &= \frac{K_I}{\sqrt{2\pi r}} \sin\frac{\theta}{2}\cos\frac{\theta}{2}\cos\frac{3}{2}\theta + \frac{K_{II}}{\sqrt{2\pi r}} \cos\frac{\theta}{2}\left(1 - \sin\frac{\theta}{2}\sin\frac{3}{2}\theta\right)
\end{aligned} \tag{177}$$

where r and θ are defined in Fig. 11 and the size of r is much smaller than the crack length. The factors K_I and K_{II} are the stress intensity factors corresponding to the opening and sliding mode, respectively. They characterize, as indicated by Irwin, the state of stress near the tip.

The leading terms for displacements near the crack tip for plane strain are

$$u_1 = \frac{K_I}{\mu}\sqrt{\frac{r}{2\pi}}\cos\frac{\theta}{2}\left(1 - 2\nu + \sin^2\frac{\theta}{2}\right) + \frac{K_{II}}{\mu}\sqrt{\frac{r}{2\pi}}\sin\frac{\theta}{2}\left(2 - 2\nu + \cos^2\frac{\theta}{2}\right)$$

$$u_2 = \frac{K_I}{\mu}\sqrt{\frac{r}{2\pi}}\sin\frac{\theta}{2}\left(2 - 2\nu - \cos^2\frac{\theta}{2}\right) + \frac{K_{II}}{\mu}\sqrt{\frac{r}{2\pi}}\cos\frac{\theta}{2}\left(-1 + 2\nu + \sin^2\frac{\theta}{2}\right) \quad (178)$$

where μ is the shear modulus and ν the Poisson's ratio.

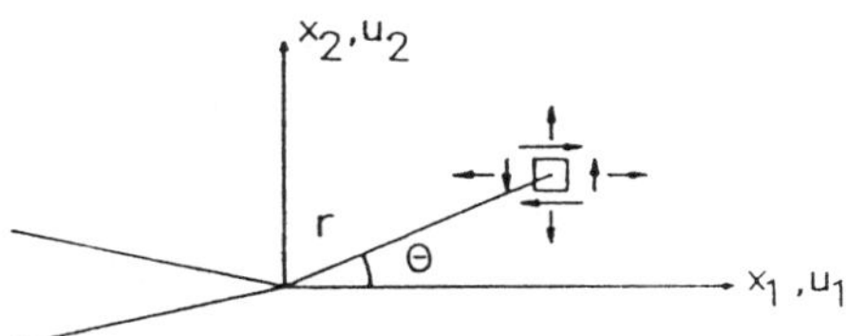

Figure 11: Coordinates near the crack tip

In spite of the existence of a zone in the immediate vicinity of the tip for which no real material remains elastic, there is a much larger elastic zone close to the tip for which eqns (177) and (178) are accurate representations of the stress and displacement fields. Irwin recognized that the fracture strength of many engineering materials can be given in terms of the critical stress intensity factor (K_{Ic}) based on these representations. This fact makes the computation of the SIF a very important matter and the main target of most linear elastic fracture mechanics problems. The SIF was related by Irwin to the previous energy release rate fracture criterium by Griffith.[38]

SIF for different geometries and load conditions can be found in the literature. Rooke and Cartwright[72] presented a collection of SIF for two-dimensional problems under static loading conditions. A more complete collection of SIF has been recently produced in Japan (Murakami[59]).

3.2 Dynamic stress intensity factors

Analytical solutions for dynamic fracture problems are limited to a small number of infinite domain problems. De Hoop[26] presented the dynamic analysis of a semi-infinite crack in an infinite solid subjected to a step pressure. Sih and Loeber,[75] and Mal[54] computed dynamic stress intensity factors of a finite crack in an infinite plane under plane harmonic waves. Mal[52,53] studied the diffraction of axisymmetric harmonic waves by a circular crack in an infinite medium. Transient dynamic stress intensity factors for a finite crack in an infinite plane for times prior to the arrival at the crack tip of the waves rediffracted by the other tip were obtained by Thau and Lu.[81] Sih, Embley and Ravera[76] studied the same problem as Thau and Lu for longer periods of time. Transient dynamic stress intensity factors for the half-plane crack in an unbounded solid were computed by Freund[34] by superposition over a fundamental solution obtained from the theory of dislocations. Dynamic stress intensity factors for cracks in a half-plane subjected to time harmonic excitation were obtained by Achenbach *et al.*,[2] Stone *et al.*,[79] Keer *et al.*,[44] Lin *et al.*,[49] Shah *et al.*,[74] and Tittmann *et al.*,[82] and for 3D problems by Angel and Achenbach.[4] Van der Hijden and Neerhoff[83] studied the same problem as Sih and Loeber[75] using a different computation approach which permits an increase in the range of frequencies of the analysis.

The aforementioned procedures for dynamic stress intensity factor computations are restricted to infinite or semi-infinite domains and include a numerical evaluation of integrals or a solution of integral equations. When the domain is finite the dynamic crack problem has been tackled by discretization of the body using a numerical method. Among the numerical methods used for dynamic SIF computations, the Finite Differences Method (FDM) and the Finite Element Method (FEM) have been well known for years. More recently the Boundary Element Method (BEM) has appeared as an efficient alternative in static and dynamic fracture mechanics. Interesting review papers of numerical solution techniques in dynamic fracture mechanics are those published by Kanninen[42] and more recently by Beskos.[11] The FDM was used by Chen[16] to study the case of a center cracked plate subjected to a step function load. A 5000 point mesh was used for this study. Finite Elements have been used in the last fifteen years by several authors. The works of Nishioka and Atluri,[65] Atluri and Nishioka,[5] Aoki *et al.*[6] and Murti and Valliapan[60] using the FEM should also be mentioned.

The direct formulation of the BEM has been used in combination with singular quarter-point (SQP) elements to compute SIF of static linear elastic problems. Blandford *et al.*[12] and Martínez and Domínguez[55] showed that very accurate results are obtained with simple BE meshes. More recently, Chirino and Domínguez[20] computed dynamic SIF in the frequency domain using the direct BEM formulation and SQP elements. They presented results for finite cracks in finite and infinite domains under harmonic and transient load conditions. The latter were obtained using the Fourier Transform. The possibilities and efficiency of the approach for static and frequency domain SIF computation were clearly established in those papers.

Transient stress intensity factors have also been computed using the step by step direct time domain B.E. formulation. The preliminary work of Nicholson and Mettu[62] in this field should be mentioned. They used two types of elements, one with constant interpolation functions in space and time and the other with interpolation functions quadratic in space and linear in time. For the first type of elements they obtained results that may be considered good, if one takes into account the kind of elements used and the way the SIF is computed from displacements given by constant elements. Those authors do not recommend the use of their quadratic elements because they obtain poor results that show an anomalous oscillatory behavior. A more precise and reliable approach than the use of constant elements to compute SIF was obviously needed. Domínguez and Gallego[29] were the first to present stable B.E. results for transient stress intensity factors. They used the direct time domain B.E. formulation in combination with the same SQP element that had been used by Martínez and Domínguez[55] and Chirino and Domínguez[20] for static and time harmonic SIF computations, respectively. A comparison of three different B.E. techiques for computation of transient dynamic SIF (harmonic analysis combined to Fourier tranform, direct step-by-step analysis, equivalent mass matrix method by the Dual Reciprocity BEM)[68,73] was performed by Chirino *et al.*[21]

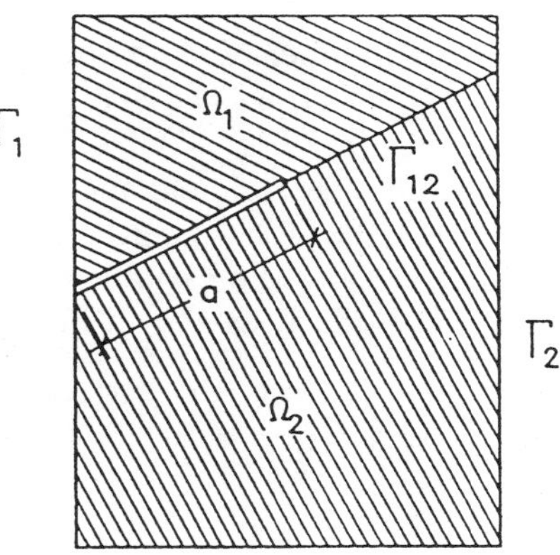

Figure 12: Substructuring

To end this subsection, reference should be made to the work carried out using B.E. formulations whose integral equations contain singularities stronger than those of the standard direct formulation. Nishimura and Kobayashi[63] and Krishnasamy *et al.*[46] have presented hypersingular formulations for static and dynamic crack problems. Some kind of regularization of the strongly singular expressions is needed. This fact makes methods based on those formulations rather involved and only solutions to simple particular problems have been presented. The hypersingular formulations permit a discretization only of the crack surfaces and the external boundaries. However, the conventional formulation, used by most authors, requires a substructuring of the body (Fig.12) which is subdivided into regions by internal boundaries including the crack line. The B.E. equations are written for each subregion and equilibrium and compatibility conditions are imposed along those parts of the internal boundaries outside the actual crack line. By using this procedure one avoids having two equal equations for each couple of nodes on the same location but on different surfaces of the crack.

3.3 Singular quarter-point element

The singular quarter-point element (SQP element) is a special type of quadratic element for two dimensional domains which is able to reproduce the spatial distribution of stresses and displacements near the crack tip. It includes the $r^{-1/2}$ singularity of the stresses shown in eqns (175) and (176).

Any displacement, traction, pair of coordinates x_1, x_2 or, in general, any variable along the boundary, can be represented in a quadratic element 'j' in terms of the value of such variable at the three nodes of the element,

$$\mathbf{f} = \boldsymbol{\Phi}\, \mathbf{f}^j \tag{179}$$

where $\mathbf{f}$ represents the interpolated quantity, $\mathbf{f}^j$ its nodal values, and $\boldsymbol{\Phi}$ the quadratic shape function matrices. A single component can be written as

$$f_i = \Phi_1\, f_i^1 + \Phi_2\, f_i^2 + \Phi_3\, f_i^3 \tag{180}$$

where the superindex stands for the number of the node, and Φ_k are the well-known quadratic shape functions: $\Phi_1 = \frac{1}{2}\xi(\xi - 1)$, $\Phi_2 = 1 - \xi^2$, and $\Phi_3 = \frac{1}{2}\xi(\xi + 1)$, being ξ the natural coordinate of the element (see Fig. 13).

For the particular case where the quadratic element has a straight-line geometry and the mid-node is placed at a quarter of the length (Fig.13) a simple relationship can be found between the coordinate ξ and the variable $\bar{r}$ along the element. Assume, for instance, that the crack tip is at node 1, as shown in Fig. 13. Then,

$$\xi = 2\sqrt{\frac{\bar{r}}{L}} - 1 \tag{181}$$

In this case eqn (180) yields

$$f_i = a_i^1 + a_i^2(\bar{r}/l)^{1/2} + a_i^3\, \bar{r}/l \tag{182}$$

where

$$\begin{aligned} a_i^1 &= f_i^1 \\ a_i^2 &= -3f_i^1 + 4f_i^2 - f_i^3 \\ a_i^3 &= 2f_i^1 + 4f_i^2 + 2f_i^3 \end{aligned} \tag{183}$$

Equation (182) ensures that for this position of the mid-point, the $r^{1/2}$ behavior of the displacement near the crack tip as given by eqns (176) and (178) is reproduced by the boundary element. This type of element is usually known as the 'quarter-point element'.

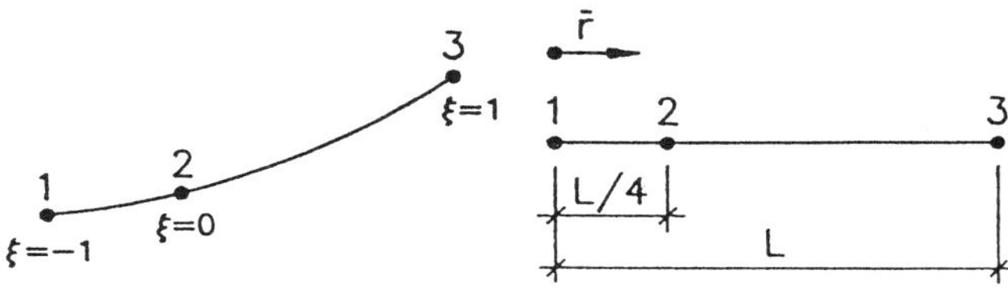

Figure 13: Quadratic and quadratic quarter point elements

Since displacements and tractions are represented independently in the BEM, a correct representation of the displacements does not mean a correct representation of the tractions. The singularity has to be included in the representation of the tractions by using modified shape functions. To this end one may write,

$$p_i = \Phi_1 \bar{p}_i \, (l/\bar{r})^{1/2} + \Phi_2 \, \bar{p}_i^2 \, (l/\bar{r})^{1/2} + \Phi_3 \, \bar{p}_i^3 \, (l/\bar{r})^{1/2} \tag{184}$$

or

$$p_i = \bar{\Phi}_1 \, \bar{p}_i^1 + \bar{\Phi}_2 \bar{p}_i^2 + \bar{\Phi}_3 \, \bar{p}_i^3 \tag{185}$$

where $\bar{\Phi}_1$, $\bar{\Phi}_2$ and $\bar{\Phi}_3$ are the modified shape functions which include the $r^{-1/2}$ singularity. Now $\bar{p}_i^j$ stands for the value of p_i at node j divided by the value of $\bar{\Phi}_i$ at that node; i.e.

$$\bar{p}_i^3 = p_i^3$$

$$\bar{p}_i^2 = p_i^2/2$$

$$\bar{p}_i^1 = \lim_{\bar{r}\to 0} p_i^1 (\bar{r}/l)^{1/2}$$

Equation (184) for p_i can now be written as

$$p_i = \bar{a}_i^1 \, (\bar{r}/l)^{-1/2} + \bar{a}_i^2 + \bar{a}_i^3 \, (\bar{r}/l)^{1/2} \tag{186}$$

where $\bar{a}_i^1 = \bar{p}_i^1$; $\bar{a}_i^2 = -\bar{p}_i^3 + 4\bar{p}_i^2 - 3\bar{p}_i^1$ and $\bar{a}_i^3 = 2\bar{p}_i^3 - 4\bar{p}_i^2 + 2\bar{p}_i^1$

Using the quarter-point element with the shape functions of eqn (185) for the tractions, both displacements and tractions are correctly represented. The element including this kind of representation is known as the traction singular quarter-point element.

The first and second mode stress intensity factors can be defined by the following limits (Fig. 11)

$$K_I = \lim_{x_1\to 0} \sqrt{2\pi x_1} \, \sigma_{22}$$

$$K_{II} = \lim_{x_1\to 0} \sqrt{2\pi x_1} \, \sigma_{12} \tag{187}$$

If the boundary discretization is done in such a way that the first interface element from the crack tip has $\theta = 0$ and this element is a singular quarter-point boundary element, then for this element, $\bar{r} \equiv x_1$, $p_1 \equiv \sigma_{12}$, $p_2 \equiv \sigma_{22}$ and the nodal values for the tractions at the tip node k are:

$$\bar{p}_1^k = \lim_{\bar{r}\to 0} p_1^k (\bar{r}/l)^{1/2} = \lim_{x_1\to 0} \sigma_{12} (x_1/l)^{1/2}$$

$$\bar{p}_2^k = \lim_{\bar{r}\to 0} p_2^k (\bar{r}/l)^{1/2} = \lim_{x_1\to 0} \sigma_{22} (x_1/l)^{1/2} \tag{188}$$

Thus, the stress intensity factors coincide with the traction nodal values except for a constant and may be computed directly by the boundary element code , i.e.

$$K_I = \bar{p}_2^k \ (2\pi l)^{1/2}$$

$$K_{II} = \bar{p}_1^k \ (2\pi l)^{1/2} \tag{189}$$

Martínez and Domínguez,[55] in their static analysis, showed how the use of the traction nodal values of the singular element at the crack tip for SIF computation (eqn 215) is substantially less sensitive to the discretization than any of the displacement correlation procedures. The same conclusion can be drawn in the context of dynamic SIF computations. An example of this will be given in subsection 3.5 where the transient dynamic K_I for a center cracked plate is computed, not only by using eqn (189), but also by using two displacement correlation formulae. One of these correlation formulae, taken from the FEM and used in the B.E. context by Blandford *et al.*,[12] is based on equating the $r^{1/2}$ term of the displacement eqn (178) and the corresponding term of the crack displacement representation given by eqn (182). By doing so, the K_I and K_{II} factors can be written in terms of the displacement values of the two nodes of the crack closer to the tip. Another correlation formula can be obtained by simple identification of the u_1 and u_2 values given by eqn (178) for the point of the crack surface at 1/4 from the tip and the corresponding computed value at the mid-node of the quarter-point element.

3.4 Time-harmonic SIF computations

3.4.1 Diffraction of waves by a crack in the complete plane

The stress and displacement fields caused by P or SV plane waves whose direction of propagation forms an angle γ with a crack (Fig. 14) in an infinite two-dimensional domain, can be represented in terms of two potentials (see subsection 1.4):

$$\phi = \phi_0 \exp\left\{ -i\frac{\omega}{c_P} \left(x_1 \ \cos\gamma + x_2 \ \sin\gamma + c_P \ t\right)\right\} \tag{190}$$

$$\Psi = 0$$

for P-waves, and

$$\phi = 0$$

$$\Psi = \Psi_0 \exp\left\{ -i\frac{\omega}{c_S} \left(x_1 \ \cos\gamma + x_2 \ \sin\gamma + c_S \ t\right)\right\} \tag{191}$$

for SV-wave; where γ is the angle between the wave propagation direction and the x_1-axis; c_P and c_S are the P and S-waves velocity, respectively; ω is the frequency and ϕ_0, Ψ_0 are amplitudes. Displacements and stresses are easily obtained by differentiation of these potentials.

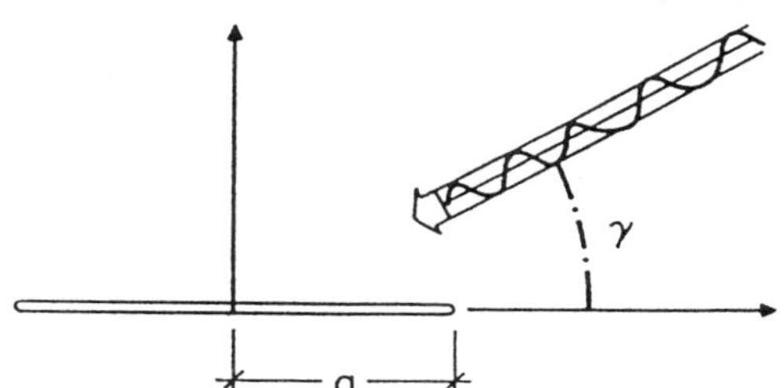

Figure 14: Waves impinging on a crack

The analysis of the diffraction of waves by a crack with length $2a$ in an infinite x_1x_2-plane is carried out by the superposition of two problems: one, the incident field in the uncracked domain; and the other, the cracked domain loaded on the crack faces by tractions equal and opposite to those that appear in those surfaces in the uncracked plane, (scattered field):

$$u_i^t = u_i^{in} + u_i \qquad \sigma_{ij}^t = \sigma_{ij}^{in} + \sigma_{ij} \tag{192}$$

where u_i^{in} and σ_{ij}^{in} are the displacement and stress components for the incident field, while u_i and σ_{ij} correspond to the scattered field. Only the second problem has to be solved. The incident field in the complete plane is known. Since there are not infinite values of the stress in the uncracked plane, the stress intensity factors of the original diffraction problem are the same as in the second problem (scattered field).

To avoid the problem of having the same B.E. equation for two different points, one on each face of the crack, a boundary that divides the domain into two parts along the x_1-axis is introduced. This boundary extends to infinity. However, the discretization is truncated at a distance to the crack tip equal to fifteen times the half-length of the crack. This can be done because elements that are far from the crack have very little effect on the solution near the tip since both the scattered field and the fundamental point load solutions satisfy the regularity conditions. Figure 15 shows the elements used in one half of the boundary. The problem is always decomposed into its symmetric and skewsymmetric parts. The two elements to which the node at the crack tip belongs are singular quarter-point elements, whereas all the others are standard quadratic elements.

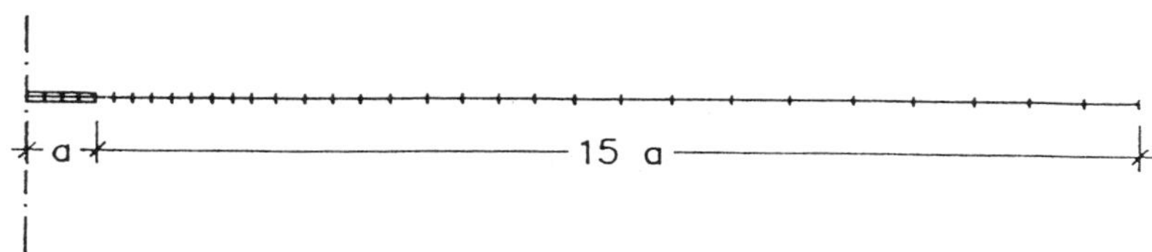

Figure 15: Boundary discretization

Stress intensity factors for P and SV waves with several angles of incidence have been computed. The values of K_I and K_{II} depend on the incident angle γ and the Poisson's ratio of the elastic medium. Figure 16 shows the moduli of the stress intensity factors for P-waves normalized with respect to their corresponding static values. Results are plotted vs. the dimensionless frequency $\omega a/c_P$ for a Poisson's ratio 0.25. Plane strain is assumed. In the figure it can be seen how K_I is maximum for $\gamma = 90^o$ and K_{II} for $\gamma = 45^o$. In all cases the values tend to rise first, reach a maximum, and then fall.

The dynamic stress intensity factor can be as much as 1.3 times the static one. The values computed using the BEM with singular quarter-point elements are compared with those obtained by Chen and Sih[18] using a method derived by Copson[22] to solve a system of integral equations. As it can be seen in the figure, the agreement is very good.

The values of K_I and K_{II} for SV-waves are shown in Fig. 17. As it could be expected, the maximum value of K_I is now for $\gamma = 45^o$ and the maximum of K_{II} for $\gamma = 90^o$. The agreement with Chen and Sih[18] is again very good.

3.4.2 Crack near a free surface

Achenbach *et al.*[2,3] and Keer *et al.*[44] studied the scattering of harmonic waves by edge or sub-surface cracks. Their formulation leads to a system of integral equations which is solved numerically. The fundamental solution of these integral equations corresponds to dislocations in the half-plane and is calculated numerically as a combination of Hankel functions and contour integrals.

In the following, the BEM with SQP elements is used to compute stress intensity factors for a horizontal subsurface crack subjected to harmonic excitation. The problem has been taken from Keer *et al.*[44] for comparison. The frequency domain dynamic formulation of the BEM with the complete space fundamental solution is used. Thus, an integral equation has to be solved numerically, but, the kernels of the integral are much simpler and easier to compute than those of the dislocations in the half-plane. On the other hand, the free surface boundary conditions are not satisfied by the fundamental solution and have to be enforced by discretization of the free surface into boundary elements for which zero tractions are prescribed. However, the elements far from the crack have very little effect on the stress field near the crack and accurate representation of the effect of the free surface can be obtained by discretizing a limited zone close to the crack. Due to the nature of the Boundary Element Method, the fact of extending the discretization only to a finite part of the free surface does not introduce undesired artificial reflections.

Figure 18 shows the boundary element discretization used for the problem at hand. The free surface discretization and the internal boundary extend to a distance from the crack of $15a$ along the x_1 axis. All the elements are quadratic and those at the crack tips are quarter-point singular. The incident field is a uniform tension suddenly applied on $x_2 = 0$ which results in uniform tension on the crack faces. The Poisson's ratio is 0.3. Figure 19 shows the values of K_I (magnitude of the ratio of the dynamic and static mode I stress intensity factors) vs. the parameters d/a and $\omega d/c_R$, with c_R as the Rayleigh wave velocity. The results are in good agreement with those taken from the figures of Keer *et al.*[44] The same comparison is made in Fig. 20 for

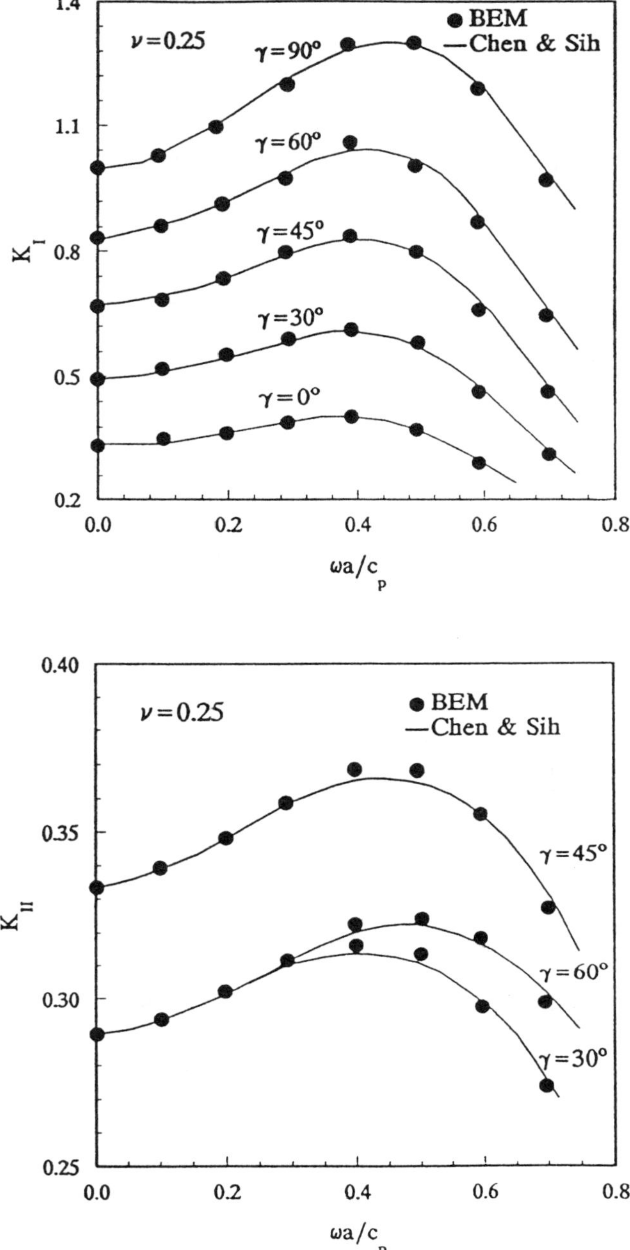

Figure 16: Ratio of dynamic and static stress intensity factors for incident P-waves

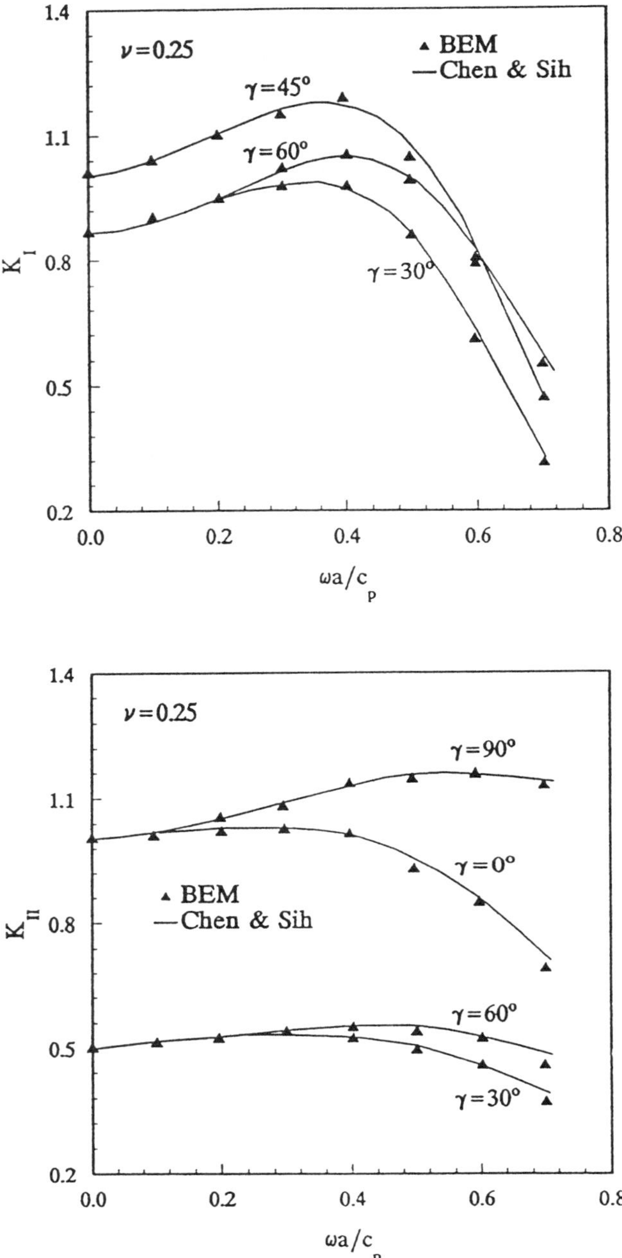

Figure 17: Ratio of dynamic and static stress intensity factors for incident SV-waves

the mode II stress intensity factors. A more extensive treatment of the computation of SIF using the frequency domain B.E. formulation in combination with the SQP element may be found in the work by Chirino[19] and Chirino and Domínguez.[20]

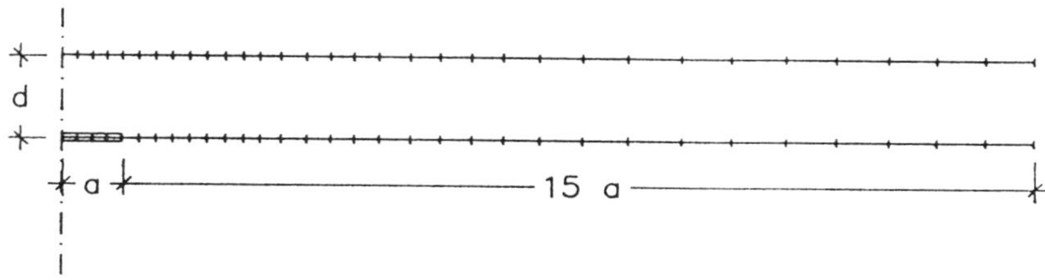

Figure 18: Boundary element discretization for a subsurface crack

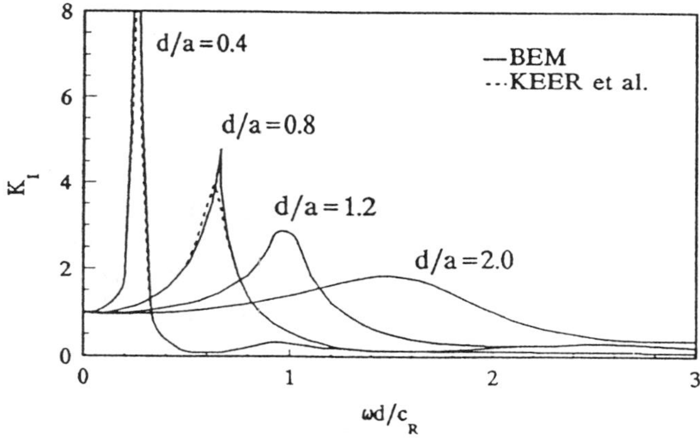

Figure 19: Mode-I SIF for a subsurface crack under vertically incident P-waves

3.5 Transient dynamic SIF computations. Step-by-step approach

The time domain formulation is now used to solve transient dynamic fracture mechanics problems. In the following, three different problems of transient dynamic fracture mechanics are studied. The results clearly show the accuracy and robustness of the approach.

3.5.1 Center cracked plate under Mode-I transient load

A rectangular plate with a central crack (see Fig. 21) is analyzed. The plate is loaded dynamically along opposite boundaries at $t = 0$ by uniform tractions σ with Heaviside-function time dependence. This problem was solved by Chen[16] using finite differences and has been frequently used as a reference to validate other methods. The material is linear elastic with properties as given by Chen: shear modulus $\mu = 76923$ GPa,

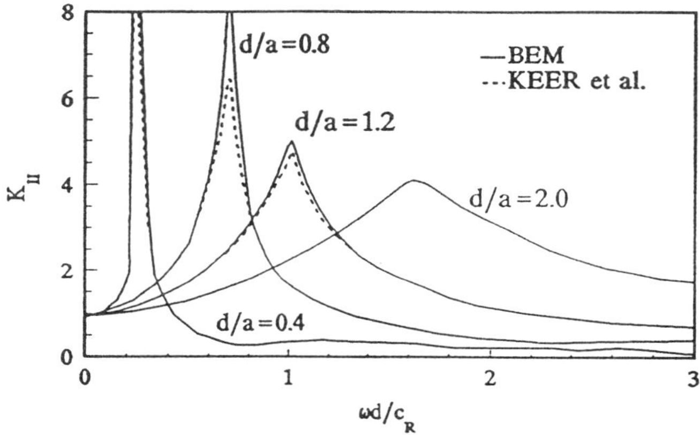

Figure 20: Mode-II SIF for a subsurface crack under vertically incident SV-waves

Poisson's ratio $\nu = 0.3$ and density $\rho = 5000$ Kg/m^3. The crack length is $2a = 4.8$ mm.

Because of the symmetry only one quarter of the plate is discretized, as shown in Fig. 21. Two equal length elements are used to discretize one half of the crack ($l/a = 0.5$). The two elements that contain the tip are SQP. The time step $\Delta t = 0.32$ μs is such that P-waves travel 2.4 mm per time step. This corresponds approximately to twice the length of the smallest elements and to one half of the largest. An analysis of the optimum time step and the effects of its changes may be found in the work of Domínguez and Gallego.[28] Figure 22(a) shows the K_I SIF normalized by $\sigma(\pi a)^{1/2}$, computed directly from the traction nodal values at the tip by using eqn (189), versus time. The results are in excellent agreement with those reached by Chen. To check the possibility of computing the SIF using a displacement correlation formula, two of those formulae were used. In the first case, the SIF is computed using the crack opening displacement (COD) of the first node in the crack. The second formula makes use of the COD of the first two nodes. The SIF computed by those two displacement correlation formulae are denoted by K_{u1s} and K_{u2s}, respectively, and that SIF computed directly from the traction nodal value at the tip is denoted by K_t. Figures 22(b) and 22(c) show K_{u1s} and K_{u2s}, respectively. The agreement with Chen's results is also very good.

In order to evaluate the spatial discretization sensitivity of the different approaches, the length of the elements adjacent to the tip was varied from $0.2a$ to $0.9a$ while the next elements were varied from $0.8a$ to $0.1a$ and the rest of the mesh remained unchanged. Figure 22(d) represents the integrals of the square of the difference between the curves computed by the three BEM approaches and Chen's curve. The integrals are represented versus the length of the tip elements. It can be seen from Fig. 22(d) that the approach that uses the traction nodal values at the tip to compute the SIF (K_t) is the less sensitive to the spatial discretization, and hence the

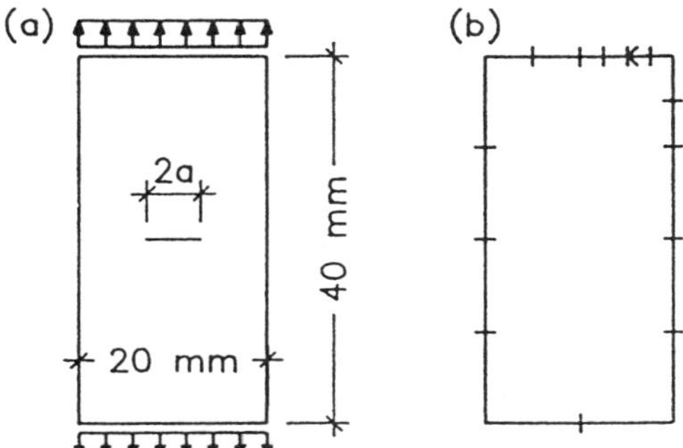

Figure 21: Center cracked plate: (a) geometry; (b) boundary discretization for one quarter of the plate.

most reliable for problems for which the solution is not known. As a reference for comparison, Figs 23(a) and 23(b) show results obtained by Murti and Valliappan[60] for the same problem using quarter-point finite elements and a similar discretization over the boundaries and crack surfaces. Consistent and lumped mass matrices were used by Murti and Valliapan to compute the results in Figs 23(a) and 23(b), respectively. The results of Fig. 23 are probably the best shown by those authors. Other FEM discretizations produced a more oscillatory behavior of the solution.

3.5.2 Rectangular plate with central inclined crack

The plate shown in Fig. 24(a) contains a central crack inclined 45°. The plate is under the effects of uniform tractions applied at $t = 0$ along opposite boundaries with a Heaviside-function time dependence. The problem was studied by Chen and Wilkins[17] using the FDM and by Murti and Valliappan[60] using the FEM. The material properties are the same as in the previous example and the BE mesh used is that shown in Fig. 24(b), where all the elements are regular quadratic elements, except those that have a node at the crack tip, which are SQP. The domain is divided into two parts as shown in Fig. 24(b). The time step is $\Delta t = 0.35\mu s$.

Figures 24(c) and 24(d) show the mode-I and mode-II SIF, respectively normalized by $\sigma(\pi a)^{1/2}$ versus time. The SIF have been computed from the SQP elements nodal values of the tractions at the crack tips. The results are compared with those of Murti and Valliappan[60] and also with the FDM results of Chen and Wilkins.[17] From observation of Fig. 24 it can be said that the BE results are in good general agreement with the FE results but not with the FD results. It is reasonable to say, in agreement with Murti and Valliappan, that the Chen and Wilkins[17] results are doubtful, in particular if one takes into account that the latter authors, using an averaging technique, computed approximate static values of K_I and K_{II} of 0.65 and -0.27 respectively, whereas the exact solutions (Murakami[58]) are 0.594 and -0.541.

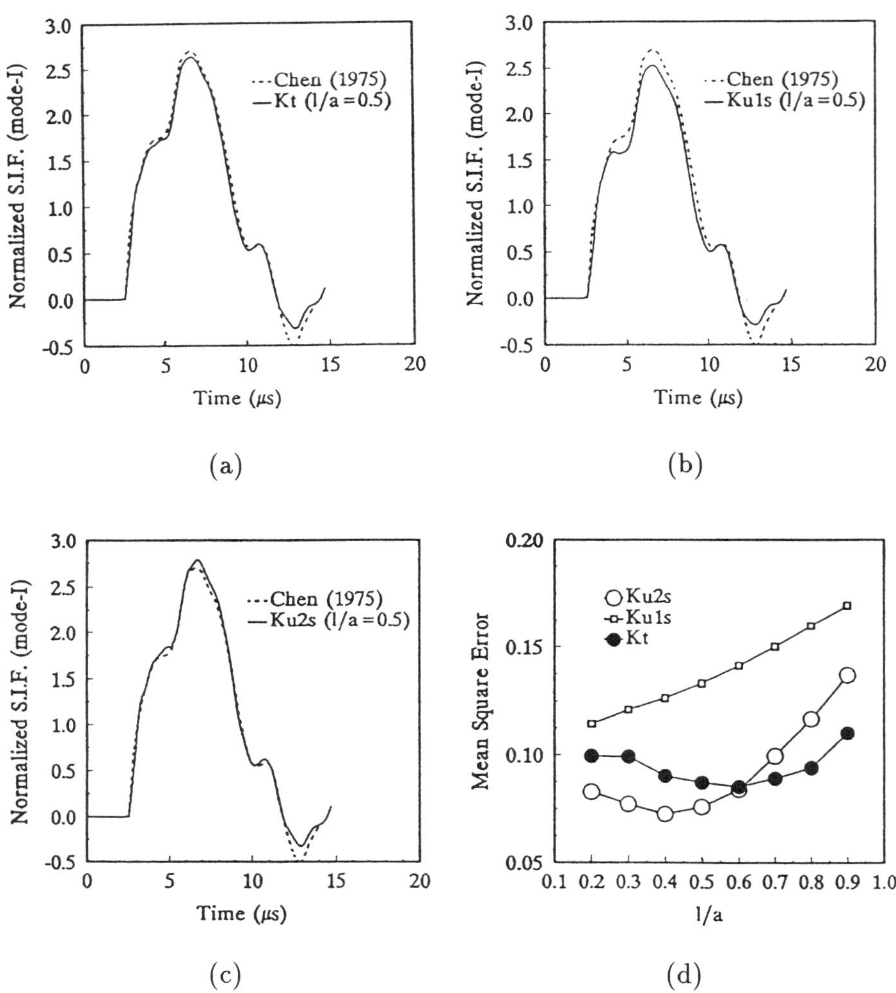

Figure 22: (a) Normalized SIF computed from the traction nodal value at the tip. (b) Normalized SIF computed from one nodal COD. (c) Normalized SIF computed from two nodal COD. (d) Integral of the square of the difference with Chen's results for the three approaches versus the length of the crack tip SQP elements.

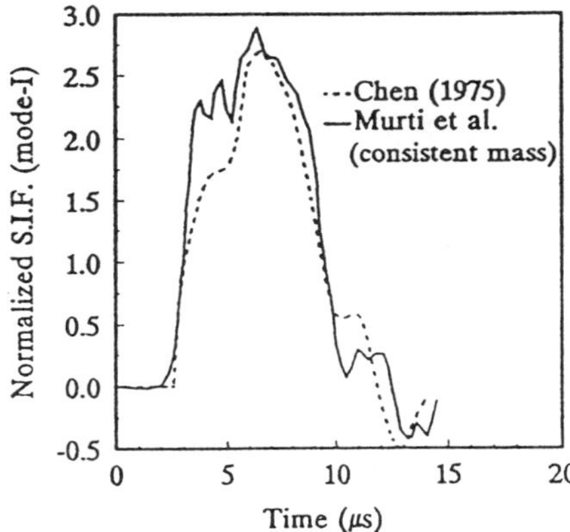

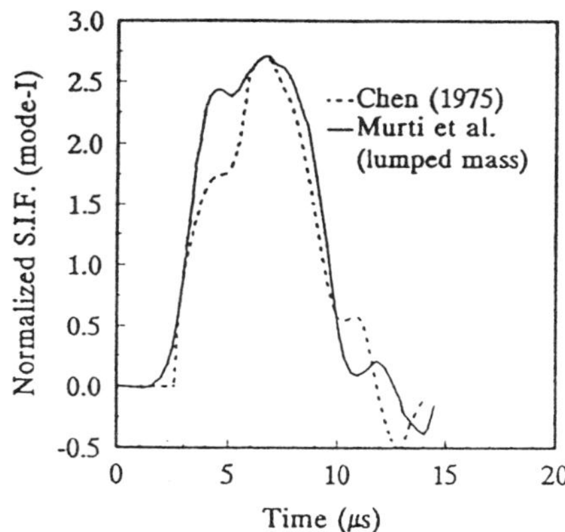

Figure 23: Finite element results for mode-I SIF of center cracked plate (from Murti and Valliappan, 1986): (a) consistent mass; (b) lumped mass.

3.5.3 Rectangular plate with an inclined surface crack

The problem shown in Fig. 25(a) contains a crack inclined 45° from the boundary of a rectangular plate loaded by a uniform tension σ applied at $t = 0$ with a Heaviside-function time dependence. This problem was studied by Aoki *et al.*[7] using singular finite elements and also by Murti and Valliappan[60] using a quarter point finite element. The material properties are: shear modulus $\mu = 29.4$ GPa, Poisson's ratio $\nu = 0.286$ and density $\rho = 2450$ Kg/m^3. Figure 25(b) shows the BE discretization, where once again the elements that have nodes at the tip are SQP. The domain is divided into two parts and the size of the time step is $\Delta t = 0.4\ \mu s$.

The computed mode-I and mode-II SIF are shown versus time in Figs 25(c) and 25(d), respectively, in comparison with those computed in the above-mentioned FE studies. As it can be seen from the figures, the BE results agree with the previous FE results. The agreement with the results obtained by Murti and Valliappan[60] is very good. The results arrived at by Aoki *et al.* show some differences for the last part of the curves and in particular do not present the peak for K_{II} at $t = 11.5\ \mu s$ that appears for the other two approaches. An extensive treatment of transient dynamic SIF computations using the time domain formulation of the BEM in combination with the SQP element may be found in the work by Gallego[35] and Domínguez and Gallego.[29]

3.6 Transient dynamic SIF computations. Frequency domain approach

Transient dynamic SIF can be computed using the frequency domain formulation as in subsection 3.4 in combination with the Fast Fourier Transform (FFT).

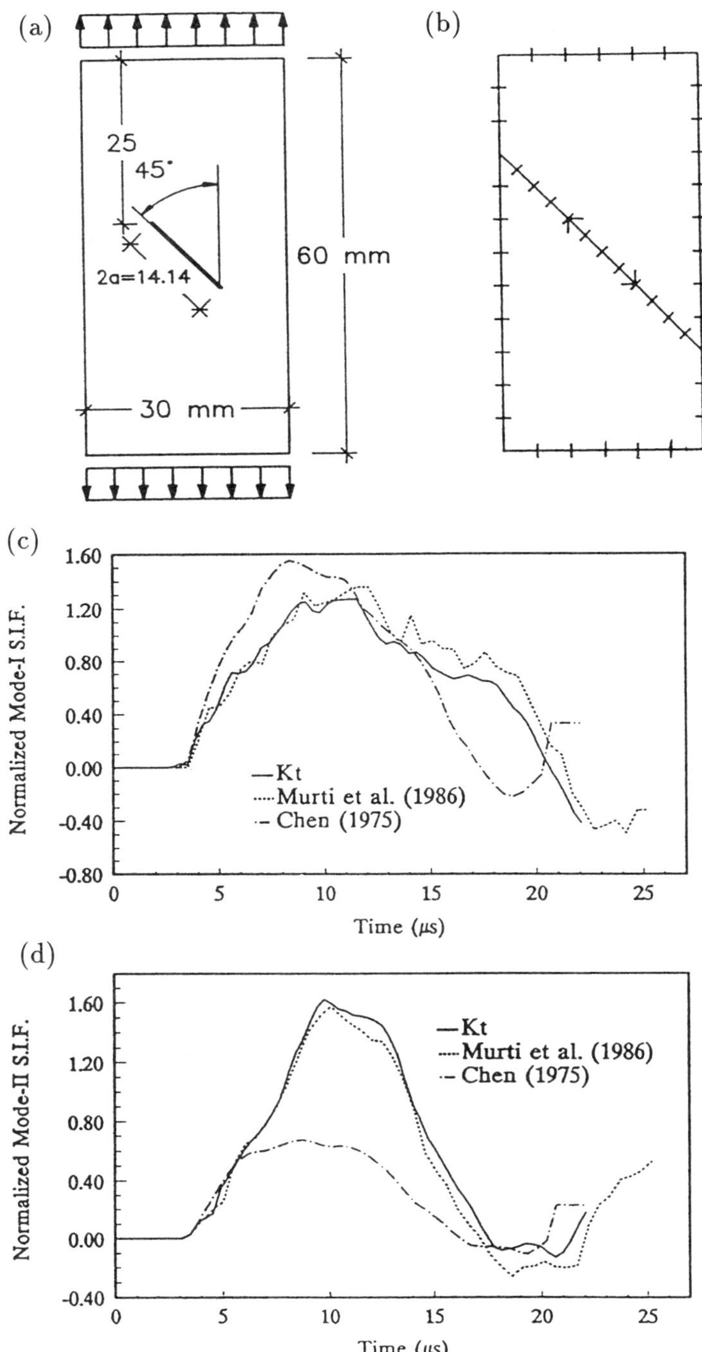

Figure 24: Plate with a central inclined crack under a uniform step load: (a) geometry; (b) boundary discretization; (c) normalized mode-I SIF versus time; (d) normalized mode-II SIF versus time.

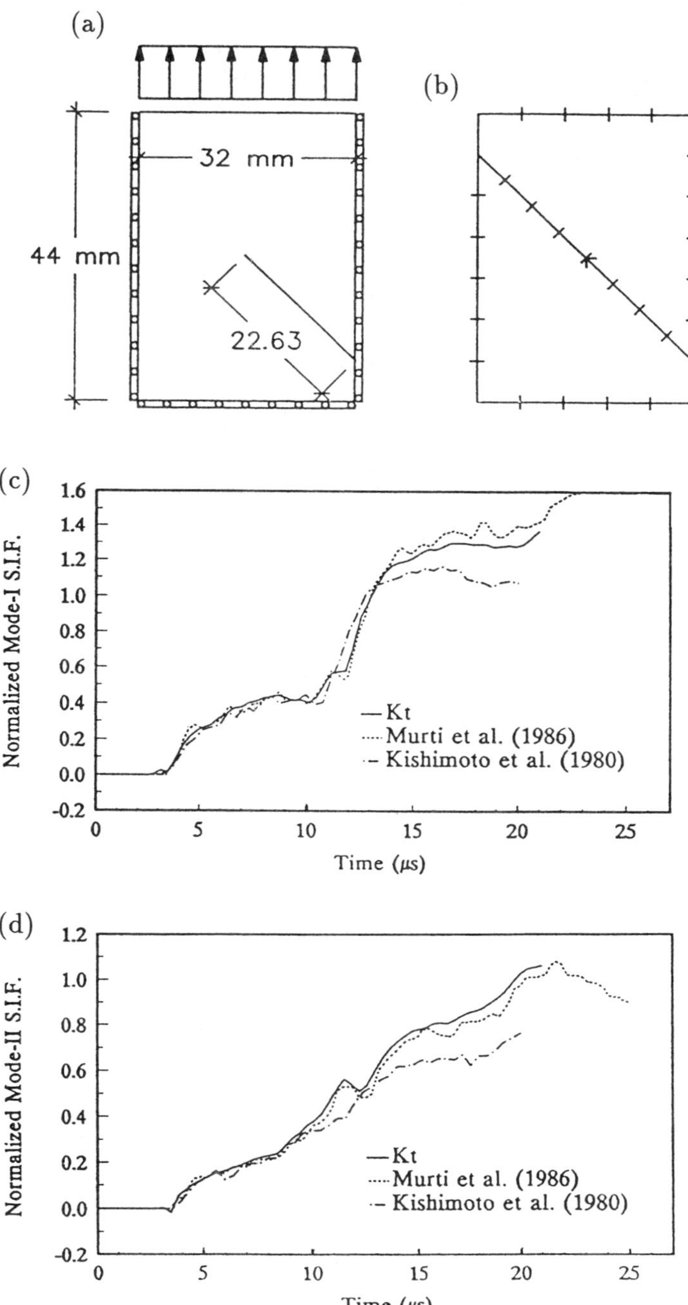

Figure 25: Plate with a crack inclined from the boundary: (a) geometry; (b) boundary discretization; (c) normalized mode-I SIF versus time; (d) normalized mode-II SIF versus time.

3.6.1 Crack in the complete plane

Going back to the problem of diffraction of waves by a crack in an infinite medium (Fig. 14), a normal incident P-wave with a step function time variation is considered. Results for transient stress intensity factors may be obtained by using the results of the frequency analysis carried out previously and the FFT algorithm. Thau and Lu[81] computed values of the stress intensity factors for this problem using a generalized Weiner-Hopf technique, but they were only able to study the problem until the time when a diffracted P-wave on reaching the opposite edge is rediffracted and comes back to the original edge. A comparison of the boundary element results with those of Thau and Lu[81] may be seen in Fig. 26 where K_I is shown vs. dimensionless time. The K_I SIF reaches a value which is 30% higher than the static one. Sih *et al.*[76] studied the same problem using integral transforms coupled with the technique of Cagniard. They obtained values of K_I for longer periods of time, but because of numerical errors they could not represent the discontinuity in the slope, shown by the BEM and by Thau and Lu, which is a consequence of the arrival of the Rayleigh waves diffracted by the other tip. A comparison of the boundary element results with those obtained by due Sih *et al.*[76] is shown in Fig. 27.

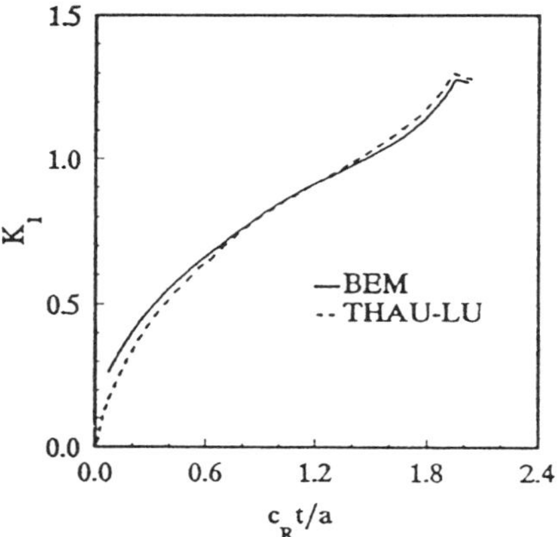

Figure 26: Ratio of transient and static mode I stress intensity factor for crack in the complete plane under a step pressure.

3.6.2 Center cracked plate under Mode-I transient load

The same centrally cracked plate under the same transient load that was studied in subsection 3.5 using the time domain formulation is now analyzed by first doing a frequency domain analysis and then a Fourier transformation. In addition to the purely elastic material, internal damping is considered by means of a complex shear modulus $\mu_c = \mu(1 + 2\beta i)$. Values of the damping factor $\beta = 0.01$, $\beta = 0.025$ and $\beta = 0.05$ are assumed. The load is a uniform tension applied on two opposite sides. Figure 28 shows one quarter of the plate with the boundary divided into three different

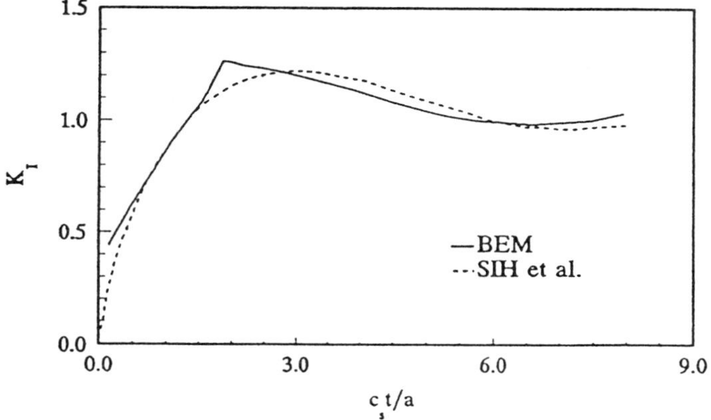

Figure 27: Ratio of transient and static mode I stress intensity factor for crack in the complete plane under a step pressure.

zones: zone A ($a + L_1 = 2a$), zone B (L_2) and zone C ($L_3 + L_4 + L_5$). The boundary discretization was made according to the following rules: the length of the elements should be $l \leq \lambda/10$ in zone A, $l \leq \lambda/5$ in zone B, and $l \leq \lambda/3$ in zone C, with λ being the length of the S-waves in the plate material.

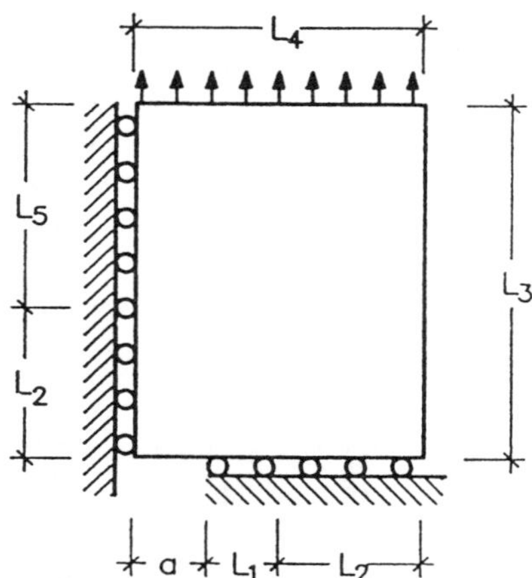

Figure 28: Model for center cracked plate

The magnitude of the ratio of the dynamic and static mode I stress intensity factor for four different values of the internal damping, is shown vs. frequency in Fig. 29. Several peaks of resonance can be seen in the figure. The peaks are shifted and damped with increasing values of β. The first peak is 20% below the first natural frequency of the uncracked plate. The transient stress intensity factor for a uniform traction on two opposite sides that varies with time as a Heaviside step function is now computed using the Fig. 29 frequency response function and the FFT algorithm. The magnitude of the mode I stress intensity factor normalized by $\sigma(\pi a)^{1/2}$ is represented for the first 12 μs in Fig. 30. Values for damping factors of 1%, 2.5% and 5% are

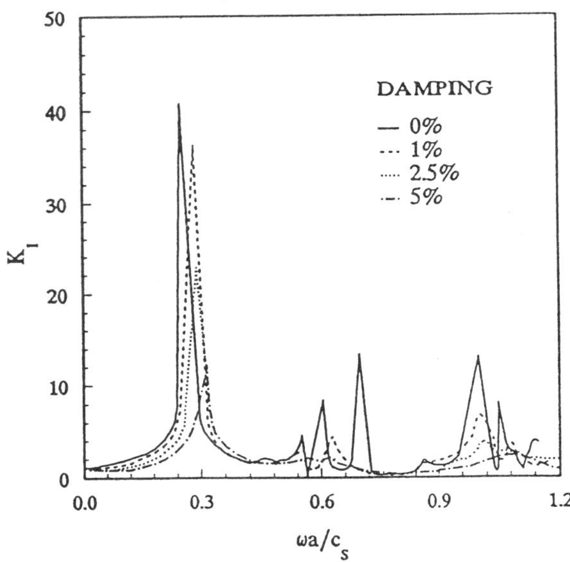

Figure 29: Ratio of dynamic and static mode I stress intensity factor for center cracked plate under uniform harmonic traction.

plotted. In this figure, results are compared with those obtained by Chen[16] using finite differences, and with Baker's[9] for the semi-infinite crack under pressure in an infinite plate. The latter are valid for times prior to the arrival of the first wave diffracted by the other tip (R_1). Both consider purely elastic material. It can be said that the results are in good agreement with Chen's and Baker's and that, as it could be expected, the existence of internal damping reduces the maximum value reached by the stress intensity factor. This reduction becomes more important as the internal damping increases. B.E. results for the purely elastic material are not shown because they present spurious oscillations that can be explained by numerical errors in the values of the peaks of the frequency response function for zero damping and in the subsequent Fourier transformation process. The use of filters could solve the problem, but, since the internal damping makes the spurious oscillations disappear, it is easier to introduce a small amount of damping in the material.

A comparison between the time domain and the frequency domain computation of transient SIF shows that only purely elastic materials can be studied with the time domain approach whereas internal damping is easily included in the frequency domain procedure. On the other hand, crack problems in purely elastic finite regions tend to produce numerical difficulties in the Fourier inversion process. The number of elements required for the frequency domain analysis is usually higher than that of the time domain because the analysis of high frequencies requires elements that can represent the corresponding short waves. However, due to the greater complexity of the time domain fundamental solution, this kind of analysis is more time-consuming than the frequency domain one when moderately large periods of time have to be studied. The same can be said about the memory requirement because of the need to store for all the matrices of the previous time steps in the time domain approach. Nevertheless, in the cases of short periods of time and rapid load variation, the time

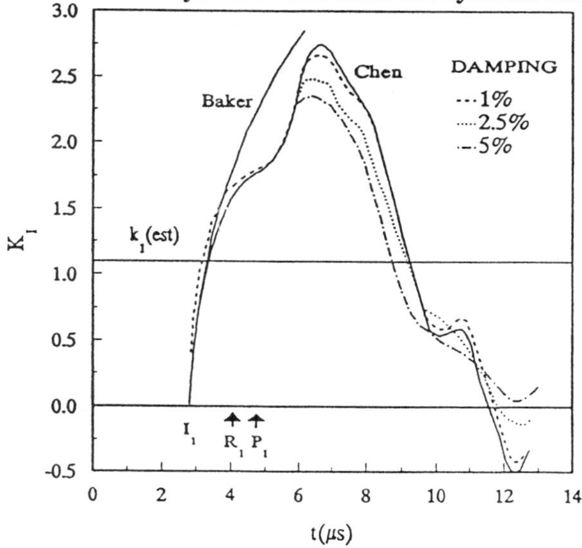

Figure 30: Ratio of transient mode I stress intensity factor and $P(t)$ $(\pi a)^{1/2}$ for center cracked plate under step uniform traction.

domain analysis can be less time-consuming.

The above results for transient SIF computed using a frequency domain analysis and the FFT have been taken from the works of Chirino[19] and Chirino and Domínguez.[20]

3.7 Transient dynamic SIF computations. Dual reciprocity approach

The so-called Dual Reciprocity Method is an alternative to the dynamic analysis of linear elastic problems.[68] This method is used here in combination with the SQP element to solve linear elastic fracture mechanics problems.[73]

Only very little changes are needed for SIF computations when one starts from the general quadratic element Dual Reciprocity formulation and implementation. The SQP elements can be included as part of the BE discretization. The only special attention for those elements is that the corresponding terms of the **G** and **M** matrices are computed using the special shape functions of those elements as done for static SIF computation as well as for the other dynamic approaches.

For the sake of comparison, the same examples already studied in the previous sections are now analyzed using the Dual Reciprocity approach.

3.7.1 Center cracked plate under Mode-I transient load

The same centrally cracked plate under the same loading conditions of subsections 3.5 and 3.6 is studied. Figure 31(a) shows the quadratic boundary element discretization of one quarter of the plate. The discretization is the same used in subsection 3.5 for the time domain analysis. The two elements next to the crack tip are SQP. Two meshes including internal points have also been used in order to check the accuracy of the results. One includes 6 regularly distributed internal points (Fig. 31b) and the other 12 (Fig. 31c). A time step of 0.19 μs has been chosen. By doing so the parameter $\beta = c_P \Delta t/l$, l being the length of the elements, remains in a range between 0.56 and 2.32 which was proved to have given stable and accurate results in previous analyses.[73]

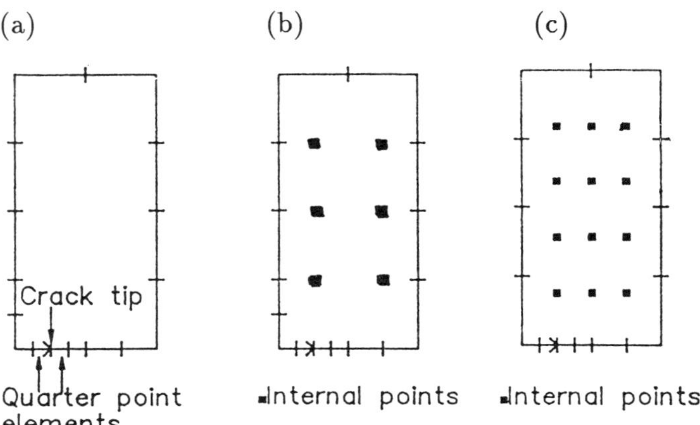

Figure 31: One quarter of a center cracked plate: (a) boundary discretization; (b) boundary discretization and six internal points; (c) boundary discretization and twelve internal points.

Figure 32 shows the normalized K_I SIF versus time. The results are compared with Chen's and with the time domain B.E. results already shown in Fig. 22. It can be seen from the figure that the computed results, including six internal points, are in very good agreement with the time domain and Chen's results. The inclusion of more internal points does not increase the accuracy. The results corresponding to the B.E. discretization without internal points show significant differences with the other solutions in the upper zone of the time range represented.

3.7.2 Rectangular plate with central inclined crack

Again the same problem of the time domain analysis is studied using the Dual Reciprocity approach. The discretization is the same as in the time domain study. Two discretizations including internal points (Fig. 33) are also tested. The time increment considered is $\Delta t = 0.3\ \mu s$ which yields values of the time step parameter β between 0.88 and 1.24.

The normalized values of K_I and K_{II} are shown versus time in Fig. 34. Results for the three meshes are compared with those obtained by Murti and Valliappan[60] using F.E. and with the time domain results of subsection 3.5. To obtain accurate results, it is necessary to include a few internal points (5 per subregion). However, an increase in the number of internal points does not produce noticeable changes in the accuracy of the computed SIF.

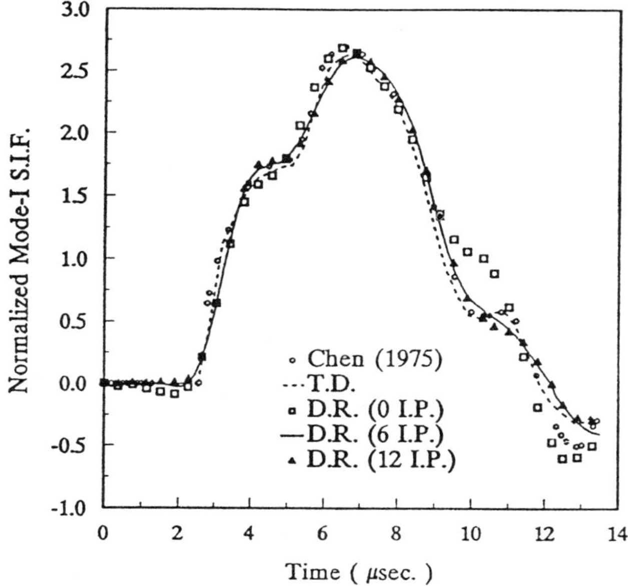

Figure 32: Center cracked plate under uniform step load. Normalized mode-I SIF

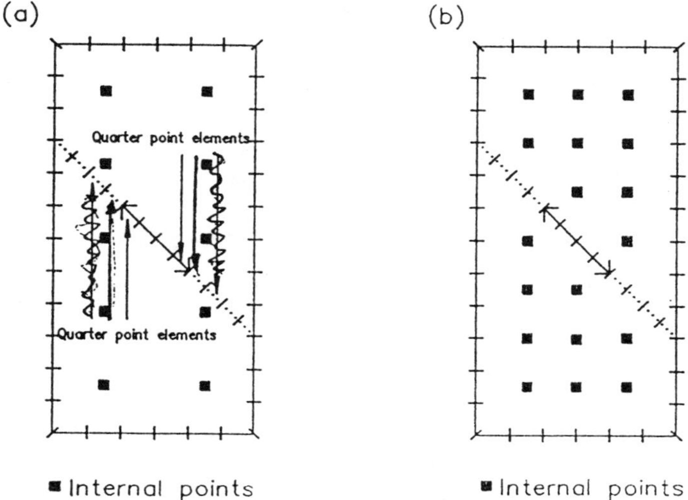

Figure 33: Plate with central inclined crack: (a) B.E. discretization with five internal points in each subregion; (b) B.E. discretization with nine internal points in each subregion.

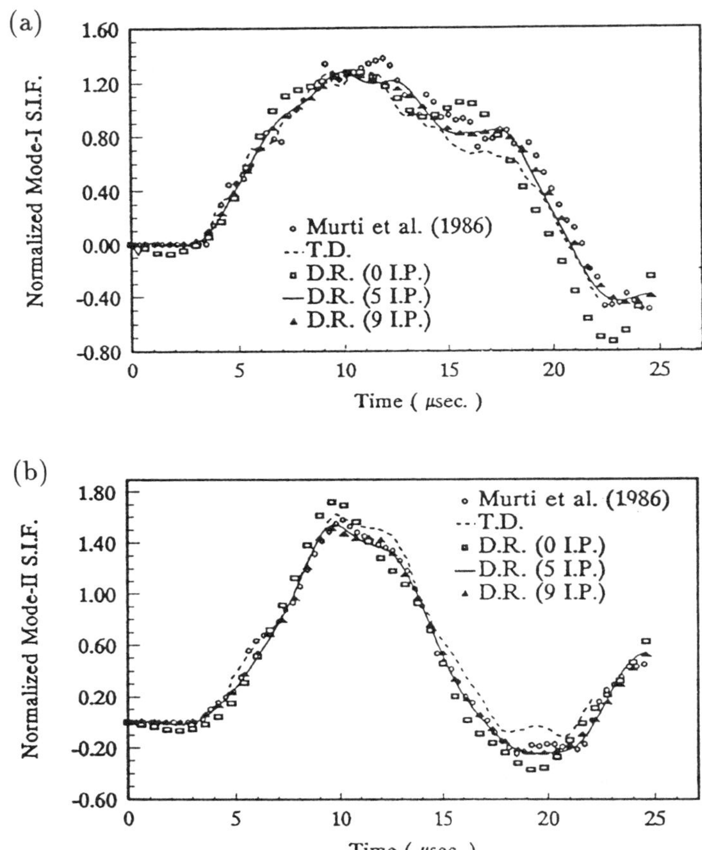

Figure 34: Plate with a central inclined crack under uniform step load: (a) Normalized mode-I SIF; (b) Normalized mode-II SIF.

3.7.3 Rectangular plate with an inclined surface crack

The problem of Fig. 25 has also been studied using the Dual Reciprocity approach. Once again the discretization is the same used for the direct time domain B.E. analysis. Two discretizations including internal points (Fig. 35) are also tested as in previous examples. The time increment considered is $\Delta t = 0.4$ μs which yields values of the parameter β between 0.79 and 1.09.

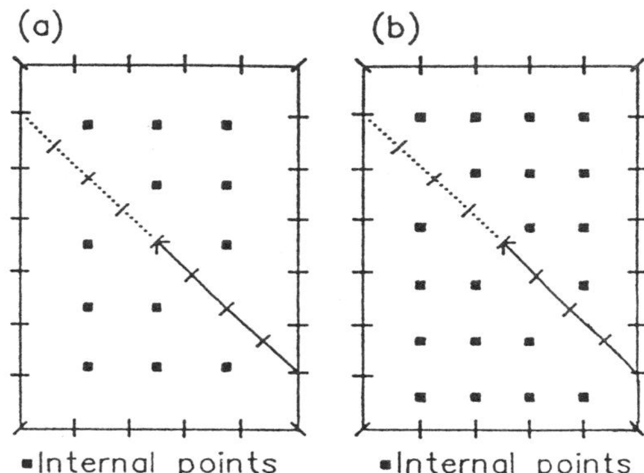

Figure 35: Plate with a crack inclined from the boundary: (a) B.E. discretization with six internal points in each subregion; (b) B.E. discretization with ten internal points in each subregion.

The computed values of K_I and K_{II} are shown versus time in Fig. 36. Results are again compared with the F.E. results reached by Murti and Valliappan[60] and with those of the previous time domain B.E. analysis. The Dual Reciprocity results again show an increase of the accuracy when some (six) internal points are considered. No substantial improvement is obtained by adding more internal points. The results of Fig. 36 indicate a reasonably good agreement between the time domain and the FEM. However, some of the quick changes in the slope and peaks shown by those results are not reproduced by the Dual Reciprocity ones.

It can be said from the comparison of the Dual Reciprocity approach and the time domain approach for transient dynamic SIF computations that the computational effort when using Dual Reciprocity is less than when using the direct time domain analysis; however, there is an additional mesh definition effort required (number and location of the internal points) and the accuracy for complicated problems and similar discretizations can be expected to be smaller when using Dual Reciprocity than when using the direct time domain approach.

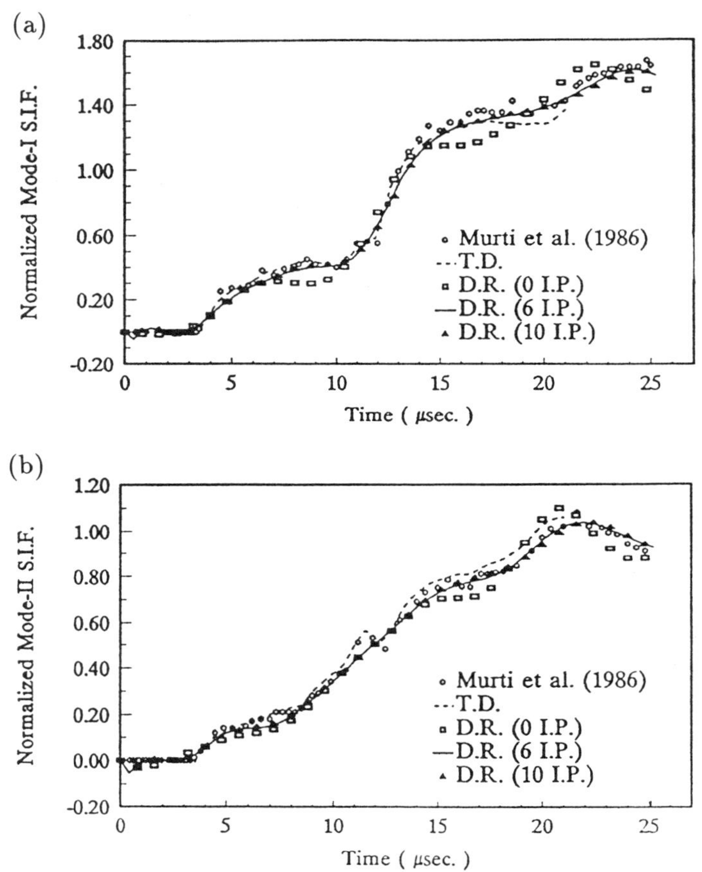

Figure 36: Plate with a crack inclined from the boundary: (a) normalized mode-I SIF; (b) normalized mode-II SIF.

3.8 Dynamic crack propagation

Dynamic crack propagation is devoted to the study of solid bodies containing cracks that propagate under conditions for which the inertia effects are important. A number of analytical solutions for crack propagation problems exist. However, these solutions are limited to simple loadings and infinite domains (Freund[31,32,33]). Apart from experimental studies, numerical methods are today the only alternative for the analysis of cracks propagating in finite bodies under general loading conditions.

In the past two decades, the application of the finite element method (FEM) to dynamic crack propagation has seen an important advance. Interesting reviews of the FEM techniques for dynamic crack propagation can be found in the papers by Aoki *et al.*,[6] Atluri and Nishioka,[5] and Beskos.[11] The different approaches are based on one of the following two concepts: a stationary mesh, which includes a 'node-release' mechanism, or a moving mesh which normally includes a singular element.

The BEM, which appeared recently as an alternative for elastic fracture mechanic problems, seems to be a better choice than the FEM for elastodynamic fracture mechanics because the discretization is restricted to the boundary surface and the concept of singular elements is simplified. In particular, when dealing with dynamic crack propagation, the remeshing process is conceptually much simpler in the BEM than in any domain technique.

In the following, the direct time-domain formulation of the BEM is used in combination with a singular element to study dynamic crack propagation in finite and infinite elastic bodies. To do so, a moving singular element and a remeshing technique developed by Gallego and Domínguez[36] is used. These ideas are applied in the general context of time domain B.E. with shape functions for space and time discretization. Integral equation formulations have been applied to problems of propagating cracks in elastic bodies by other authors. However, all these studies, directed towards the simulation of earthquake sources, are limited to infinite domains and use either the BEM in conjunction with a node release mechanism (Das,[24] Das and Kostrow[25]) or are restricted to particular formulations related to the BEM (Burridge,[14] Bergkvist [10]).

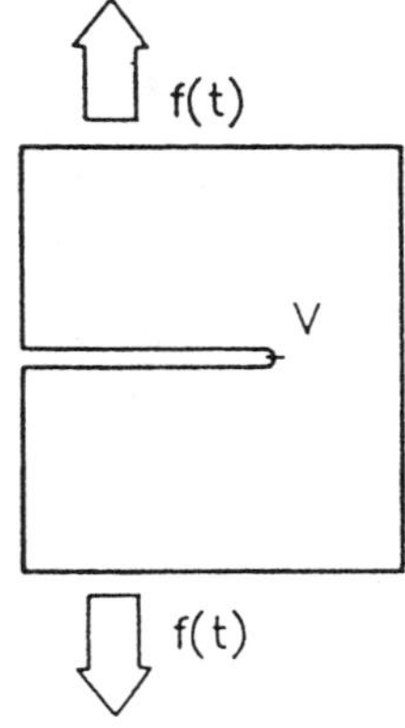

Figure 37: Mode-I dynamic crack propagation example

3.9 Moving singular element

Consider a mode-I dynamic crack propagation problem as shown in Fig. 37. A boundary is introduced along the crack, following the known direction of propagation. Because of the symmetry, only one half of the domain is analyzed. Assume that the crack propagates at a certain speed c, and that at time t the position of the crack tip and the discretization of the part of the boundary that contains one side of the crack is as shown in Fig. 38(a). Two elements, one before and one after the tip, are SQP. After a time increment Δt, the crack tip has moved $c\Delta t$ (Fig. 38b). As the crack tip advances from left to right, the size of the elements to the left is increased and that of the elements to the right is decreased. The SQP elements are excluded from this adjustment and their size remains the same along the crack propagation process. The translation of the crack tip for each time step can take any value and is not related to the assumed discretization. After some time, the elements on the left-hand side would be much bigger than those on the right-hand side. To avoid this, when the ratio of the sizes is greater than 5/3 (Fig. 38c), the number of elements to the left is increased by one and the number of elements to the right is decreased by one. The element discretization of both sides is then redefined (Fig. 38d).

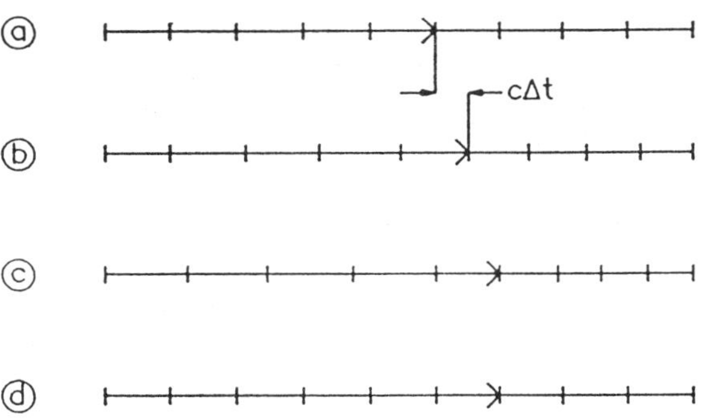

Figure 38: Movement of the elements and remeshing process

The B.E. matrix equation for time domain elastodynamic at step n,

$$\sum_{m=1}^{n} \mathbf{H}^{nm}\mathbf{u}^{m} = \sum_{m=1}^{n} \mathbf{G}^{nm}\mathbf{p}^{m} \tag{193}$$

can be written in a subdivided form as

$$\sum_{m=1}^{n} \begin{pmatrix} H_{ff}^{nm} & H_{fc}^{nm} \\ H_{cf}^{nm} & H_{cc}^{nm} \end{pmatrix} \begin{pmatrix} u_f^m \\ u_c^m \end{pmatrix} = \sum_{m=1}^{n} \begin{pmatrix} G_{ff}^{nm} & G_{fc}^{nm} \\ G_{cf}^{nm} & G_{cc}^{nm} \end{pmatrix} \begin{pmatrix} p_c^m \\ p_f^m \end{pmatrix} \tag{194}$$

where the subindex 'f' stands for the nodes whose position remains fixed during the remeshing process and 'c' for those that change position during that process. The vectors $\mathbf{u}^m$ and $\mathbf{p}^m$ contain the unknowns on the boundary at time step m, and the elements of matrices $\mathbf{H}^{nm}$ and $\mathbf{G}^{nm}$ are the integral along the boundary elements of the tractions and displacements of the fundamental solution evaluated at time step n when the load is applied at time step m. In the 'stationary element' formulation those matrices do not depend on the particular values of m and n but on their difference $n-m$. However, It is important to note that when eqn (194) is written for a new time step n, the submatrices corresponding to changing collocation points and/or to changing integration elements have to be calculated anew for all the previous steps m. The translation property of the fundamental solution cannot be used in those cases because the position of the nodes at time step n is not the same as before. Thus, H_{ij}^{nm} and G_{ij}^{nm} $(i,j = c \text{ or } f)$ depend on the n and m values and not only on the difference $n-m$.

If the general time domain B.E. discretization is used for moving elements, the results are poor because the space and time dependence of the variables are not uncoupled. The space interpolation functions of the variables move with the elements and therefore these functions are not only space dependent but also time dependent. Displacements and tractions are represented as follows:

$$u_j = \sum_{q=1}^{Q}\sum_{m=1}^{n} \varphi^q(r,\tau)\eta^m(\tau)\, u_j^{mq}$$

$$p_j = \sum_{q=1}^{Q}\sum_{m=1}^{n} \Psi^q(r,\tau)\mu^m(\tau)\, p_j^{mq} \tag{195}$$

where Q is the number of boundary nodes and n the current time step.

The approximation for velocities is now

$$\dot{u}_j = \sum_{q=1}^{Q}\sum_{m=1}^{n} \left[\varphi^q(r,\tau)\dot{\eta}^m(\tau) + \dot{\varphi}^q(r,\tau)\eta^m(\tau)\right] u_j^{mq} \tag{196}$$

and the integral representation, after substitution of the above approximations, can be written as

$$c_{ij}^p\, u_j^{np} = \sum_{q=1}^{Q}\sum_{m=1}^{n} \left\{ \left[\int_{\Gamma_q}\int_{\tau_{m-1}}^{\tau_m^+} u_{ij}^* \mu^m \Psi^q \, \mathrm{d}\tau \mathrm{d}\Gamma\right] p_j^{mq} \right.$$

$$\left. - \left[\int_{\Gamma_q}\int_{\tau_{m-1}}^{\tau_m^+} \left(z_{ij}^*\eta^m\varphi^q - w_{ij}^*\dot{\eta}^m\varphi^q - w_{ij}^*\eta^m\dot{\varphi}^q\right) \mathrm{d}\tau \mathrm{d}\Gamma\right] u_j^{mq} \right\} \tag{197}$$

In the regular formulation of the time domain BEM, the space shape functions are taken away from the time integrals which are done analytically. This is not possible when Ψ^q and φ^q are time dependent. In order to do the time integration, an approximation of the space shape functions time dependence is made. Two terms of their series expansion are taken:

$$\Psi^q(r,\tau) = \Psi^q(r,\tau_{m-1}) + (\tau - \tau_{m-1})\dot{\Psi}^q(r,\tau_{m-1})$$

$$\phi^q(r,\tau) = \phi^q(r,\tau_{m-1}) + (\tau - \tau_{m-1})\dot{\phi}^q(r,\tau_{m-1}) \tag{198}$$

By substitution of eqn (198) into eqn (197), the time integrals can be separated from the space integrals and eqn (197) can be written as the typical B.E. equation, where now the coefficients of the system are

$$\begin{aligned}
G_{ij}^{nmpq} &= \int_{\Gamma_q}\left[\int_{\tau_{m-1}}^{\tau_m^+} u_{ij}^{*}\mu^m \,\mathrm{d}\tau\right]\Psi^q(r,\tau_{m-1})\,\mathrm{d}\Gamma \\
&\quad + \int_{\Gamma_q}\left[\int_{\tau_{m-1}}^{\tau_m^+} (\tau-\tau_{m-1})u_{ij}^{*}\mu^m \,\mathrm{d}\tau\right]\dot{\Psi}^q(r,\tau_{m-1})\,\mathrm{d}\Gamma \\
H_{ij}^{nmpq} &= \int_{\Gamma_q}\left[\int_{\tau_{m-1}}^{\tau_m^+} (z_{ij}^{*}\eta^m - w_{ij}^{*}\dot{\eta}^m)\,\mathrm{d}\tau\right]\varphi^q(r,\tau_{m-1})\,\mathrm{d}\Gamma \\
&\quad + \int_{\Gamma_q}\left\{\int_{\tau_{m-1}}^{\tau_m^+} [(z_{ij}^{*}\eta^m - w_{ij}^{*}\dot{\eta}^m)(\tau-\tau_m) \right. \\
&\quad \left. - w_{ij}^{*}\eta^m]\,\mathrm{d}\tau\right\}\dot{\varphi}^q(r,\tau_{m-1})\,\mathrm{d}\Gamma
\end{aligned} \tag{199}$$

The first term of each coefficient is the same as in the stationary elements formulation. The computation of the second term requires a time integration followed by a space integration.

The time integration is done analytically. Explicit expressions for those integrals may be found in the work by Gallego.[35] The space integration first requires the computation of the time derivatives of the shape functions φ^q and Ψ^q.

Assuming that for each time step the elements move with a constant speed c, the shape functions can be written as

$$\varphi(r,t) = \varphi(y) \tag{200}$$

where $y = r - C\,t$. Then,

$$\dot{\varphi}(r,\tau_{m-1}) = -C\left.\frac{\mathrm{d}\varphi}{\mathrm{d}y}\right|_{y=y_{m-1}} \tag{201}$$

The expressions or the shape functions $\varphi(y)$ are those of the usual space-shape functions and their derivatives are easily obtained in terms of the natural coordinate ξ. In this formulation, at least one of the moving elements is singular of the $r^{-1/2}$ kind. Its derivative is singular of the $r^{-3/2}$ kind, which is too strong for a one-dimensional integration. Nevertheless, the singularity appears multiplied by the Heaviside function and the existence of the integral can be demonstrated in this case (Jones[41]). Once the shape function has been differentiated, the resulting singular kernels of the second terms of eqn (199) are integrated numerically. This integration can be done because the Heaviside function makes only the computation of the finite part of the integrals necessary (Kutt[47]).

The numerical formulae given by Kutt[47] for finite part computations have been used. A ten-point integration scheme produces accurate results for the finite parts of the integrals containing $r^{-3/2}$ singularities.

3.10 Dynamic crack propagation examples

3.10.1 Half-plane crack propagating in an unbounded domain

To check the accuracy of the present approach, a problem with known analytical solution is solved first. The problem is that of a semi-infinite crack that extends at constant speed in an unbounded domain under the action of a uniform and time-independent normal pressure applied on the crack faces from $t = 0$. This problem was solved by Baker[9] and its solution was extended by Freund[31] to general loading and a time delay prior to crack growth. The properties of the elastic region are: Poisson's ratio, $\nu = 0.3$; shear modulus, $\mu = 2.24\ 10^{10}$Pa and density, $\rho = 2450$ Kg/m^3. The resulting wave speeds are $c_1 = 5880$ m/s and $c_2 = 3143$ m/s. Because of the symmetry, only one half of the problem is studied (Fig. 39).

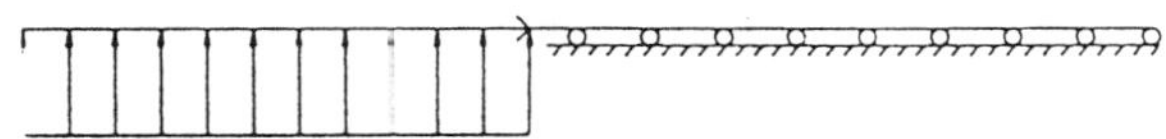

Figure 39: Half-plane crack in unbounded domain

The boundary, which divides the domain into two equal half-planes, is modeled by a boundary element mesh containing 26 elements; 13 to each side of the tip. The two elements ending at the crack tip are moving SQP elements 3.2 mm long. The rest of the elements are space quadratic elements whose initial length increases uniformly with the distance to the tip until a or the maximum value of 5 mm. The time step $\Delta t = 0.27\ \mu s$ is such that the parameter $\beta = c_1 \Delta t / L$, l being the distance between two consecutive nodes, varies between 0.99 and 0.64. The problem is studied for two different values of the crack propagation velocity $c/c_2 = 0.2$ and 0.4.

The two SQP elements move with the tip and their size does not change with time. The rest of the elements change size in accordance with the remeshing technique described in the previous section. At each time step, the SIF is computed as usual by eqn (189) from the crack-tip nodal value of the traction p_2.

Figures 40 and 41 show the computed values of the mode-I SIF normalized by $\sigma(0.06\pi)^{1/2}$ versus time for crack propagation velocities $C = 0.2\ c_2$ and $0.4\ c_2$, respectively. The B.E. results are compared with the analytical solutions for $0.2\ c_2$ and $0.4\ c_2$, denoted in the figures by $K(t, 0.2\ c_2)$ and $K(t, 0.4\ c_2)$, respectively. It can be seen in the figures that the agreement between the B.E. and the analytical solutions is very good. For the sake of completeness, the analytical solution for a stationary crack $K(t, 0)$ is also plotted in the figures.

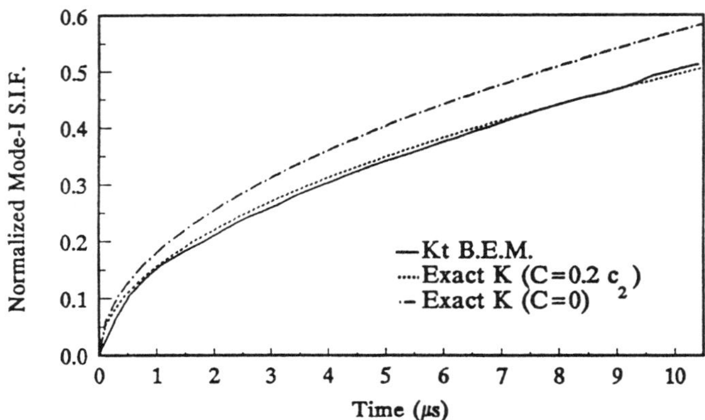

Figure 40: Time dependence of dynamic SIF for a half-plane crack propagating with constant speed $c = 0.2\, c_2$.

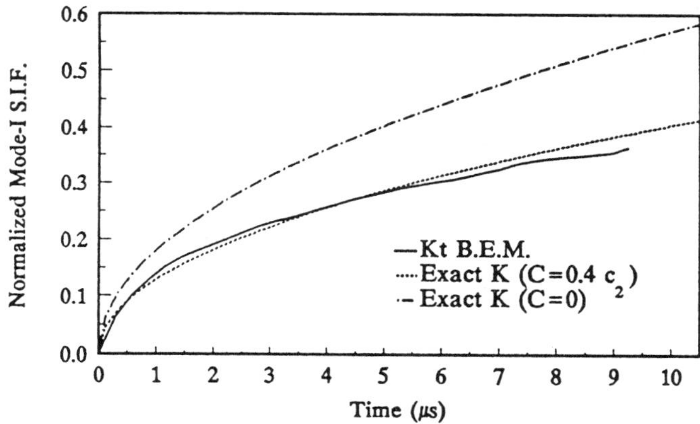

Figure 41: Time dependence of dynamic SIF for a half-plane crack propagating with constant speed $c = 0.4\, c_2$.

3.10.2 Crack propagation at constant speed in a finite body

In order to evaluate the application or the present approach to crack propagation in finite bodies under dynamic loading conditions a problem of this kind, previously studied by Nishioka and Atluri[64] using a different numerical procedure, is analyzed. The problem is that of a rectangular plane domain with a central crack (Fig. 42). A uniformly distributed traction, with a Heaviside-function time dependence, is applied at the two sides parallel to the crack. The material properties are: shear modulus, $\mu = 2.94\ 10^{10}$ Pa; Poisson's ratio $\nu = 0.286$, and density $\rho = 2450$ Kg/m^3. The crack, with an initial length $2a = 24$ mm, remains stationary until a time $t_0 = 4.4\ \mu s$ and then propagates with a constant velocity $c = 1000$ m/s.

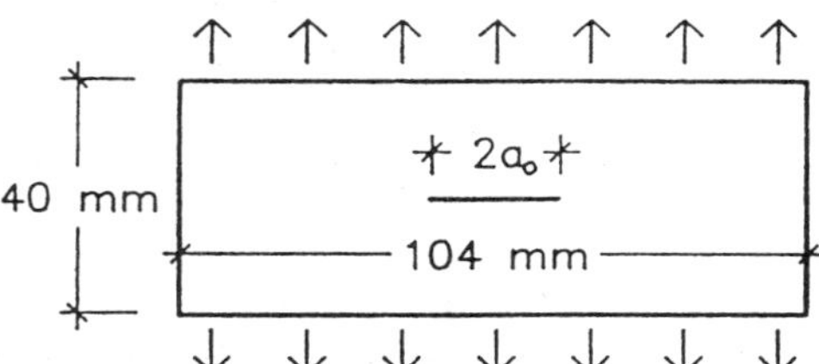

Figure 42: Center-cracked rectangular plate subjected to a step-function normal stress

Due to symmetry, only one quarter of the plate is discretized by boundary elements. The discretization at the initial crack length is shown in Fig. 43. The two elements containing the crack tip are moving SQP elements and the rest are standard space quadratic elements. The boundary elements discretization is the same that results from Nishioka and Atluri's[64] finite element discretization of the boundary. In modeling the crack propagation, the sizes of the two moving SQP do not change. The regular quadratic elements are readjusted in accordance with the aforementioned criterion. The time step $\Delta t = 0.34\ \mu s$ is such that the parameter β for the equal size regular elements that model most of the boundary is $\beta = c_1\ \Delta t/L = 1.08$.

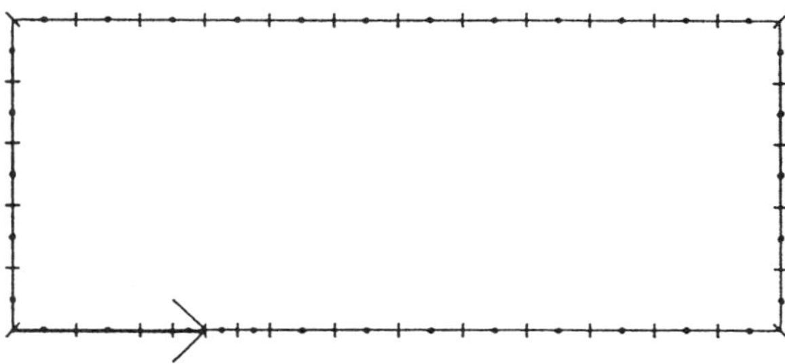

Figure 43: Boundary element mesh for a center-cracked rectangular plate

The computed values of the mode-I SIF normalized by $\sigma(\pi a)^{1/2}$ are shown versus time in Fig. 44, along with analytical results by Freund[33] and the finite element results by Nishioka and Atluri.[64] The boundary element results are in excellent agreement with the half-plane crack in infinite domain results by Freund,[33] until interaction of waves coming from one crack tip with the other crack tip takes place. The present

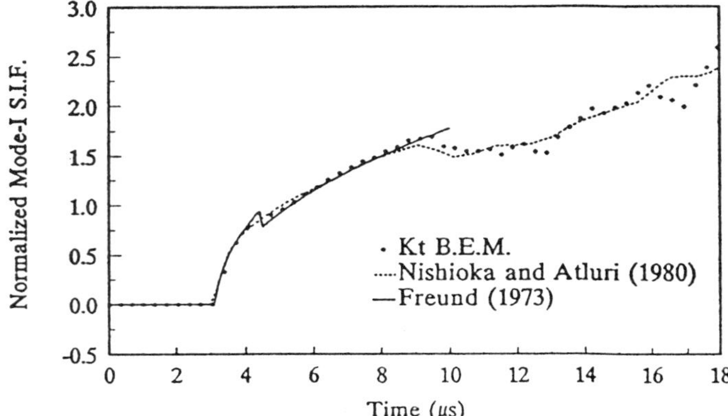

Figure 44: Time dependence dynamic SIF for a central crack propagating with $C = 1000$ m/s in a rectangular plate.

results are also in good agreement with those computed by finite elements by Nishioka and Atluri,[64] for times after the time for which the infinite domain solution may be considered valid.

References

1. Achenbach, J.D. *Wave Propagation in Elastic Solids*, North-Holland, Amsterdam, 1973.

2. Achenbach, J.D., Keer, L.M. & Mendelsohn, D.A. Elastodynamic analysis of an edge crack, *J. Appl. Mech.*, 1980, **47**, 551-556.

3. Achenbach, J.D. & Brind, R.J. Scattering of surface waves by a sub-surface crack, *J. Sound Vibration*, 1981, **76**, 43-56.

4. Angel, Y.C. & Achenbach, J.D. Stress intensity factor for 3-D dynamic loading of a cracked halfspace, *J. Elasticity*, 1985, **15**, 89-102.

5. Atluri, S.N. & Nishioka, T. Numerical studies in dynamic fracture mechanics, *Int. J. Fract.*, 1985, **27**, 245-261.

6. Aoki, S., Kishimoto, K., Kondo, H. & Sakata, M. Elastodynamic analysis of cracks by finite element method using singular elements, *Int. J. Fract.*, 1978, **14**, 59-68.

7. Aoki, S., Kishimoto, K., Izumihara, Y. & Sakata, M. Dynamic analysis of cracked linear viscoelastic solids by finite element method using singular element, *Int. J. Fract.*, 1980, **16**, 97-109.

8. Bachelor, G.K. *An Introduction to Fluid Mechanics*, Cambridge Univ. Press., Cambridge, 1967.

9. Baker, B.R. Dynamic stresses created by a moving crack, *ASME J. App. Mech.*, 1969, **29**, 449-454.

10. Bergkvist, H. Brittle crack propagation at non-uniform velocity, *Num. Meth. in Fracture Mechanics*, A.R. Luxmoore & D.R.J. Owen (ed), 1978, Quadrant, Swansea, pp. 685-695.

11. Beskos, D.E. Numerical methods in dynamic fracture mechanics, Research Report EUR 11300 EN, Ispra Establishment, European Joint Research Centre, 1987.

12. Blandford, G.E., Ingraffea, A.R. & Ligget, J.A. Two-dimensional stress intensity factor computations using the boundary element method, *Int. J. Num. Meth. Eng.*, 1981, **17**, 387-404.

13. Broek, D. *Elementary Engineering Fracture Mechanics*, Martinus Nijhoff Publishers, Dordrecht, The Netherlands, 1986.

14. Burridge, R. The numerical solution of certain integrals equations with non-integrable kernels arising in the theory of crack propagation and elastic wave diffraction, *Proc. Roy. Soc.*, London, 1969, **A265**, pp. 353-381.

15. Cagniard, L. *Réflexion et réfraction des ondes seismiques progresive*, Gauthier-Villiard, Paris, 1939. *Reflection and refraction of progressive seismic waves*, McGraw-Hill, New York, 1962.

16. Chen, Y.M. Numerical computation of dynamic stress intensity factor by Lagrangian finite-difference method, *Engng. Fracture Mech.*, 1975, **7**, 653-660.

17. Chen, Y.M. & Wilkins, M.L. Stress analysis of crack problems with a three dimensional time dependent computer program, *Int. J. Fract.*, 1976, **2**, 607-617.

18. Chen, E.P. & Sih, G.C. Scattering waves about stationary and moving cracks, in *Mechanics of Fracture: Elastodynamic Crack Problems*, G.C.Sih (ed), Noordhoff, Leyden, 1977, pp.119-212.

19. Chirino, F. Stress intesity factor computations by the BEM (In Spanish), Ph. D. Thesis, Univ. de Las Palmas, Spain, 1987.

20. Chirino, F. & Domínguez, J. Dynamic analysis of cracks using boundary element method, *Eng. Fract. Mech.*, 1989, **34**, 1051-1061.

21. Chirino, F., Gallego, R., Sáez, A. & Domíguez, J., A comparative study of three boundary element approaches to transient dynamic crack problems, *Engrg. Anal. Boundary Element*, in press.

22. Copson, E.T. On certain dual integral equations, *Proc. Glasgow Math. Assoc.*, 1961, **5**, 19-24.

23. Cruse, T.A. *Boundary Element Analysis in Computational Fracture Mechanics*, Kluwer Academic Publishers, Dordrecht, The Netherlands, 1988.

24. Das, S. A numerical method for determination of source time functions for general three-dimensional rupture propagation, *Geophys. J. Roy. Astr. Soc.*, 1980, **62**, 591-604.

25. Das, S. & Kostrov, B.V. On the numerical boundary integral equation method for three-dimensional dynamic shear crack problems, 1987, **54**, 99-104.

26. De Hoop, *Representation Theorems for the Displacement in an Elastic Solid and their Applications to Elastodynamics Diffraction Theory*, Thesis, Technische Hogeschool, Delft, 1959.

27. Domínguez, J. *Boundary Elements in Dynamics*, CMP-Elsevier, Southampton, 1993.

28. Domínguez, J. & Gallego, R. The time domain boundary element method for elastodynamic problems, *Math. Comp. Modelling*, 1991, **15**, 119-129.

29. Domínguez, J. & Gallego, R. Time domain boundary element method for dynamic stress intensity factor computations, *Int. J. Numer. Meth. Eng.*, 1992, **33**, 635-647.

30. Eringen, A.C. & Suhubi, E.S. Linear theory, *Elastodynamics*, Vol. 2, Academic Press, New York, 1975.

31. Freund, L.B. Crack propagation in an elastic solid subjected to general loading-I: constant rate of extension, *J. Mech. Phys. Solids*, 1972, **20**, 129-140.

32. Freund, L.B. Crack propagation in an elastic solid subjected to general loading-II: non-uniform rate of extension, *J. Mech. Phys. Solids*, 1972, **20**, 141-152.

33. Freund, L.B. Crack propagation in an elastic solid subjected to general loading-III: stress wave loading, *J. Mech. Phys. Solids*, 1973, **21**, 47-61.

34. Freund, L.B. The stress intensity factor due to normal impact loading of the faces of a crack, *Int. J. Engng. Sci.*, 1974, **12**, 179-189.

35. Gallego, R. *Numerical Studies of Elastodynamic Fracture Mechanics Problems*, Ph.D. Thesis, Universidad de Sevilla (in Spanish), 1990.

36. Gallego, R. & Domínguez, J. Dynamic crack propagation analysis by moving singular boundary elements, *ASME J. of App. Mech.*, 1992, **59**, S158-S162.

37. Graffi, D. Sul teorema di reciprocitá nella dinamica dei corpi elastici, *Mem. Accad. Sci.*, 1946, **4**, Bologna, Ser.10, pp. 103-111.

38. Griffith, A.A. The phenomena of rupture and flow in solids, *Phil. Trans. Roy. Soc., Ser.A.*, 1920, **221**, 163-198.

39. Iacovache, M. O extindere a metodei lui Galerkin pentru sistemul ecuatiilar elasticitatii, *Bul. Stiint. Acod. Rep. Pop. Romane. Ser.A.*, 1949, **1**, 593-596.

40. Irwin, G.R. Analysis of stresses and strains near the end of a crack traversing a plate, *ASME J. App. Mech.*, 1957, **24**, 361-364.

41. Jones, D.S. *Generalized Functions*, McGraw-Hill, London, 1966.

42. Kanninen, M.F. A critical appraisal of solution techniques in dynamic fracture mechanics, in *Num. Meth. in Fracture Mechanics*, A.R. Luxmore & D.R.J. Owen (ed), 1978, Quadrant, Swansea, pp. 612-634.

43. Kanninen, M.F. & Popelar, C.H. *Advanced Fracture Mechanics*, Oxford University Press, New York, 1986.

44. Keer, L.M., Lin, W. & Achenbach, J.D. Resonance effects for a crack near a free surface, *ASME J. Appl. Mech.*, 1984, **51**, 65-70.

45. Kishimoto, K., Aoki, S. & Sakata, M. Dynamic stress intensity factors using J-integral and finite element method, *Eng. Fract. Mech.*, 1980, **13**, 387-394.

46. Krishnasamy, G., Schmerr, L.W., Rudolphi, T.J. & Rizzo, F.J. Hypersingular boundary integral equations: some applications in acoustic and elastic wave scattering, *ASME J. Appl. Mech.*, 1990, **57**, 404-414.

47. Kutt, H.R. Quadrature formulae for finite-part integrals, CSIR Special Report, WISK 178, National Research Institute for Mathematical Sciences, Pretoria, Republic of South Africa, 1975.

48. Lamé, G. *Leçons sur la Théorie Mathématique de l'Elasticité des Corps Solides*, Bachelier, Paris, 1852.

49. Lin, W., Keer, L.M. & Achenbach, J.D. Dynamic stress intensity factors for an inclined subsurface crack, *ASME J. Appl. Mech.*, 1984, **51**, 773-779.

50. Love, A.E.H. The propagation of wave-motion in an isotropic elastic medium, *Proc. London Math. Soc.*, Ser. 2, 1904, pp. 291.

51. Love, A.E.H. *Some Problems of Geodynamics*, Cambridge University Press, London, 1911.

52. Mal, A.K. Diffraction of elastic waves by a penny-shaped crack, *Q. Appl. Math.*, 1968, **26**, 231-238.

53. Mal, A.K. Interaction of elastic waves with a Griffith crack, *Int. J. Engng. Sci.*, 1970, **8**, 763-776.

54. Mal, A.K. Interaction of elastic waves with a penny-shaped crack, *Int. J. Engng. Sci.*, 1970, **8**, 381-389.

55. Martínez, J. & Domínguez, J. On the use of quarter-point boundary elements for stress intensity factor computations, *Int. J. Num. Meth. Eng.*, 1984, **20**, 1941-1950.

56. Mettu, S.R. & Nicholson, J.W. Computation of dynamic stress intesity factors by the time domain boundary integral equation method-II examples, *Eng. Fract. Mech.*, 1988, **31**, 769-782.

57. Milkowitz, J. *The Theory of Elastic Waves and Wave Guides*, North-Holland, Amsterdam, 1977.

58. Murakami, Y. Analysis of mixed-mode stress intensity factors by body force method, in *Num. Meth. Fracture Mechanics*, D.R.J. Owen & A.R. Luxmore (ed), Swansea, 1980, pp. 145-147.

59. Murakami, Y. (ed), *Stress Intesity Factores Handbook*, Pergamon Press, Oxford, 1987.

60. Murti, V. & Valliappan, S. The use of quarter point element in dynamic crack analysis, *Engng. Fracture Mech.*, 1986, **23**, 585-614.

61. Neuber, H. Ein neuer ansatz zur losung Raumlicher probleme der elastizitatstheorie, *Z. Augew. Math-Mech.*, 1934, **14**, 203.

62. Nicholson, J.W. & Mettu, S.R. Computation of dynamic stress intensity factors by the time domain integral equation method I. Analysis, *Eng. Fract. Mech.*, 1988, **31**, 759-767.

63. Nishimura, N. & Kobayashi, S. Regularised BIEs for miscellaneous elasticity problems, in *Boundary Element Methods in Engineering*, B.S. Annigeri & K.Tseng (ed), Springer-Verlag, Berlin, 1990.

64. Nishioka, T. & Atluri, S.N. Numerical modelling of dynamic crack propagation in finite bodies by moving singular element, Part I: formulation; Part II: results, *ASME J. App. Mech.*, 1980, **47**, 570-576; 577-582.

65. Nishioka, T. & Atluri, S.N. A numerical study of the use of path independent integrals in elastodynamic crack propagation, *Engng. Fracture Mech.*, 1983, **18**, 23-33.

66. Pao, Y.H. & Mow, C.C. *Diffraction of Elastic Waves in Dynamic Stress Concentrations*, Crane Russak, New York, 1973.

67. Papkovich, P.F. The representation of the general integral of the fundamental equations of elasticity theory in terms of harmonic functions (in Russian), *Izv.Akad. Nauk SSSR, Phys.-Math.Ser.*, 1932, **10**, 1425.

68. Partidge, P.W., Brebbia, C.A. & Wrobel, L.C. *The Dual Reciprocity Boundary Element Method*, CMP-Elsevier, Southampton, 1992.

69. Poisson, S.D. Mémoire sur l'équilibre et le mouvement des corps élastiques, *Mém. Acad-Sci.*, Paris, 1829, **8**, 357-570.

70. Poisson, S.D. Addition au mémoire sur l'équilibre et le mouvement des corps élastiques, *Mém. Acad-Sci.*, Paris, 1829, **8**, 623.

71. Rayleigh, L. On waves propagated along the plane surface of an elastic solid, *Proc. London Math. Soc.*, 1885, Vol. 17, pp. 4-11.

72. Rooke, D.P. & Cartwright, D.J. *Compedium of Stress Intensity Factors*, Her Majesty's Stationary Office, London, 1976.

73. Sáez, A. Application of the BEM to elastodynamic fracture mechanic problem: the equivalent mass matrix approach, Diploma Project, Escuela Superior Ingenieros Industriales, Sevilla (In Spanish), 1992.

74. Shah, A.H., Wong, K.C. & Datta, S.K. Surface displacements due to elastic wave scattering by buried planar and non-planar cracks, *Wave Motion*, 1980, **7**, 319-333.

75. Sih, G.C. & Loeber, J.F. Wave propagations in an elastic solids with a line of discontinuity or finite crack, *Q. Appl. Math.*, 1969, **27**, 193-213.

76. Sih, G.C., Embley, G.T. & Ravera, R.S. Impact response of a finite crack in plane extension, *Int. J. Solids Struct.*, 1972, **8**, 977-993.

77. Sternberg, E. On the integration of the equations of motion in the classical theory of elasticity, *Arch. Rat. Mech. Anal.*, 1960, **6**, 34-50.

78. Stokes, G.G. On the dynamical theory of diffraction, *Trans. Camb. Phil. Soc.*, 1849, **9**, 1-62.

79. Stone, S.F., Ghosh, M.L. & Mal, A.K. Diffraction of antiplane shear waves by an edge crack, *ASME J. Appl. Mech.*, 1980, **47**, 359-362.

80. Stoneley, R. Elastic waves at the surface of separation of two solids, *Proc. Roy. Soc.*, 1924, Ser. A, Vol. 106, pp. 416-428.

81. Thau, S.A. & Lu, T.H. Transient stress intensity factors for a finite crack in an elastic solid caused by a dilatational wave, *Int. J. Solids Struct.*, 1971, **7**, 731-750.

82. Tittmann, B.R., Ahlberg, L.A. & Mal, A.K. Rayleigh wave diffraction from surface-breaking discontinuities, *Appl. Phys. Lett.*, 1986, **20**, 1333-1335.

83. Van Der Hijden, J.H.M.T. & Neerhoff, F.L. Scattering of elastic waves by a plane crack of finite width, *ASME J. Appl. Mech.*, 1984, **51**, 646-651.

84. Wheeler, L.T. & Sternberg, E. Some theorems in classical elastodynamics, *Arch. Rat. Mech. Anal.*, 1968, **31**, 51-90.

85. Williams, M.L. Stress singularities resulting from various boundary conditions in angular corners of plates in extension, *ASME J. App. Mech.*, 1952, **19**, 526-528.

Chapter 6

Cracking of strain-softening materials

A. Carpinteri

Politecnico di Torino, Department of Structural Engineering, 10129 Torino, Italy

Abstract

Progressive cracking in structural elements of concrete is considered. Two simple models are applied which, even though different, lead to similar predictions for the fracture behaviour. Both the Virtual Crack Propagation Model and the Cohesive Limit Analysis show a trend toward brittle behavior and catastrophical events for large structural sizes. Such a trend is fully confirmed by more refined finite element investigations and by experimental testing on plain and reinforced concrete members.

A cohesive crack model is proposed aiming at describing the size effects of fracture mechanics, i.e. the transition from ductile to brittle structure behaviour by increasing the size scale and keeping the geometrical shape unchanged.

For extremely brittle cases (e.g. initially uncracked specimens, large and/or slender structures, low fracture toughness, high tensile strength, etc.) a snap-back instability in the equilibrium path occurs and the load-deflection softening branch assumes a positive slope. Both load and deflection must decrease to obtain a slow and controlled crack propagation (whereas in normal softening only the load must decrease). If the loading process is deflection-controlled, the loading capacity presents a discontinuity with a negative jump. It is proved that such a catastrophical event tends to reproduce the classical LEFM-instability ($K_{\mathrm{I}} = K_{\mathrm{IC}}$) for small fracture toughnesses and/or for large structure sizes. In these cases, the plastic zone does not develop nor does slow crack growth occur before the unstable crack propagation.

The loss of symmetry of initially symmetrical systems is analyzed as a bifurcation problem in the final part of the paper.

Introduction

Cementitious, ceramic and rock materials present a tensile softening behaviour after the peak-load, i.e. the load sustained by the material decreases when the deformation is increased.[1–8] If the loading process is deformation-controlled, the material behaves in a stable manner and the descending load vs. deformation law can be obtained experimentally.

When strain-softening is involved in the mechanical behaviour of a structural member not homogeneous or not homogeneously loaded, it is accompanied by strain-localization.[1,4,9,10] The suitable constitutive law to describe the mechanical behaviour in the localized damage zone appears to be 'stress vs. displacement', so that energy is dissipated on a crack area rather than in a material volume. Fracture energy is defined as the energy dissipated on a unit crack area.[4,11–14]

Whereas in classical plasticity and damage theory[15,16] geometrically similar structures behave in the same way as only energy dissipation per unit volume is allowed, when energy dissipation per unit area is also contemplated (strain or curvature localization) the global brittleness becomes scale-dependent.

Size-scale and slenderness are demonstrated to have fundamental influence on the global structural behaviour, which can range from ductile to brittle when strain softening and strain localization are taken into account. The brittle behaviour coincides with a snap-back instability[2,3,11–13,17–27] in the load vs. deflection path, which shows a positive slope in the softening branch. Such a virtual branch may be revealed only if the loading process is controlled by a monotonically increasing function of time (e.g. the crack mouth opening displacement). Otherwise, the loading capacity will present a discontinuity with a negative jump.

A general explanation to the well-known decrease in apparent strength and increase in fracture toughness by increasing the member sizes is given in terms of Dimensional Analysis. Due to the different physical dimensions of tensile strength $[F\ L^{-2}]$ and fracture energy $[F\ L^{-1}]$, the true values of such two intrinsic material properties may be found precisely only with comparatively large members.

The cohesive crack model is a representative model when the plastic zone is confined to a very narrow band. The plastic stress field is represented by restraining forces which close the crack tip faces. These forces are non-increasing functions of the distance between the crack surfaces.

Such a model was originally proposed by Barenblatt,[28] who considered the cohesive forces confined in an interaction zone of constant size, with the shape of the terminal crack region being fixed even if translating. On the other hand, Dugdale[29] considered a similar model with vanishing singularity at the crack tip and an interaction zone of variable size, spreading into the entire ligament at the condition of general yielding.

Over the following years the cohesive crack model was reconsidered, with some modifications, by several authors: Bilby, Cottrell & Swinden,[30] Rice,[31] Wnuk,[32] Hillerborg, Modeer & Petersson,[4] etc.

In the present paper the cohesive crack model is applied to analyze stable vs. unstable crack propagation in elastic-softening materials. The crack propagation in real structures often presents a transition from slow to fast rate, and vice-versa. In other cases the crack propagation is only slow or fast. This happens depending on material properties, structural geometry, loading condition and external constraints.

For extremely brittle cases (e.g. initially uncracked specimens, large and/or slender elements, low fracture toughness, high tensile strength, etc.) a snap-back instability appears in the load-displacement curve, which presents a softening branch with positive slope. This means that both load and deflection must decrease to obtain a slow and controlled crack propagation, whereas in normal softening (negative slope) only the load must decrease. If, at the snap-back instability, the loading process is displacement-controlled, the load vs. displacement path will present a negative jump onto the lower branch with negative slope.

The accuracy of the numerical description of cohesive crack propagation is also investigated. It is shown that, when the finite element mesh is too coarse – i.e. when the cohesive forces are too far from one another – the cohesive model is unable to describe fracture process and mechanical behaviour regularly. In other words, when the structure is very large or the fracture toughness very small, the plastic or cohesive zone at the crack tip becomes relatively small, and the finite element mesh must be refined, so that such a zone and the LEFM-stress-singularity may be properly reproduced and the snap-back instability described in the post-peak and post-catastrophical stage.

The snap-back instability was studied in the past by several Authors: Maier,[10] Bazant,[1] Carpinteri,[2] Schreyer,[33] Rots *et al.*[8] etc. On the other hand, the object of the present paper is to put it in connection with the LEFM-instability. From an experimental point of view, the snap-back instability was originally detected by Fairhurst *et al.*[34] in the compressive behaviour of rocks and more recently by Rokugo *et al.*[35] and Biolzi *et al.*[36] in the bending behaviour of concrete.

The last part of the paper analyzes the loss of symmetry of initially symmetrical systems, due to unbalanced damage initiation and propagation.

1 Ductile-brittle transition and size-effects in fracture mechanics

1.1 Application of dimensional analysis

Due to the different physical dimensions of ultimate tensile strength, σ_u, and fracture toughness, K_{IC}, scale effects are always present in the usual fracture testing of common engineering materials. This means that, for the usual size-scale of the laboratory specimens, the ultimate strength collapse or the plastic collapse at the ligament tends to anticipate and obscure the brittle crack propagation. Such a competition between collapses of a different nature can easily be shown by considering the ASTM formula[37] for the three point bending test evaluation of fracture toughness (Fig. 1)

$$K_I = \frac{P\ell}{tb^{3/2}} f\left(\frac{a}{b}\right) \tag{1}$$

with

$$f\left(\frac{a}{b}\right) = 2.9\left(\frac{a}{b}\right)^{1/2} - 4.6\left(\frac{a}{b}\right)^{3/2} + 21.8\left(\frac{a}{b}\right)^{5/2} - 37.6\left(\frac{a}{b}\right)^{7/2} + 38.7\left(\frac{a}{b}\right)^{9/2}$$

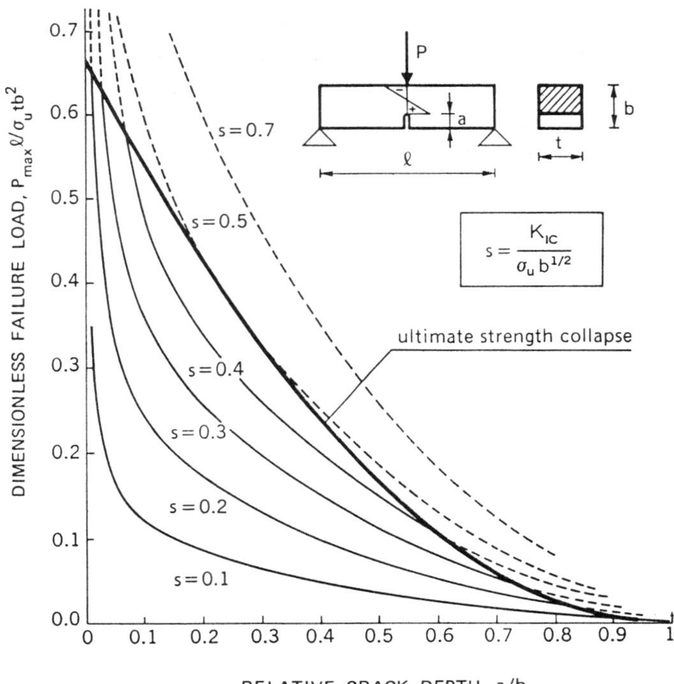

Figure 1: Dimensionless load of crack instability versus relative crack depth

At the crack propagation eqn (1) becomes

$$K_{IC} = \frac{P_{max}\ell}{tb^{3/2}} f\left(\frac{a}{b}\right) \tag{2}$$

where P_{max} is the external load of brittle fracture.
If both members of eqn (2) are divided by $\sigma_u b^{1/2}$ we obtain

$$\frac{K_{IC}}{\sigma_u b^{1/2}} = s = \frac{P_{max}\ell}{\sigma_u t b^2} f\left(\frac{a}{b}\right) \tag{3}$$

where s is the *brittleness number* defined by Carpinteri.[17,18] Rearranging eqn (3) gives

$$\frac{P_{max}\ell}{\sigma_u t b^2} = \frac{s}{f\left(\frac{a}{b}\right)} \tag{4}$$

On the other hand, it is possible to consider the non-dimensional load of ultimate strength in a beam of depth $(b-a)$

$$\frac{P_{max}\ell}{\sigma_u t b^2} = \frac{2}{3}\left(1 - \frac{a}{b}\right)^2 \tag{5}$$

Equations (4) and (5) are plotted in Fig. 1 as functions of the crack depth a/b. While the former produces a set of curves by varying the brittleness number s, the latter is represented by a unique curve. It is evident that the ultimate strength collapse at the ligament precedes crack propagation for each initial crack depth, when the brittleness number s is higher than the limit-value $s_0 = 0.50$.

For lower s numbers, ultimate strength collapse anticipates crack propagation only for crack depths external to a certain interval. This means that a true LEFM collapse occurs only for comparatively low fracture toughnesses, high tensile strengths and/or large structure sizes. The individual values of K_{IC}, σ_u and b do not determine the collapse nature, but only their function s does so, see eqn (3).

1.2 Virtual propagation of a brittle fracture

The flexural behaviour of the beam in Fig. 1 will be analyzed. The deflection due to the elastic compliance of the uncracked beam is

$$\delta_e = \frac{P\ell^3}{48EI} \tag{6}$$

where I is the inertial moment of the cross-section. On the other hand, the deflection due to the local crack compliance is[38]

$$\delta_c = \frac{3}{2}\frac{P\ell^2}{tb^2E} g\left(\frac{a}{b}\right) \tag{7}$$

with

$$g\left(\frac{a}{b}\right) = \left(\frac{a/b}{1-\frac{a}{b}}\right)^2 \left\{5.58 - 19.57\left(\frac{a}{b}\right) + 36.82\left(\frac{a}{b}\right)^2 - 34.94\left(\frac{a}{b}\right)^3 + 12.77\left(\frac{a}{b}\right)^4\right\} \tag{8}$$

The superposition principle provides

$$\delta = \delta_e + \delta_c$$

and, in non-dimensional form

$$\frac{\delta \ell}{\epsilon_u b^2} = \frac{P\ell}{\sigma_u t b^2} \left[\frac{1}{4} \left(\frac{\ell}{b} \right)^3 + \frac{3}{2} \left(\frac{\ell}{b} \right)^2 g \left(\frac{a}{b} \right) \right] \tag{9}$$

where $\epsilon_u = \sigma_u / E$. The term within the square brackets is the dimensionless compliance, which is a function of the beam slenderness, ℓ/b, as well as of the crack depth, a/b. Some linear load-deflection diagrams are represented in Fig. 2, by varying the crack depth a/b and for the fixed ratio $\ell/b = 4$.

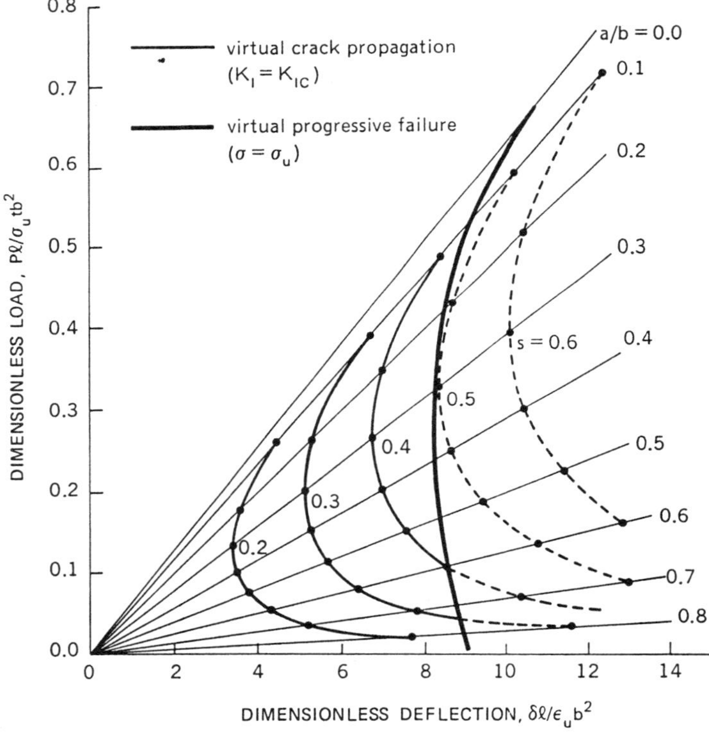

Figure 2: Dimensionless load of crack instability versus dimensionless deflection

Through eqns (4) and (5), it is possible to determine the point of crack propagation as well as the point of ultimate strength on each linear plot in Fig. 2. Whereas the former depends on the brittleness number s, the latter is unique. The set of the crack propagation points with s = constant and by varying the crack depth represents a virtual load-deflection path, where point by point the load is always that producing crack instability.

When the crack grows, the load of instability decreases and the compliance increases, so that the product at the right member of eqn (9) may result as being either decreasing or increasing. The diagrams in Fig. 2 show the deflection decreasing (with the load) up to the crack depth $a/b \cong 0.3$ and then increasing (in discordance with the load). Therefore, whereas for $a/b \gtrsim 0.3$ the $P - \delta$ curve presents the usual softening course with negative derivative, for $a/b \lesssim 0.3$ it presents positive derivative. Such a branch could not be revealed by deflection-controlled testing, and the representative point would jump from the positive to the negative branch with a behaviour discontinuity.

The set of the ultimate strength points, by varying the crack depth, is represented by the thick line in Fig. 2. Such a line intersects the virtual crack propagation curves with $s \leq s_0 = 0.50$, analogous to what is shown in Fig. 1, and presents a slight indentation with $\mathrm{d}P/\mathrm{d}\delta > 0$.

The crack mouth opening displacement w_1 is a function of the specimen geometry and of the elastic modulus[38]

$$w_1 = \frac{6P\ell a}{tb^2E}\, h\left(\frac{a}{b}\right) \tag{10}$$

with

$$h\left(\frac{a}{b}\right) = 0.76 - 2.28\left(\frac{a}{b}\right) + 3.87\left(\frac{a}{b}\right)^2 - 2.04\left(\frac{a}{b}\right)^3 + \frac{0.66}{\left(1-\frac{a}{b}\right)^2} \tag{11}$$

In non-dimensional form eqn (10) becomes

$$\frac{w_1\ell}{\epsilon_u b^2} = \frac{P\ell}{\sigma_u t b^2}\left[6\left(\frac{\ell}{b}\right)\left(\frac{a}{b}\right) h\left(\frac{a}{b}\right)\right] \tag{12}$$

The term within square brackets is the dimensionless compliance which, also in this case, depends on beam slenderness and crack depth.
Some linear load-crack mouth opening displacement diagrams are reported in Fig. 3, by varying the crack depth a/b and for $\ell/b = 4$.

The set of the crack propagation points with s = constant and by varying the crack depth represents a virtual process even in this case. When the crack grows, the product at the right member of eqn (12) always increases, the compliance increase prevailing over the critical load decrease for each value of a/b. The $P - w_1$ curve always presents negative derivative, and the crack mouth opening displacement w_1 increases even when load and deflection both decrease ($\mathrm{d}P/\mathrm{d}\delta > 0$) in the catastrophic $P - \delta$ branch. If the crack mouth opening displacement is controlled – i.e. if w_1 increases monotonically without jumping – it would be possible to go along the virtual $P - \delta$ path with positive slope.

Such a theoretical statement was confirmed in Ref. 36 through an experimental investigation on high strength concrete beams. The mechanical response of the specimens with deep cracks appeared stable (Fig. 4a).

Both load-deflection and load-CMOD curves showed the same shape with a softening branch of negative slope. By decreasing the relative crack depth, such a branch becomes steeper with an increase in the brittleness of the system. At the same time, obviously, loading capacity and stiffness increase.

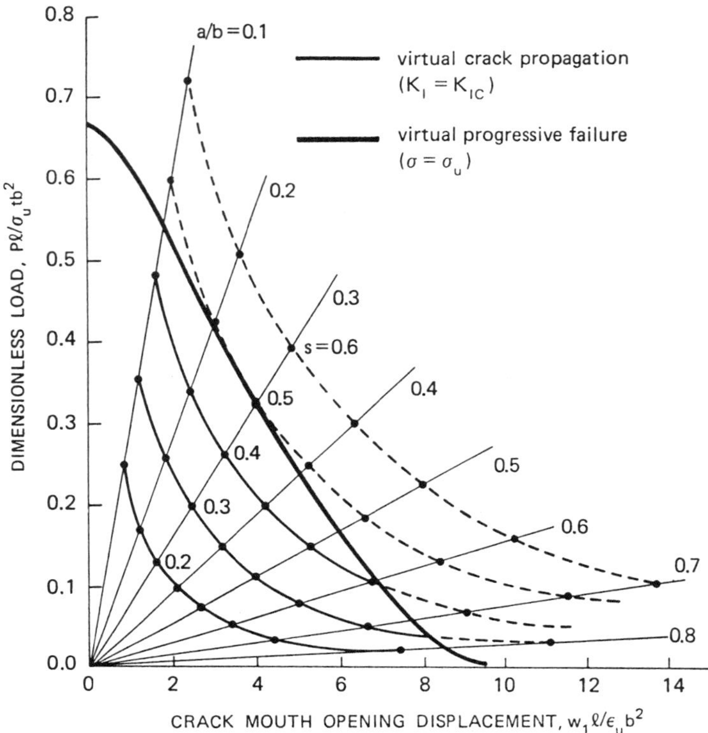

Figure 3: Dimensionless load of crack instability versus dimensionless crack mouth opening displacement

The specimens with shallow cracks (Fig. 4b), on the contrary, presented a very unstable behaviour. Whereas the load-CMOD curves have a softening tail with negative slope, the load-deflection curves are characterized by a snap-back softening instability with a softening branch of partial positive slope. More precisely, the case $a_0 = 30$ mm shows an almost vertical drop in the loading capacity when the maximum load is achieved. This experimental finding confirms the theoretical result in Fig. 2. In fact, the relative crack depth $a_0/b = 0.3$ is the critical condition between stability and instability for deflection-controlled loading processes. If the loading process had been deflection-controlled, then, once it reached the bifurcation point of the loading path the load would have presented a negative jump down to the lower softening branch with a negative slope. Therefore, it is evident that, although the process is unstable in nature, it can develop in a stable manner if CMOD-controlled.

All the diagrams in Fig. 4 converge towards the same asymptotic tail, the limit situation being independent of the initial crack length.

The previous theoretical and experimental analysis emphasizes that the (brittle or ductile) structural behaviour is connected with a geometrical feature, as is the case of the crack depth. More generally, all the geometrical features of the specimen influence the global brittleness (or ductility), and particularly slenderness and size-scale.

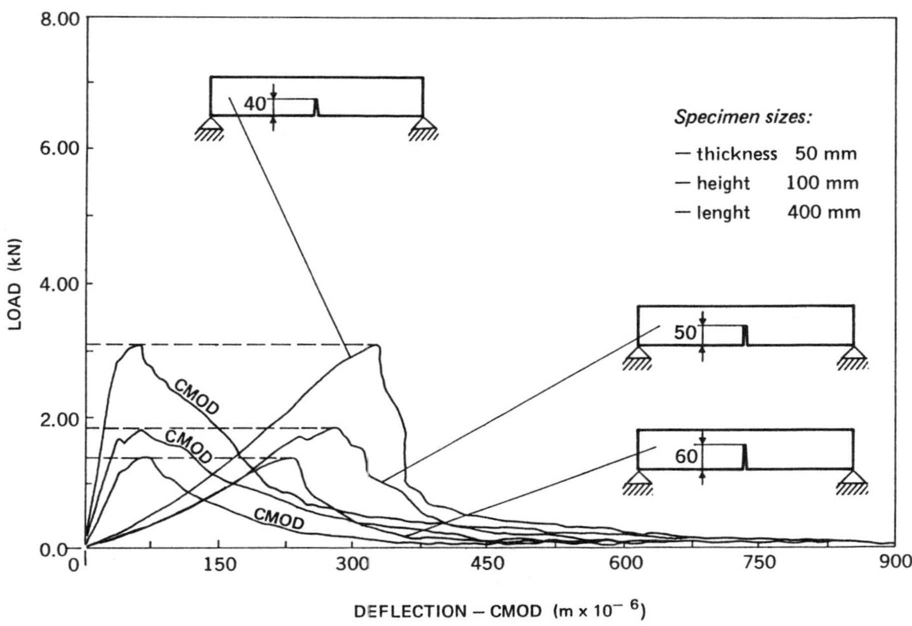

Figure 4a: Experimental load-deflection and load-CMOD diagrams ($a_0/b = 0.4, 0.5, 0.6$)

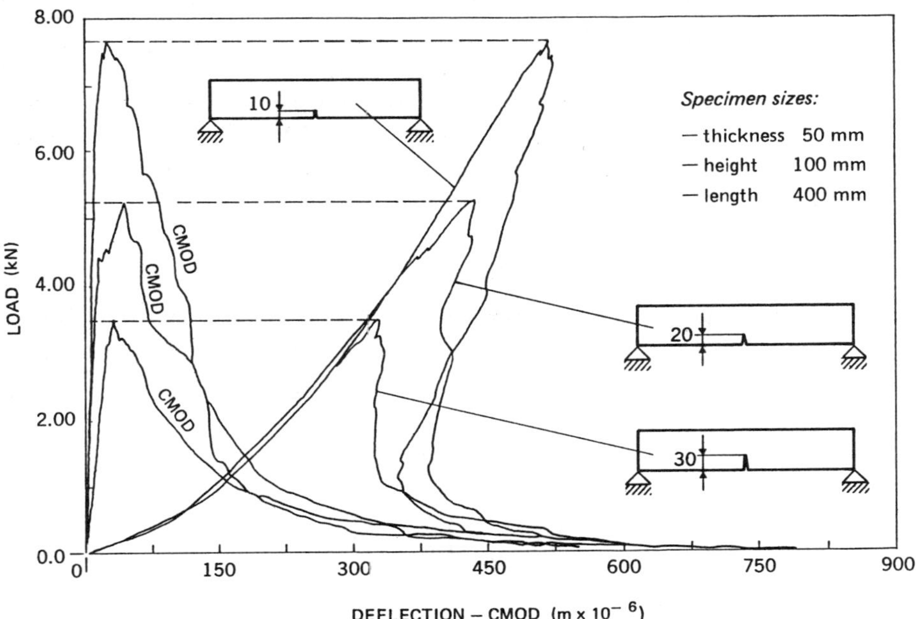

Figure 4b: Experimental load-deflection and load-CMOD diagrams ($a_0/b = 0.1, 0.2, 0.3$)

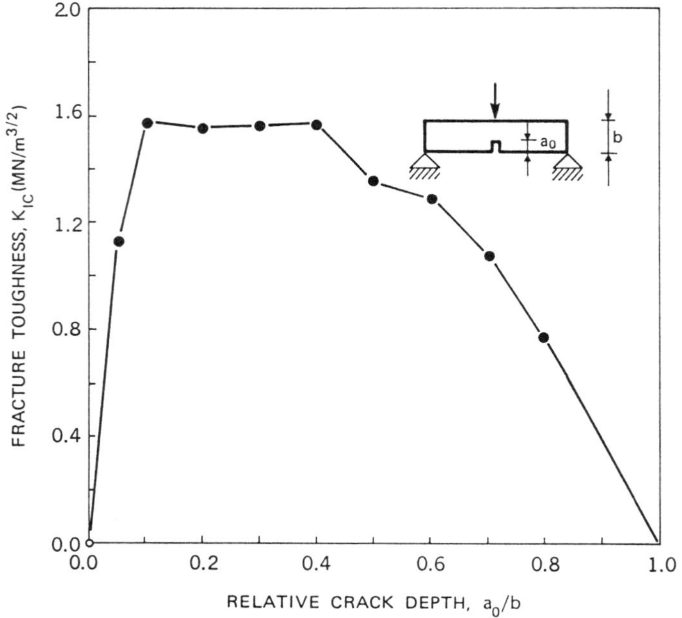

Figure 5: Fictitious fracture toughness versus relative crack depth

The values of K_{IC}, according to eqn (1), are reported in Fig. 5 against the relative crack depth. They appear nearly constant for $0.1 < a_0/b < 0.4$, and then decrease with the crack depth. Even the fictitious K_{IC} related to the crack depths $a_0/b = 0.05$ and $a_0/b = 0.7$ and 0.8 are reported. They present values much lower than the others since in these cases the ultimate tensile strength collapse at the ligament decidedly precedes the LEFM instability (Figs 1 and 2, $s = 0.52$). This fracture toughness decrease for extreme relative crack depths (tending to zero or to unity) was discussed in detail in Refs 17 and 18.

1.3 Uniaxial tensile loading of slabs

Let us consider an elastic-softening material with a double constitutive law: (a) tension σ vs. dilation ϵ, and (b) tension σ vs. crack opening displacement w, after reaching the ultimate tensile strength σ_u or strain $\epsilon_u = \sigma_u/E$ (Fig. 6)

$$\sigma = E\,\epsilon \qquad \text{for } \epsilon \leq \epsilon_u \tag{13a}$$

$$\sigma = \sigma_u \left(1 - \frac{w}{w_c}\right) \qquad \text{for } w \leq w_c \tag{13b}$$

$$\sigma = 0 \qquad \text{for } w > w_c \tag{13c}$$

According to eqn (13c), the cohesive interaction between the crack surfaces vanishes for distances larger than the critical opening w_c.

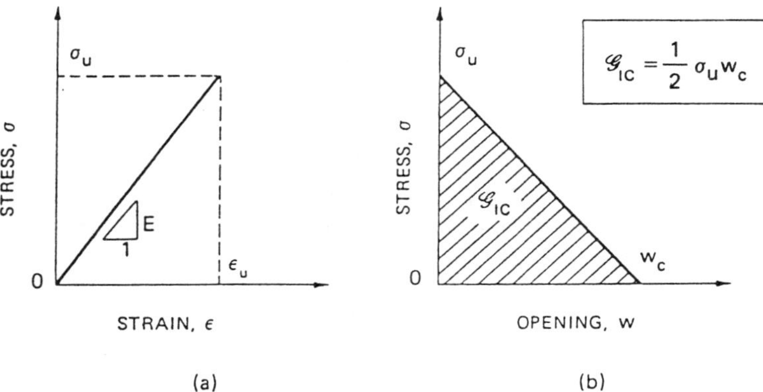

Figure 6: (a) Stress vs. strain elastic law; (b) stress vs. crack opening displacement cohesive law.

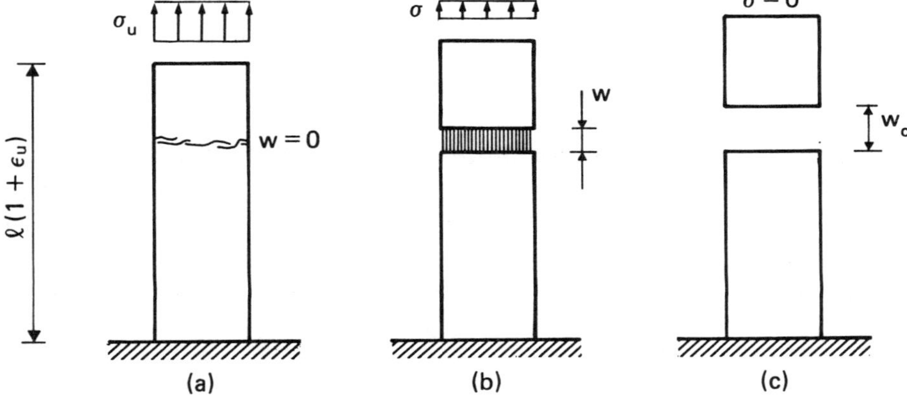

Figure 7: Three different stages of the deformation history: (a) no damage; (b) strain localization; (c) separation.

If a plane slab is increasingly loaded, the deformation history will undergo three different stages.

(i) The slab behaves elastically without damage or fracture zones (Fig. 7a). The displacement of the upper edge is

$$\delta = \frac{\sigma}{E}\ell \qquad \text{for } \epsilon \leq \epsilon_u \tag{14}$$

(ii) After reaching the ultimate tensile strength σ_u, a fracture cohesive zone develops in the weakest section of the slab. Observe that, as the stress field is homogeneous, another cause of inhomogeneity must be assumed for strain-localization. The slab behaves elastically only outside the fracture zone (Fig. 7b). The displacement of the upper edge is

$$\delta = \frac{\sigma}{E}\ell + w \qquad \text{for } w \leq w_c \tag{15}$$

Recalling eqn (13b), eqn (15) gives

$$\delta = \frac{\sigma}{E}\,\ell + w_c\left(1 - \frac{\sigma}{\sigma_u}\right) \quad \text{for } w \leq w_c \tag{16}$$

While the fracture zone opens, the elastic zone shrinks at progressively decreasing stresses. At this stage, the loading process may be stable only if it is displacement-controlled, i.e. if the external displacement δ is imposed. But this is only a necessary and not sufficient condition for stability.

(iii) When $\delta \geq w_c$ the reacting stress σ vanishes, the cohesive forces disappear and the slab is completely separated into two pieces (Fig. 7c).

Rearranging eqn (14) gives

$$\sigma = E\frac{\delta}{\ell} \quad \text{for } \delta \leq \epsilon_u \ell \tag{17}$$

while the condition of complete separation (stage (iii)) reads

$$\sigma = 0 \qquad \text{for } \delta \geq w_c \tag{18}$$

When $w_c > \epsilon_u \ell$, the softening process is stable only if displacement-controlled, since the slope $\mathrm{d}\sigma/\mathrm{d}\delta$ at stage (ii) is negative (Fig. 8a). When $w_c = \epsilon_u \ell$, the slope $\mathrm{d}\sigma/\mathrm{d}\delta$ is infinite and a drop in the loading capacity occurs, even if the loading is displacement-controlled (Fig. 8b). Eventually, when $w_c < \epsilon_u \ell$, the slope $\mathrm{d}\sigma/\mathrm{d}\delta$ becomes positive (Fig. 8c) and the same negative jump occurs as that shown in Fig. 8b.

Rearranging eqn (16) provides

$$\delta = w_c + \sigma\left(\frac{\ell}{E} - \frac{w_c}{\sigma_u}\right) \tag{19}$$

The same conditions just obtained from a geometrical point of view (Fig. 8) may also be given by the analytical derivation of eqn (19).

Normal softening occurs for $\mathrm{d}\delta/\mathrm{d}\sigma < 0$

$$\left(\frac{\ell}{E} - \frac{w_c}{\sigma_u}\right) < 0 \tag{20}$$

whereas catastrophical softening (or snap-back) for $\mathrm{d}\delta/\mathrm{d}\sigma \geq 0$

$$\left(\frac{\ell}{E} - \frac{w_c}{\sigma_u}\right) \geq 0 \tag{21}$$

Equation (21) may be rearranged in the following form

$$\frac{(w_c/2b)}{\epsilon_u(\ell/b)} \leq \frac{1}{2} \tag{22}$$

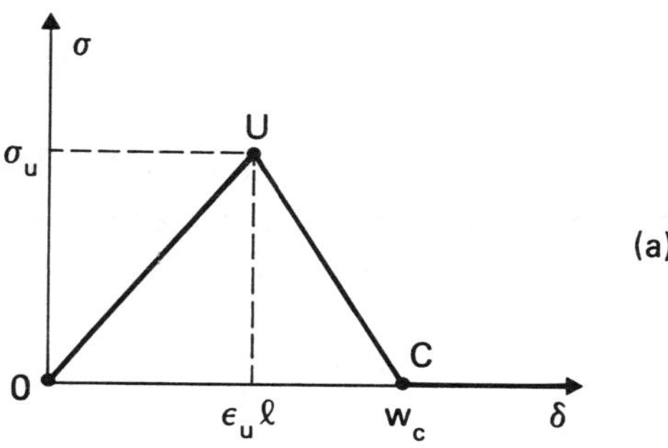

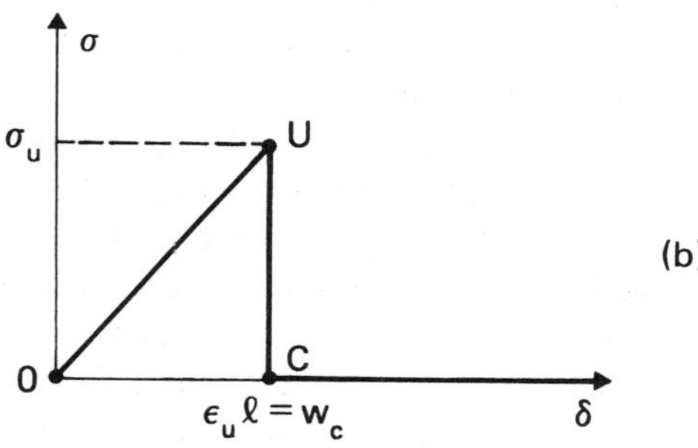

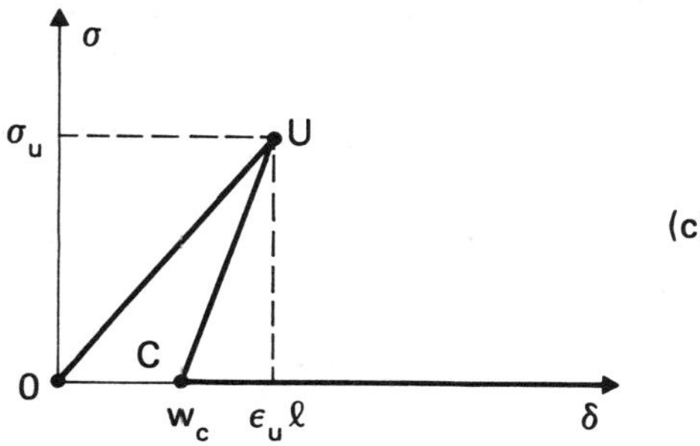

Figure 8: Stress-displacement response: (a) externally unstable; (b) and (c) internally unstable.

where b is the slab width.

The ratio $(w_c/2b)$ is a dimensionless number, which is a function of material properties and structural size-scale[2]

$$s_E = \frac{w_c}{2b} = \frac{\mathcal{G}_{IC}}{\sigma_u b} \tag{23}$$

$\mathcal{G}_{IC} = 1/2\sigma_u w_c$ being the fracture energy of the material (Fig. 6). The energy brittleness number s_E describes the scale effects of fracture mechanics, i.e. the ductile-brittle transition when the size-scale is increased. Equation (22) may be presented in the following final form

$$\frac{s_E}{\epsilon_u \lambda} \leq \frac{1}{2} \tag{24}$$

with λ = slenderness = ℓ/b.

When the size-scale and slab slenderness are relatively large and the fracture energy relatively low, the global structural behaviour is brittle. Not the single values of parameters s_E, ϵ_u and λ, but only their combination $B = s_E/\epsilon_u\lambda$ is responsible for the global brittleness or ductility of the structure considered.

When $B \leq 1/2$, the plane rectangular slab of Fig. 7 shows a mechanical behaviour which can be defined as *brittle* or *catastrophic*. A *bifurcation* or *snap-back* of the global equilibrium occurs since, if point U in Fig. 8c is reached and then the imposed external displacement δ is decreased by a very small amount $\mathrm{d}\delta$, the global unloading may occur along two alternative paths: the elastic UO or the virtual softening UC.

The global brittleness of the slab can be defined as the ratio of the ultimate elastic energy contained in the body to the energy dissipated by fracture

$$\text{Brittleness} = \frac{\frac{1}{2}\frac{\sigma_u^2}{E} \times (\text{Area}) \times \ell}{\mathcal{G}_{IC} \times (\text{Area})} = \frac{1}{2B} \tag{25}$$

Such a ratio is higher than unity when eqn (21) is verified and a catastrophical softening instability occurs.

Recalling the relationship between energy brittleness number s_E (see eqn 23) and stress brittleness number[17–20]

$$s = \frac{K_{IC}}{\sigma_u b^{1/2}} \tag{26}$$

which reads as follows

$$s_E = \epsilon_u s^2 \tag{27}$$

it is possible to relate the brittleness ratio in eqn (25) to the stress brittleness number, through the slab slenderness λ

$$\text{Brittleness} = \frac{\lambda}{2s^2} \tag{28}$$

1.4 Three point bending of beams[11,12]

The linear elastic behaviour of a three point bending, initially uncracked beam may be represented by the following dimensionless equation

$$\tilde{P} = \frac{4}{\lambda^3}\tilde{\delta} \tag{29}$$

where the dimensionless load and central deflection are, respectively, given by

$$\tilde{P} = \frac{P\ell}{\sigma_u t b^2} \tag{30}$$

$$\tilde{\delta} = \frac{\delta\ell}{\epsilon_u b^2} \tag{31}$$

with ℓ = beam span, b = beam depth, t = beam thickness.

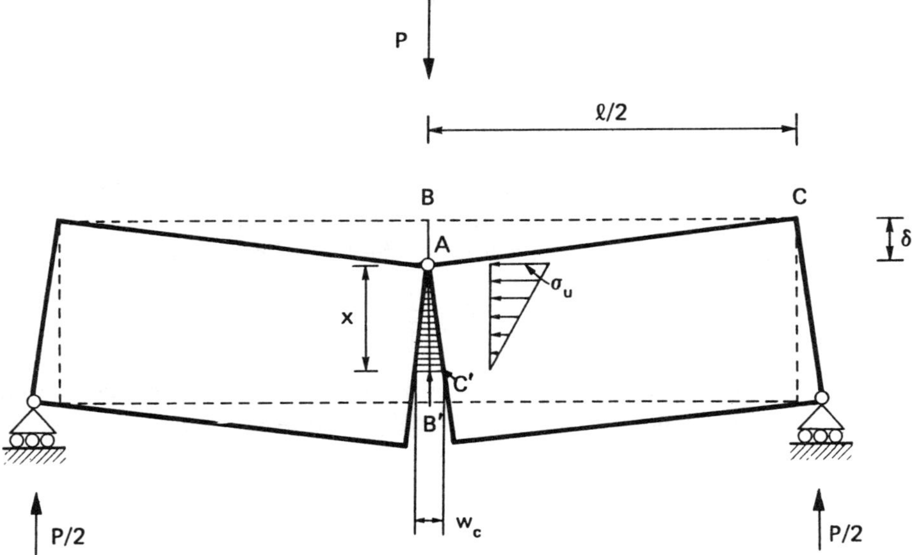

Figure 9: Limit-situation of complete fracture with cohesive forces

Once the ultimate tensile strength σ_u is achieved at the lower beam edge, a fracturing process in the central cross-section is supposed to start. Such a process admits a limit-situation like that in Fig. 9. The limit stage of the fracturing and deformation process may be considered that of two rigid parts connected by the hinge A in the upper beam edge. The equilibrium of each part is ensured by the external load, the support reaction and the closing cohesive forces. The latter depend on the distance between the two interacting surfaces: increasing the distance w, the cohesive forces decrease till they vanish for $w \geq w_c$.

The geometrical similitude of the triangles ABC and $AB'C'$ in Fig. 9 provides

$$\frac{\delta}{\ell/2} = \frac{w_c/2}{x} \tag{32}$$

where x is the extension of the triangular distribution of cohesive forces. Equation (32) can be rearranged as

$$x = \frac{w_c \ell}{4\delta} \tag{33}$$

The rotational equilibrium round point A is possible for each beam part only if the moments of the support reaction and cohesive forces, respectively, are equal

$$\frac{P}{2}\frac{\ell}{2} = \frac{\sigma_u x t}{2}\frac{x}{3} \tag{34}$$

Recalling eqn (33), the relation between load and deflection may be obtained

$$P = \frac{\sigma_u t \ell w_c^2}{24}\frac{1}{\delta^2} \tag{35}$$

Equation (35) can be put into dimensionless form

$$\tilde{P} = \frac{1}{6}\left(\frac{s_E \lambda^2}{\epsilon_u \tilde{\delta}}\right)^2 \tag{36}$$

While the linear eqn (29) describes the elastic behaviour of the beam, initially uncracked, the hyperbolic eqn (36) represents the asymptotical behaviour of the same beam, totally cracked. Equation (29) is valid only for load values lower than that producing the ultimate tensile strength σ_u at the lower beam edge

$$\tilde{P} \leq \frac{2}{3} \tag{37}$$

On the other hand, eqn (36) is valid only for deflection values higher than that producing a cohesive zone of extension x equal to the beam depth b

$$x \leq b \tag{38}$$

From eqns (33) and (38) it follows

$$\tilde{\delta} \geq \frac{s_E \lambda^2}{2\epsilon_u} \tag{39}$$

The bounds (37) and (39), upper for load and lower for deflection respectively, can be transformed into two equivalent bounds, upper for both deflection and load. Equations (29) and (37) provide

$$\tilde{\delta} \leq \frac{\lambda^3}{6} \tag{40}$$

whereas eqns (36) and (39)

$$\tilde{P} \leq \frac{2}{3} \tag{41}$$

Conditions (37) and (41) are identical. Therefore, a stability criterion for elastic-softening beams may be obtained by comparing eqns (39) and (40). When the two

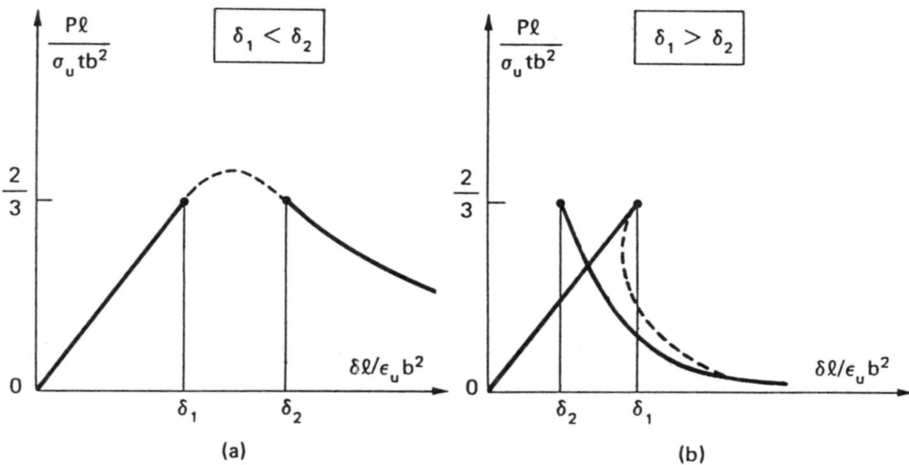

Figure 10: Load deflection diagrams: (a) ductile and (b) brittle condition. $\delta_1 = \lambda^3/6$; $\delta_2 = s_E\lambda^2/2\epsilon_u$.

domains are separated, it is presumable that the two $P - \delta$ branches – linear and hyperbolic – are connected by a regular curve (Fig. 10a). On the other hand, when the two domains are partially overlapped, it is well-founded to suppose them as connected by a curve with highly negative or even positive slope (Fig. 10b).

Unstable behaviour and catastrophical events (snap-back) may be possible for

$$\frac{s_E\lambda^2}{2\epsilon_u} \leq \frac{\lambda^3}{6} \tag{42}$$

and the brittleness condition for the three point bending geometry becomes

$$\frac{s_E}{\epsilon_u\lambda} \leq \frac{1}{3} \tag{43}$$

Even in this case, the system is brittle for low brittleness numbers s_E, high ultimate strains ϵ_u and large slendernesses λ. Observe that the same dimensionless number $B = s_E/\epsilon_u\lambda$ also appears in eqn (24), where the upper bound for brittleness is equal to 1/2.

It is therefore evident that the relative brittleness for a structure is dependent on loading condition and external constraints, in addition to material properties, size-scale and slenderness. For instance, uniaxial tension is more unstable than three point bending (Fig. 11).

As previously discussed, the global brittleness of the beam can be defined as the ratio of the ultimate elastic energy contained in the body to the energy dissipated by fracture

$$\text{Brittleness } \frac{\frac{1}{2}P_u\delta_u}{\mathcal{G}_{IC} \times (\text{Area})} = \frac{\frac{1}{18}\sigma_u\epsilon_u bt\ell}{\mathcal{G}_{IC}bt} = \frac{1}{18B} \tag{44}$$

Such a ratio is higher than unity when

$$\frac{s_E}{\epsilon_u\lambda} \leq \frac{1}{18} \tag{45}$$

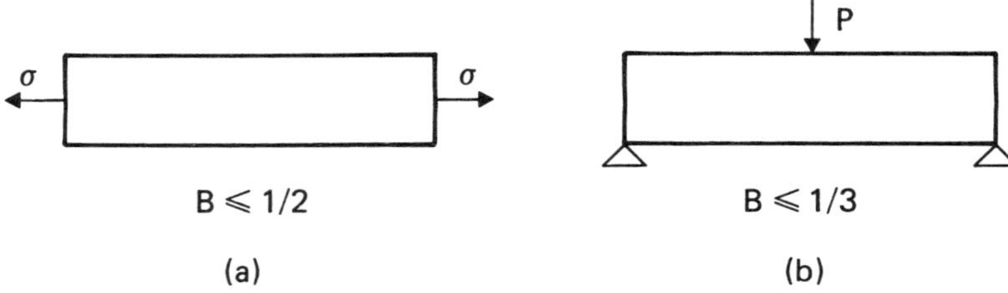

Figure 11: Bounds to relative brittleness for (a) uniaxial tension and (b) three point bending geometry. $B = s_E/\epsilon_u\lambda$.

Equation (45) represents a stricter condition for global structural brittleness compared with eqn (43).

1.5 Three point bending of slabs

When shear forces cannot be neglected (deep beams) and the Poisson ratio is negligible, eqn (29) is replaced by

$$\tilde{\delta} = \tilde{P}\left(\frac{1}{4}\lambda^3 + \frac{3}{5}\lambda\right) \tag{46}$$

whereas eqn (40) becomes

$$\tilde{\delta} \leq \left(\frac{\lambda^3}{6} + \frac{2}{5}\lambda\right) \tag{47}$$

Snap-back is then expected for

$$\frac{s_E\lambda^2}{2\epsilon_u} \leq \left(\frac{\lambda^3}{6} + \frac{2}{5}\lambda\right) \tag{48}$$

or

$$\frac{s_E}{\epsilon_u} \leq \left(\frac{\lambda}{3} + \frac{4}{5}\frac{1}{\lambda}\right) \tag{49}$$

The system is brittle for low brittleness numbers s_E and high ultimate strains ϵ_u, whereas low slendernesses $\lambda(\lambda \lesssim 1.55)$ produce a clear trend towards unstable behaviour (Fig. 12). One can observe that, below the ratio $s_E/\epsilon_u \cong 1.03$, instability is always predicted.

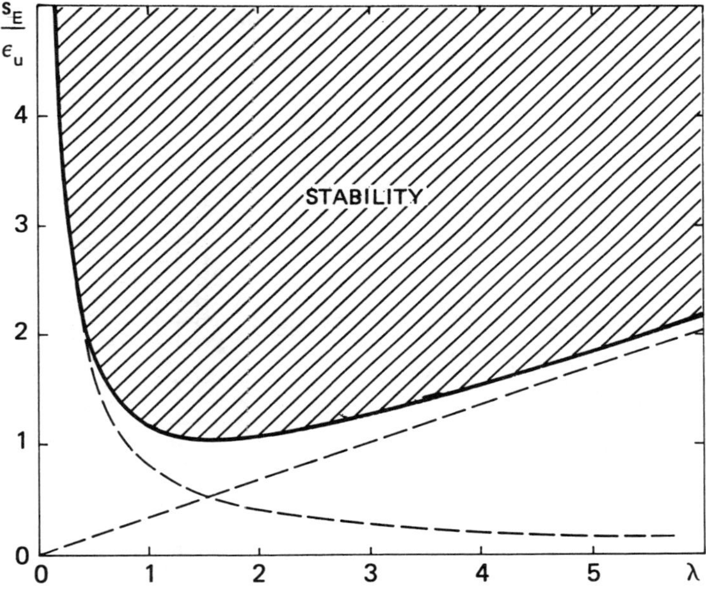

Figure 12: Size-scale vs. slenderness locus of snap-back instability

2 Cohesive crack modelling of strain-softening materials - opening mode

2.1 Cohesive crack model and ductile-brittle transition

The cohesive crack model is based on the following assumptions.[2,4]

- The cohesive fracture zone (plastic or process zone) begins to develop when the maximum principal stress achieves the ultimate tensile strength σ_u (Fig. 6a).
- The material in the process zone is partially damaged but still able to transfer stress. Such a stress is dependent on the crack opening displacement w (Fig. 6b).

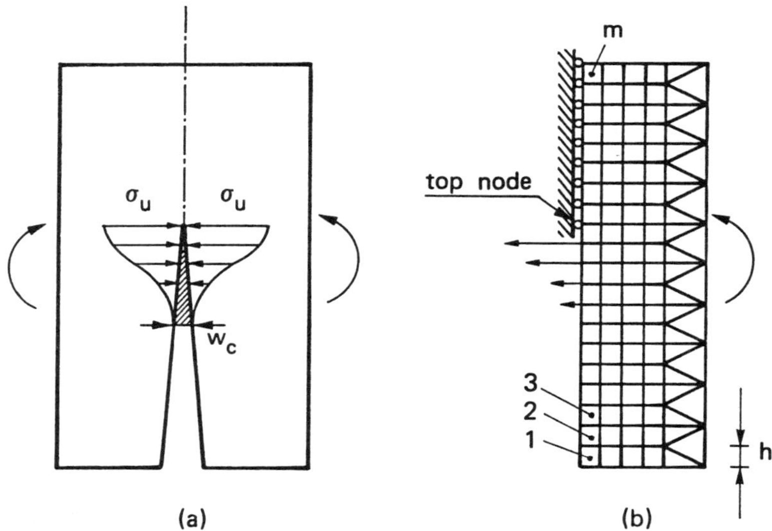

Figure 13: Stress distribution across the cohesive zone (a) and equivalent nodal forces in the finite element mesh (b).

The *real crack tip* is defined as the point where the distance between the crack surfaces is equal to the critical value of crack opening displacement w_c and the normal stress vanishes (Fig. 13a). On the other hand , the *fictitious crack tip* is defined as the point where the normal stress attains the maximum value σ_u and the crack opening vanishes (Fig. 13a).

The closing stresses acting on the crack surfaces (Fig. 13a) can be replaced by nodal forces (Fig. 13b). The intensity of these forces depends on the opening of the fictitious crack w, according to the $\sigma - w$ constitutive law of the material (Fig. 6b). When the tensile strength σ_u is achieved at the fictitious crack tip (Fig. 13b), the top

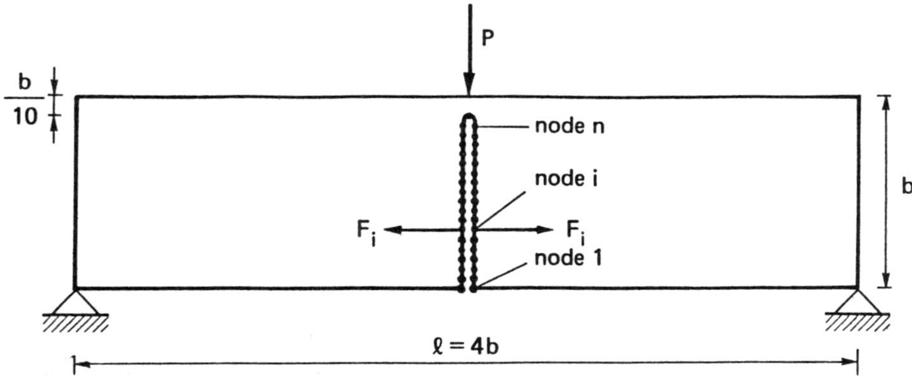

Figure 14: Finite element nodes along the potential fracture line

node is opened and a cohesive force starts acting across the crack, while the fictitious crack tip moves to the next node.

With reference to the three point bending test (TPBT) geometry in Fig. 14, the nodes are distributed along the potential fracture line.

The coefficients of influence in terms of node openings and deflection are computed by a finite element analysis where the fictitious structure in Fig. 14 is subjected to $(n+1)$ different loading conditions. Consider the TPBT in Fig. 15a with the initial crack of length a_0 and tip in the node k. The crack opening displacements at the n fracture nodes may be expressed as follows

$$\mathbf{w} = \mathbf{K}\,\mathbf{F} + \mathbf{C}\,P + \boldsymbol{\Gamma} \tag{50}$$

being

$\mathbf{w}$ = vector of the crack opening displacements,
$\mathbf{K}$ = matrix of the coefficients of influence (nodal forces),
$\mathbf{F}$ = vector of the nodal forces,
$\mathbf{C}$ = vector of the coefficients of influence (external load),
P = external load,
$\boldsymbol{\Gamma}$ = vector of the crack opening displacements due to the specimen weight.

On the other hand, the initial crack is stress-free and therefore

$$F_i = 0 \qquad \text{for } i = 1, 2, ..., (k-1) \tag{51a}$$

while at the ligament there is no displacement discontinuity

$$w_i = 0 \qquad \text{for } i = k, (k+1), ..., n \tag{51b}$$

Equations (50) and (51) constitute a linear algebraical system of $2n$ equations and $2n$ unknowns, i.e. the elements of vectors $\mathbf{w}$ and $\mathbf{F}$. If load P and vector $\mathbf{F}$ are known, it is possible to compute the beam deflection δ

$$\delta = \mathbf{C}^T\mathbf{F} + D_{\mathrm{P}}P + D_{\gamma} \tag{52}$$

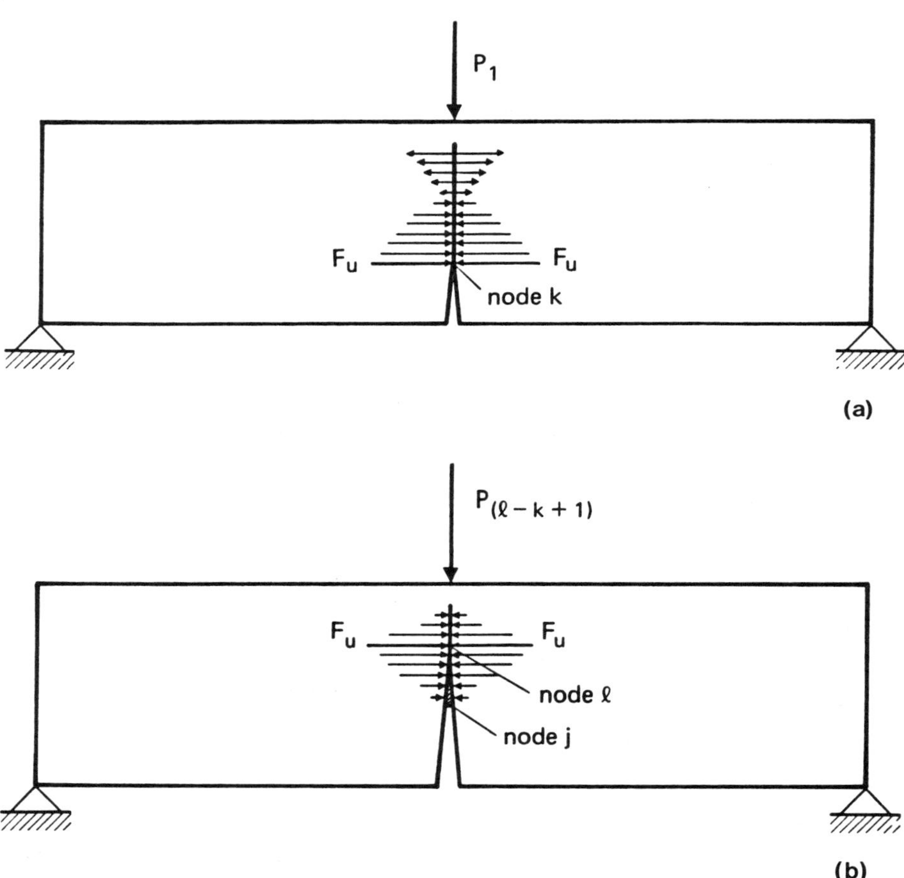

Figure 15: Cohesive crack configurations at the first (a) and $(\ell - k + 1)$th (b) crack growth increment.

where D_P is the deflection for $P = 1$ and D_γ is the deflection due to the specimen weight.

After the first step, a cohesive zone forms in front of the real crack tip (Fig. 15b), say between nodes j and ℓ. Then eqns (51) are replaced by

$$F_i = 0 \qquad \text{for } i = 1, 2, ..., (j-1) \tag{53a}$$

$$F_i = F_u\left(1 - \frac{w_i}{w_c}\right) \qquad \text{for } i = j, (j+1), ..., \ell \tag{53b}$$

$$w_i = 0 \qquad \text{for } i = \ell, (\ell+1), ..., n \tag{53c}$$

where F_u is the ultimate strength nodal force

$$F_u = b\sigma_u/m \tag{54}$$

Equations (50) and (53) constitute a linear algebraic system of $(2n+1)$ equations and $(2n+1)$ unknowns, i.e. the elements of vectors $\mathbf{w}$ and $\mathbf{F}$ and the external load P.

The present numerical program simulates a loading process where the controlling parameter is the fictitious crack depth. On the other hand, real (or stress-free) crack depth, external load and deflection are obtained at each step after an iterative procedure.

The three point bending beam in Fig. 14 is considered herein, with the constant geometrical proportions: span $= \ell = 4b$, thickness $= t = b$. The scale factor is therefore represented by the beam depth b.

As is shown in Fig. 13b, m finite elements are adjacent to the central line, whereas only $n = 0.9m$ nodes can be untied during the crack growth (Fig. 14). The finite element size h (Fig. 13b) is then connected with the beam depth b through the simple relation: $b = mh$.

The three different finite element meshes in Fig. 16 are considered. Mesh (a) presents 20 elements and 18 fracture nodes, mesh (b) 40 elements and 36 fracture nodes, mesh (c) 80 elements and 72 fracture nodes.

The load-deflection response of the three point bending beam in Fig. 14 is represented in Figs 17a, b, and c, for $m = 20$, 40 and 80 respectively. The initial crack depth is assumed $a_0/b = 0.0$, while the ultimate tensile strain $\epsilon_u = \sigma_u/E$ is 8.7×10^{-5} and the Poisson ratio $\nu = 0.1$. The diagrams are plotted in non-dimensional form by varying the brittleness number s_E . The simple variation in this dimensionless number reproduces all the cases related to the independent variations in $\mathcal{G}_{IC}$, b and σ_u. It is not the individual values of $\mathcal{G}_{IC}$, b and σ_u which are responsible for the structural behaviour, which can range from ductile to brittle, but rather their function s_E – see eqn (23). Specimens with high fracture toughness are then ductile, as well as small specimens and/or specimens with low tensile strength. Vice-versa, brittle behaviours are predicted for low fracture toughnesses, large specimens and/or high tensile strengths.

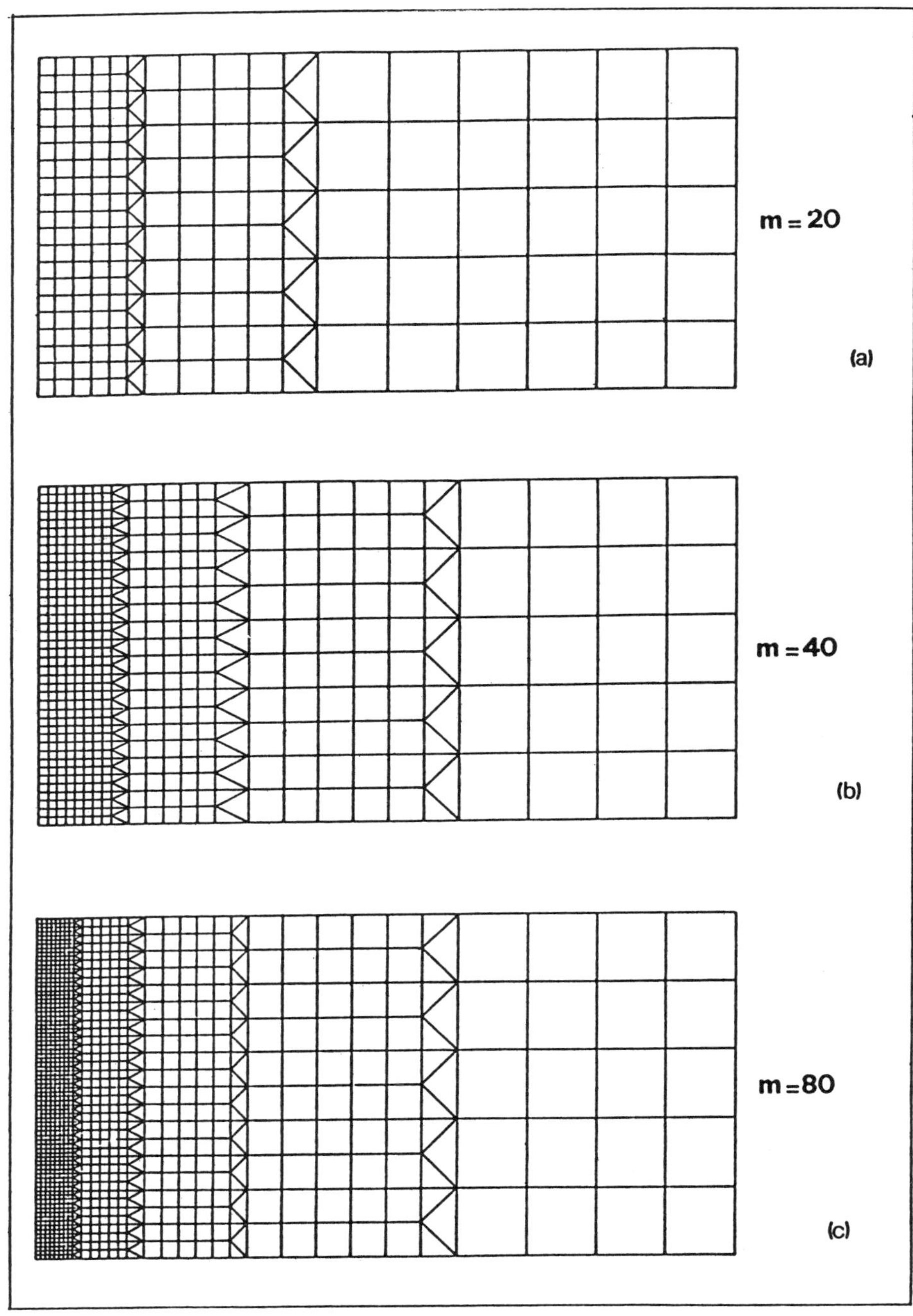

Figure 16: Refinement of the finite element mesh

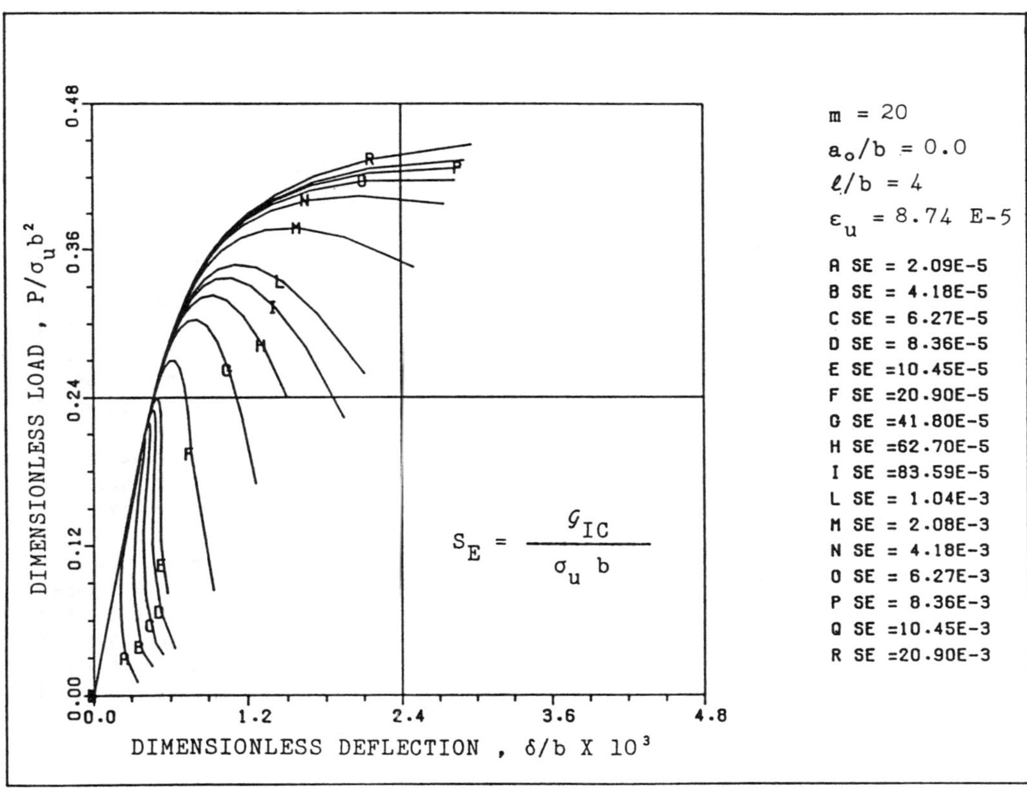

Figure 17a: Dimensionless load-deflection response of an initially uncracked specimen, by varying the brittleness number, $s_E = \mathcal{G}_{IC}/\sigma_u b = w_c/2b$, between 2×10^{-5} and 2×10^{-2}. $m = 20$.

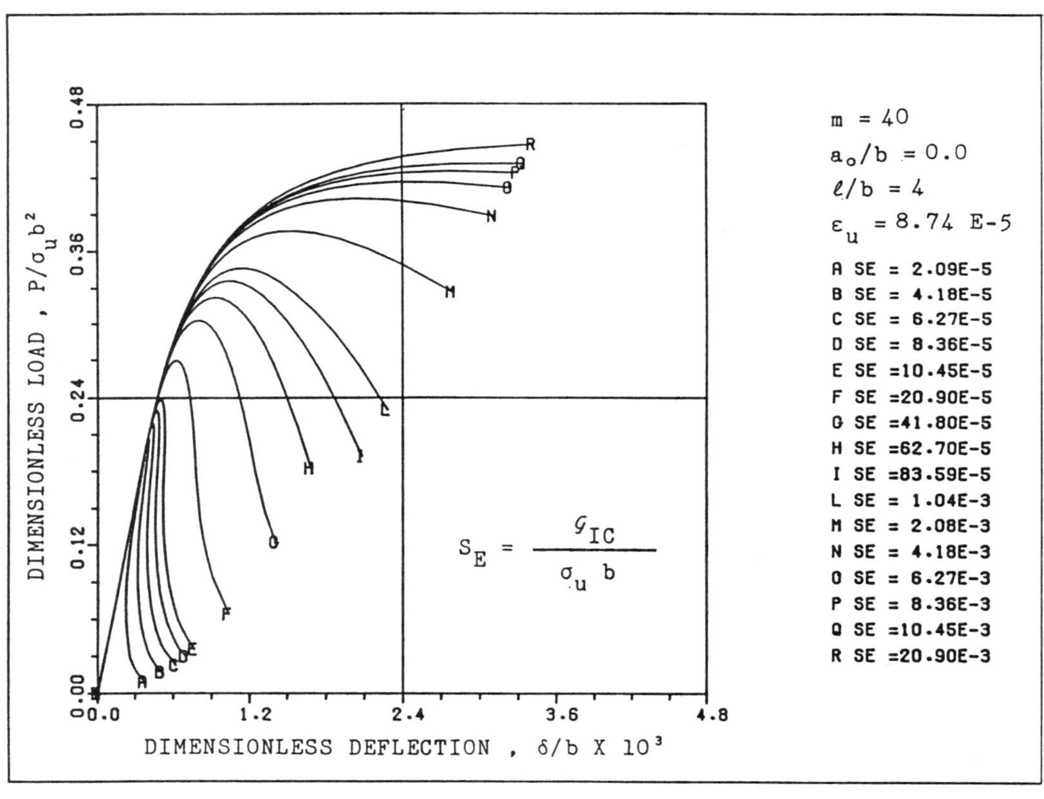

Figure 17b: Dimensionless load-deflection response of an initially uncracked specimen, by varying the brittleness number, $s_E = \mathcal{G}_{IC}/\sigma_u b = w_c/2b$, between 2×10^{-5} and 2×10^{-2}. $m = 40$.

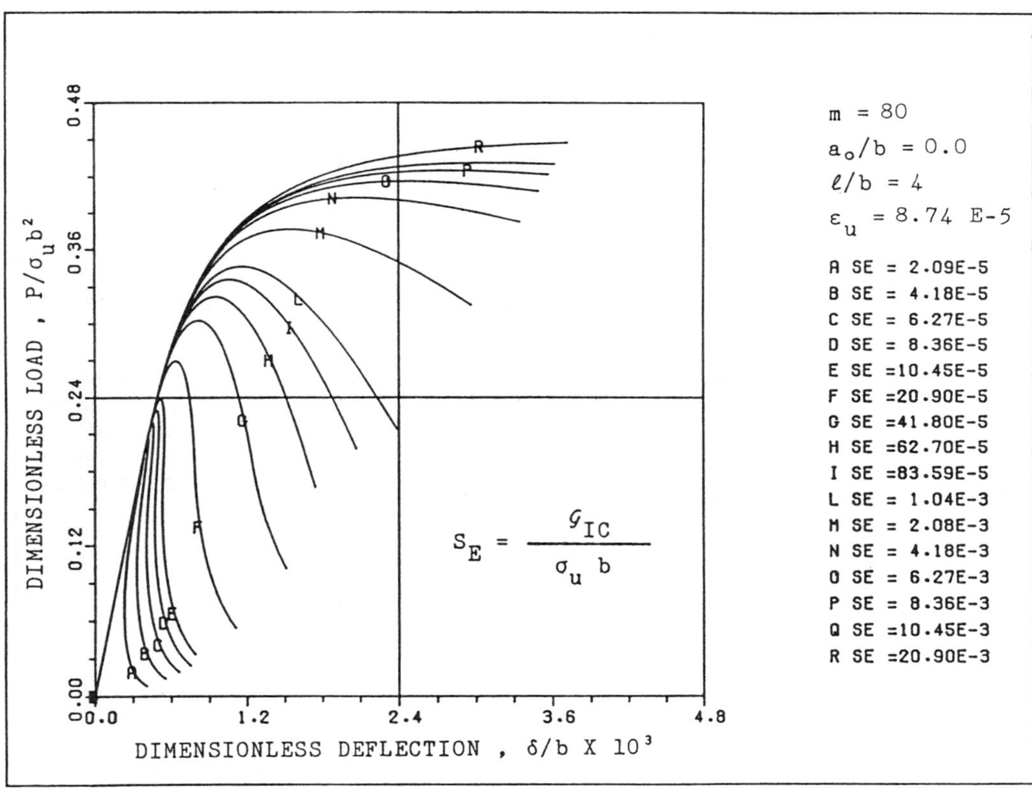

Figure 17c: Dimensionless load-deflection response of an initially uncracked specimen, by varying the brittleness number, $s_E = \mathcal{G}_{IC}/\sigma_u b = w_c/2b$, between 2×10^{-5} and 2×10^{-2}. $m = 80$.

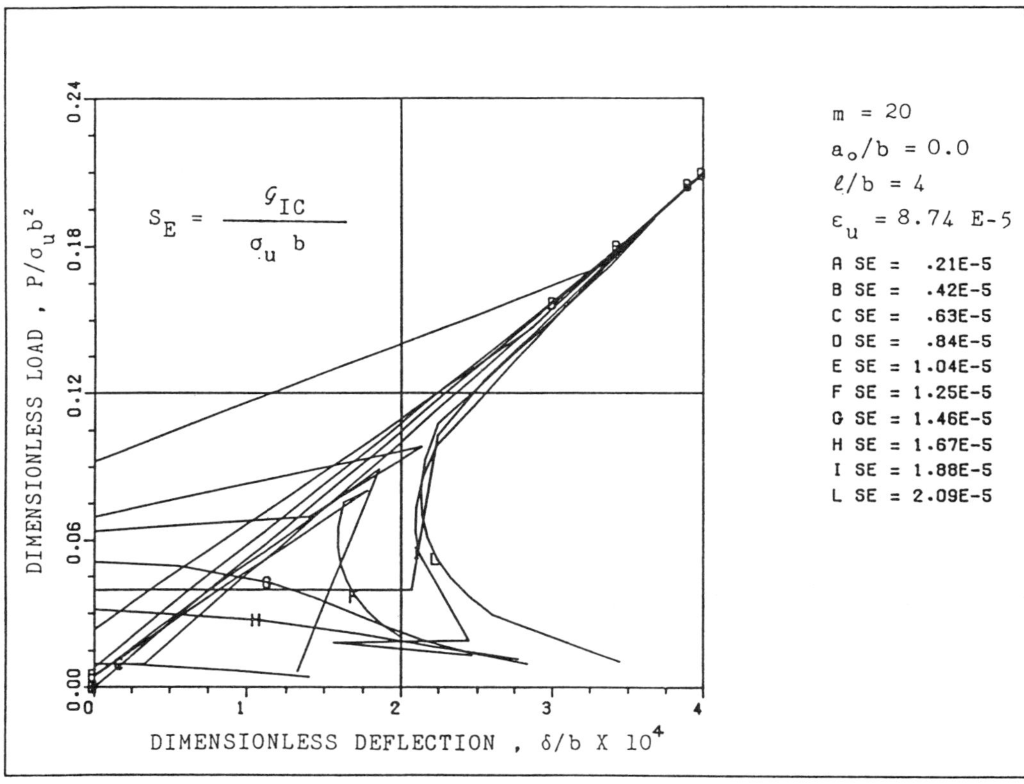

Figure 18a: Dimensionless load-deflection response of an initially uncracked specimen, by varying the brittleness number, $s_{\mathrm{E}} = \mathcal{G}_{\mathrm{IC}}/\sigma_{\mathrm{u}}b = w_{\mathrm{c}}/2b$, between 2×10^{-6} and 2×10^{-5}. m=20.

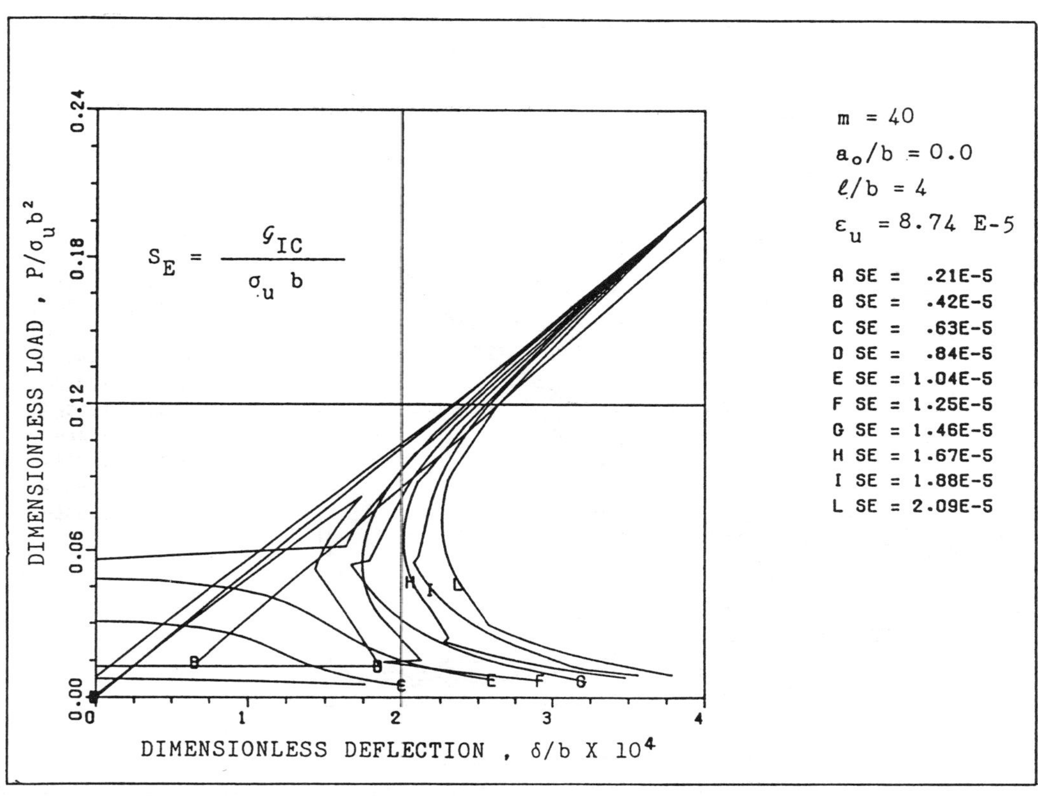

Figure 18b: Dimensionless load-deflection response of an initially uncracked specimen, by varying the brittleness number, $s_E = \mathcal{G}_{IC}/\sigma_u b = w_c/2b$, between 2×10^{-6} and 2×10^{-5}. m=40.

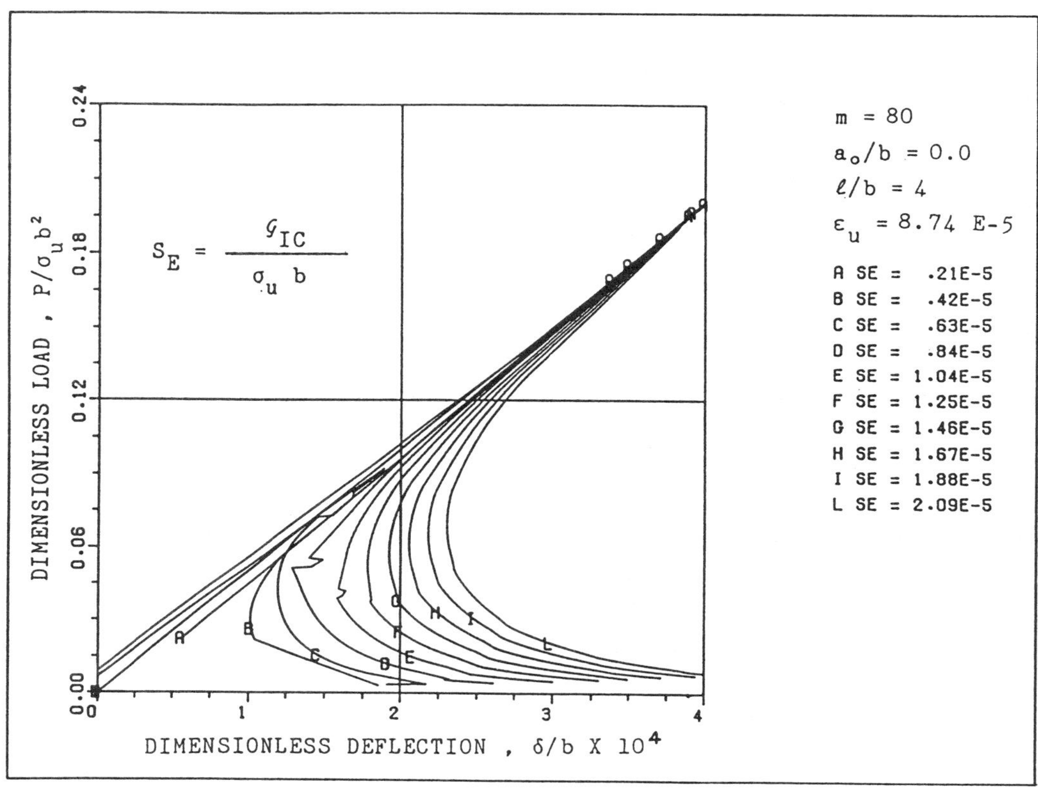

Figure 18c: Dimensionless load-deflection response of an initially uncracked specimen, by varying the brittleness number, $s_E = \mathcal{G}_{IC}/\sigma_u b = w_c/2b$, between 2×10^{-6} and 2×10^{-5}. m=80.

The influence of the variation in the number s_E is investigated over four orders of magnitude in Figs 17, from 2×10^{-2} to 2×10^{-5}. The results reported in Figs 17a, b, and c appear very similar. Of course, the diagrams for $m = 20$ (Fig. 17a) are slightly less regular than those for $m = 80$ (Fig. 17c), and present some weak cuspidal points specially for low s_E numbers.

When still lower s_E numbers are contemplated, the $P - \delta$ diagrams lose their regularity, from a mathematical point of view, and their resolution, from a graphical point of view. The influence of the variation in the s_E number is further analyzed over one order of magnitude in Figs 18, from 2×10^{-5} to 2×10^{-6}. The results reported in Figs 18a, b, and c appear much less uniform than those in Figs 17a, b, and c. The diagrams for $m = 20$ (Fig. 18a) are lacking in mathematical regularity, graphical resolution and physical meaning. The diagrams present a slightly better regularity and resolution for $m = 40$ (Fig. 18b), whereas for $m = 80$ (Fig. 18c) they appear sufficiently regular, especially for brittleness numbers that are not too low ($10^{-5} \lesssim s_E \lesssim 2 \times 10^{-5}$). If a better resolution is requested for $2 \times 10^{-6} \lesssim s_E \lesssim 10^{-5}$, the mesh must be refined, i.e. the number m increased. On the other hand, it is evident that the mesh must be refined – i.e. the cohesive forces must be closer – for relatively large structures and/or for relatively brittle materials, where the cohesive zone is confined to a relatively small crack tip region.

From the cases shown in Figs 17 and 18, the s_E -threshold below which the results are unacceptable is approximately

$$s_E = \frac{w_c}{2mh} \cong \frac{80}{m} \times 10^{-5} \tag{55}$$

The lower bound to s_E can be regarded as an upper bound to the finite element size h

$$h \lesssim 600 w_c \tag{56}$$

For a concrete-like material with maximum aggregate size of 2 cm, it is approximately $w \cong 0.1$ mm, and then eqn (56) provides: $h \lesssim 6$ cm.

The load-deflection response shows the same trends even when an initial crack is present in the lower edge of the three point bending beam. The initial crack depth is considered to be $a_0/b = 0.3$ (Figs 19 and 20). The deeper the initial crack is, the more ductile the beam behaviour results as being.

In addition to the slenderness $\ell/b = 4$ considered thus far, the ratios $\ell/b = 8$ and 16 are then contemplated. For initially uncracked specimens ($a_0/b = 0.0$), Fig. 17b ($\ell/b = 4$) is to be compared with Fig. 21a ($\ell/b = 8$) and Fig. 21b ($\ell/b = 16$). The brittleness increase with the slab slenderness is manifest.

Such a trend is due to the variation in the elastic compliance of the non-damaged zone. An increase of this compliance produces an increase of brittleness in the system. In the softening stage, in fact, the elastic recovery prevails over the localized increase of deformation, so that a snap-back instability occurs.

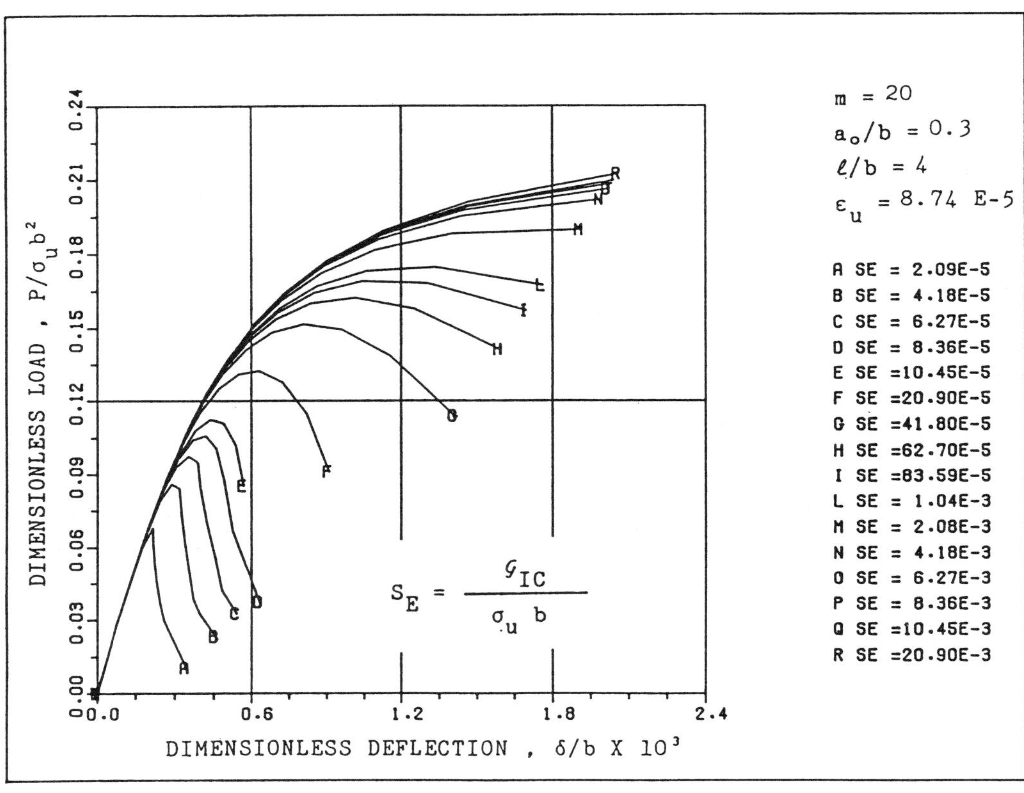

Figure 19a: Dimensionless load-deflection response of an initially cracked specimen ($a_0/b = 0.3$), by varying the brittleness number, $s_E = \mathcal{G}_{IC}/\sigma_u b = w_c/2b$, between 2×10^{-5} and 2×10^{-2}. $m = 20$.

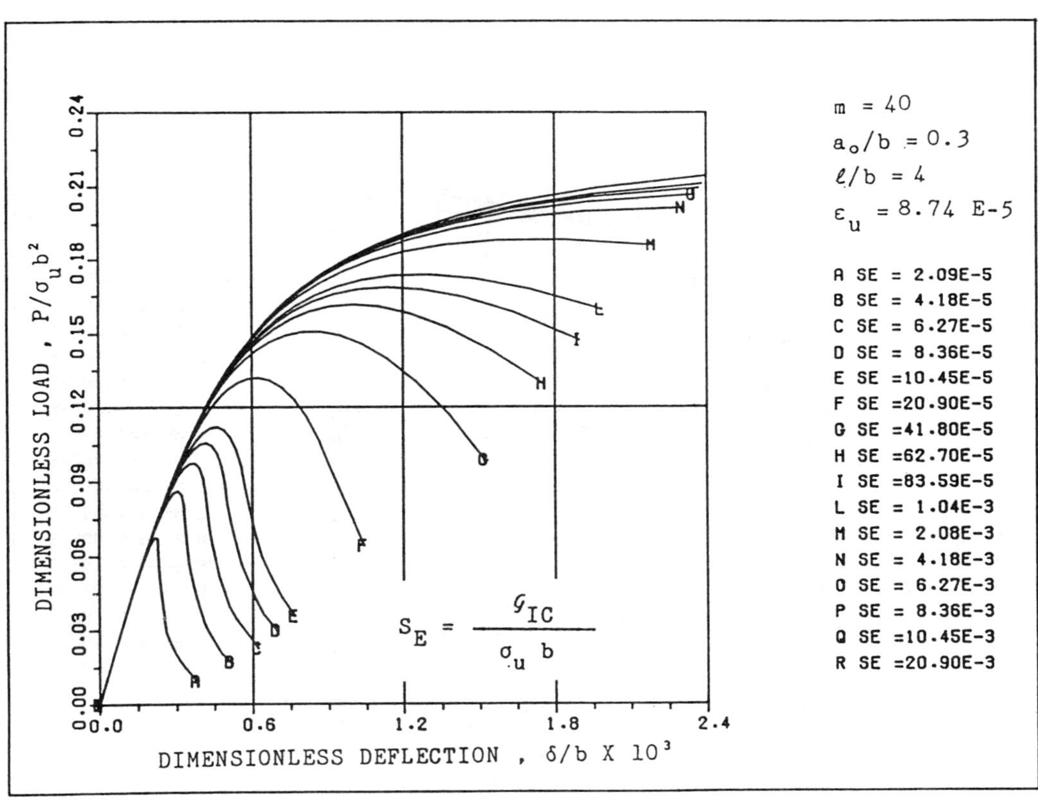

Figure 19b: Dimensionless load-deflection response of an initially cracked specimen ($a_0/b = 0.3$), by varying the brittleness number, $s_E = \mathcal{G}_{IC}/\sigma_u b = w_c/2b$, between 2×10^{-5} and 2×10^{-2}. $m = 40$.

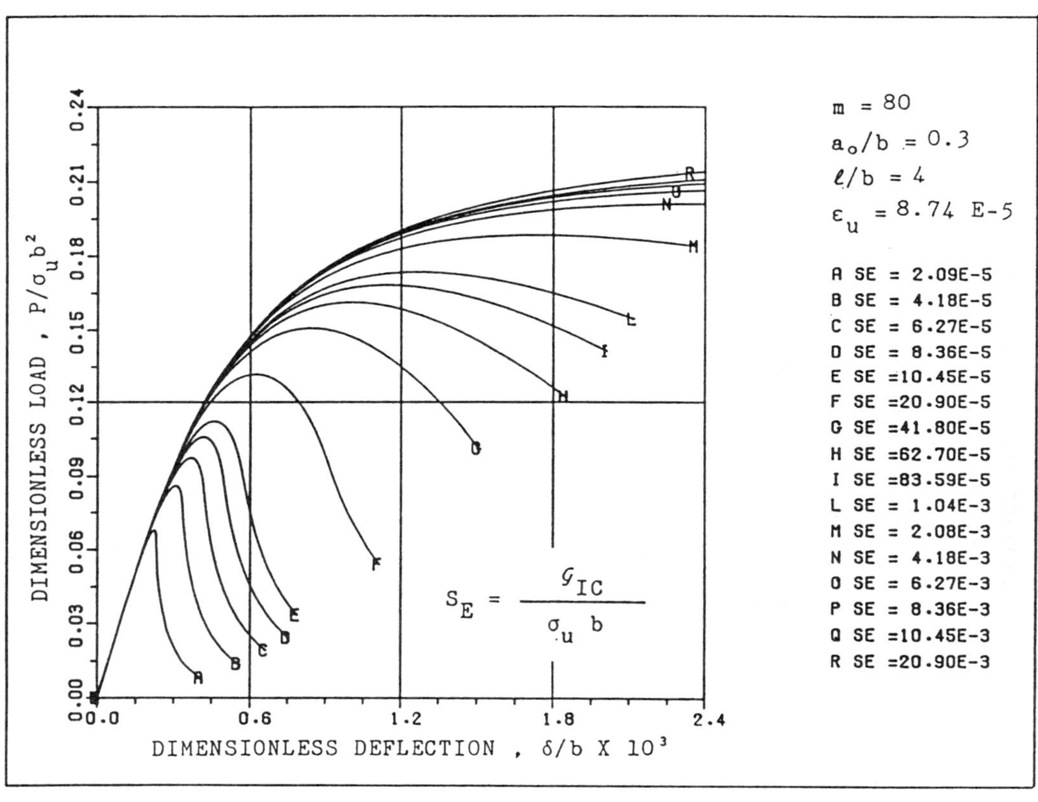

Figure 19c: Dimensionless load-deflection response of an initially cracked specimen ($a_0/b = 0.3$), by varying the brittleness number, $s_{\rm E} = \mathcal{G}_{\rm IC}/\sigma_{\rm u}b = w_{\rm c}/2b$, between 2×10^{-5} and 2×10^{-2}. $m = 80$.

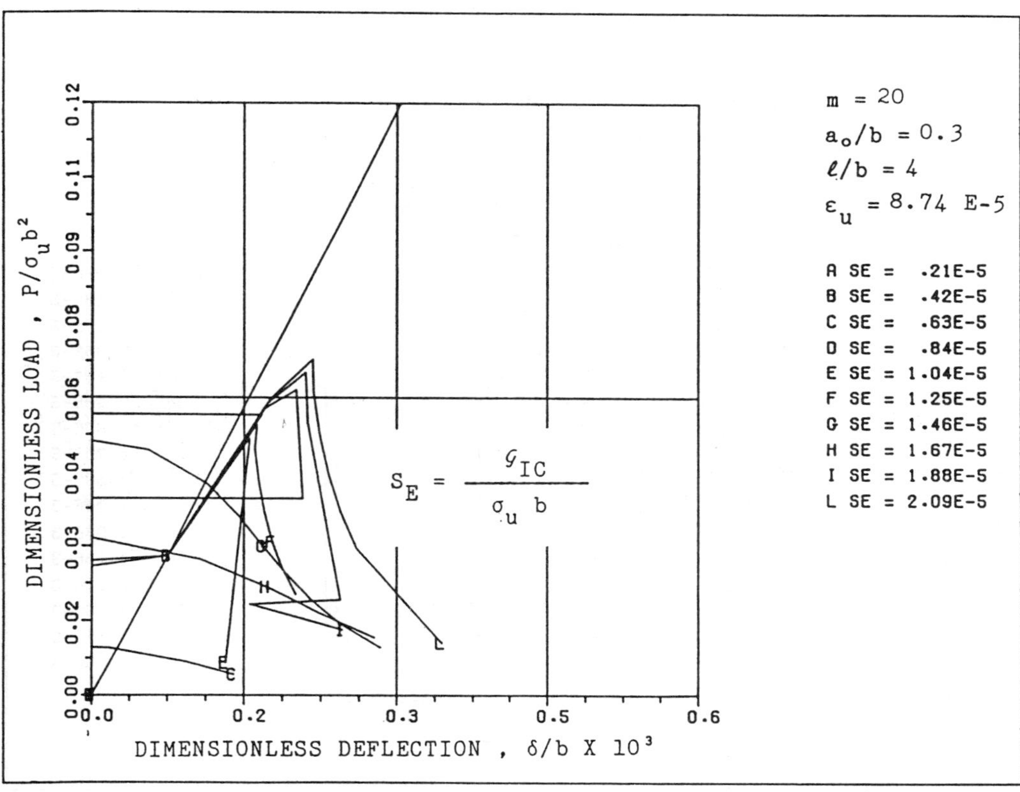

Figure 20a: Dimensionless load-deflection response of an initially cracked specimen ($a_0/b = 0.3$), by varying the brittleness number, $s_E = \mathcal{G}_{IC}/\sigma_u b = w_c/2b$, between 2×10^{-6} and 2×10^{-5}. m= 20.

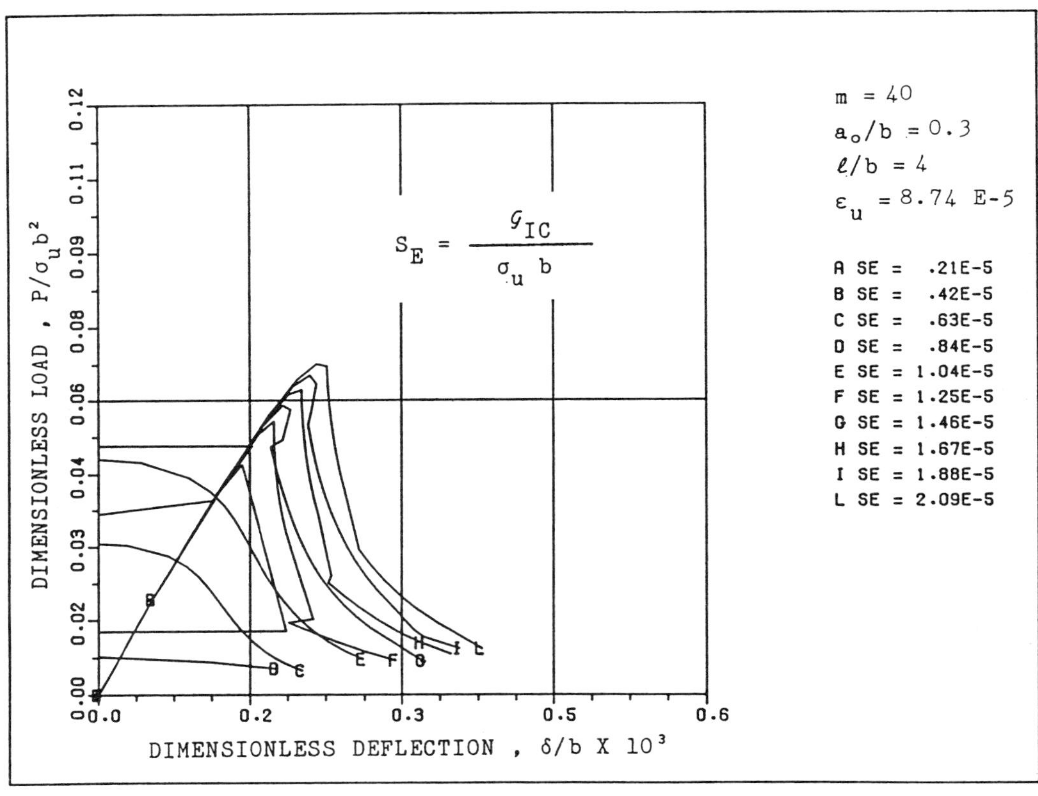

Figure 20b: Dimensionless load-deflection response of an initially cracked specimen (a_0/b = 0.3), by varying the brittleness number, $s_E = \mathcal{G}_{IC}/\sigma_u b = w_c/2b$, between 2×10^{-6} and 2×10^{-5}. m= 40.

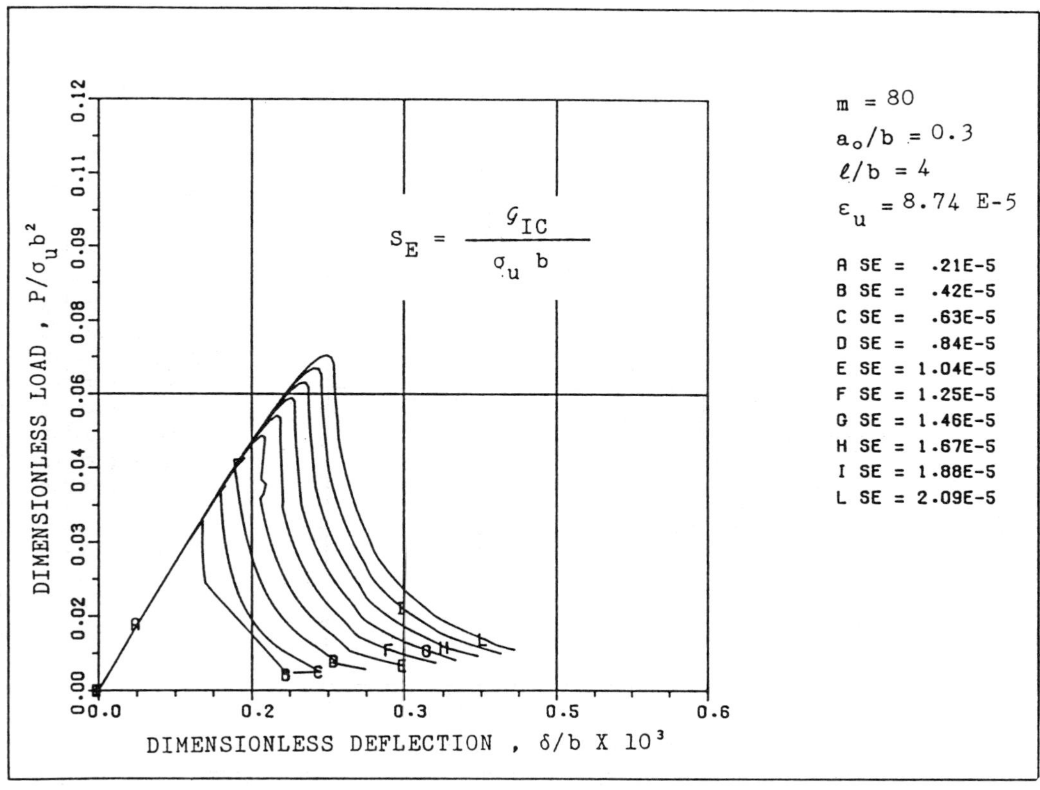

Figure 20c: Dimensionless load-deflection response of an initially cracked specimen ($a_0/b = 0.3$), by varying the brittleness number, $s_E = \mathcal{G}_{IC}/\sigma_u b = w_c/2b$, between 2×10^{-6} and 2×10^{-5}. $m = 80$.

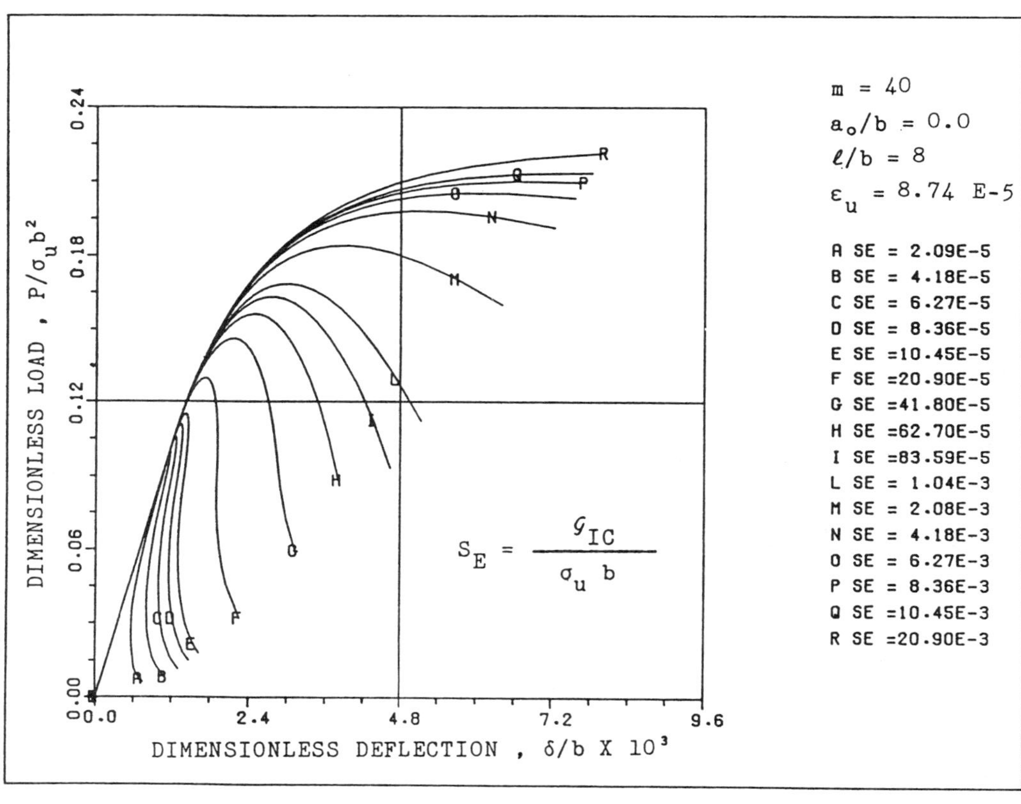

Figure 21a: Dimensionless load-deflection response of an initially uncracked specimen, by varying the brittleness number, $s_E = \mathcal{G}_{IC}/\sigma_u b = w_c/2b$, between 2×10^{-5} and 2×10^{-2}. $\ell/b = 8$.

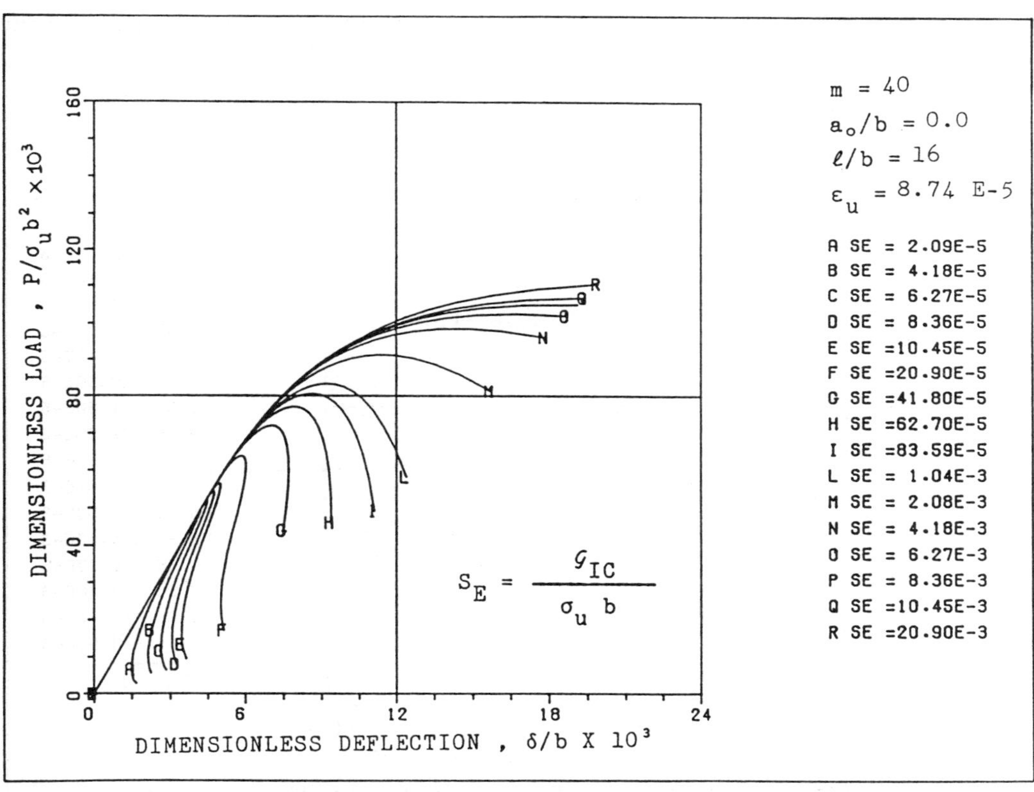

Figure 21b: Dimensionless load-deflection response of an initially uncracked specimen, by varying the brittleness number, $s_E = \mathcal{G}_{IC}/\sigma_u b = w_c/2b$, between 2×10^{-5} and 2×10^{-2}. $\ell/b = 16$.

2.2 Size-scale effects: decrease in apparent strength and increase in fictitious fracture toughness

The maximum loading capacity $P^{(1)}_{\mathbf{max}}$ of initially uncracked specimens with $\ell = 4b$ is obtained from Figs 17 and 18. On the other hand, the maximum load $P^{(3)}_{\mathbf{max}}$ of ultimate strength is given by

$$P^{(3)}_{\mathbf{max}} = \frac{2}{3}\frac{\sigma_{\mathrm{u}} t b^2}{\ell} \tag{57}$$

The values of the ratio $P^{(1)}_{\mathbf{max}}/P^{(3)}_{\mathbf{max}}$ may also be regarded as the ratio of the apparent tensile strength σ_{f} (given by the maximum load $P^{(1)}_{\mathbf{max}}$ and applying eqn (57)) to the true tensile strength σ_{u} (considered as a material constant).

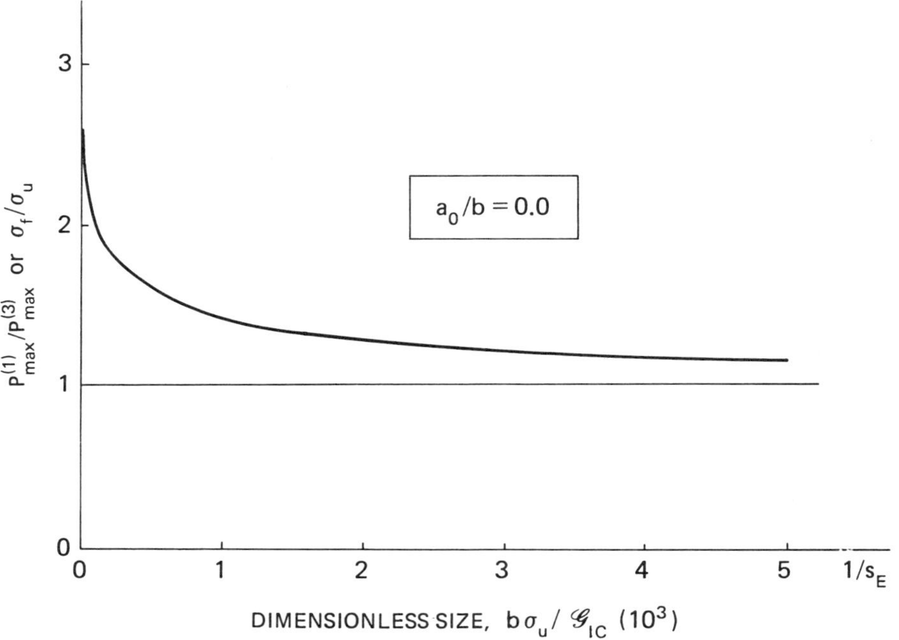

Figure 22: Decrease in the apparent ultimate tensile strength σ_{f} by increasing the specimen size.

It is evident from Fig. 22 that the results of the cohesive crack model tend to those of the ultimate strength analysis for low s_{E} values

$$\lim_{s_{\mathrm{E}}\to 0} P^{(1)}_{\mathbf{max}} = P^{(3)}_{\mathbf{max}} \tag{58}$$

Therefore, only for comparatively large specimen sizes can the tensile strength σ_{u} be obtained as $\sigma_{\mathrm{u}} = \sigma_{\mathrm{f}}$. With the usual laboratory specimens, an apparent strength higher than the true one is always found.

The maximum loading capacity $P^{(1)}_{\mathbf{max}}$ of initially cracked specimens according to the cohesive crack model is obtained from the $P - \delta$ diagrams in Figs 19 and 20.

On the other hand, the maximum loading capacity $P^{(2)}_{max}$ according to LEFM can be derived from the following formula

$$P^{(2)}_{max} = \frac{K_{IC} t b^{3/2}}{\ell f(a_0/b)} \tag{59}$$

with the shape-function f given by eqn (1), and the critical value of stress-intensity factor K_{IC} computed according to the well-known relationship

$$K_{IC} = \sqrt{\mathcal{G}_{IC} E} \tag{60}$$

Eventually, a simple ultimate strength analysis on the center-line with the assumption of a butterfly stress variation through the ligament, provides

$$P^{(3)}_{max} = \frac{2}{3} \frac{\sigma_u t (b - a_0)^2}{\ell} \tag{61}$$

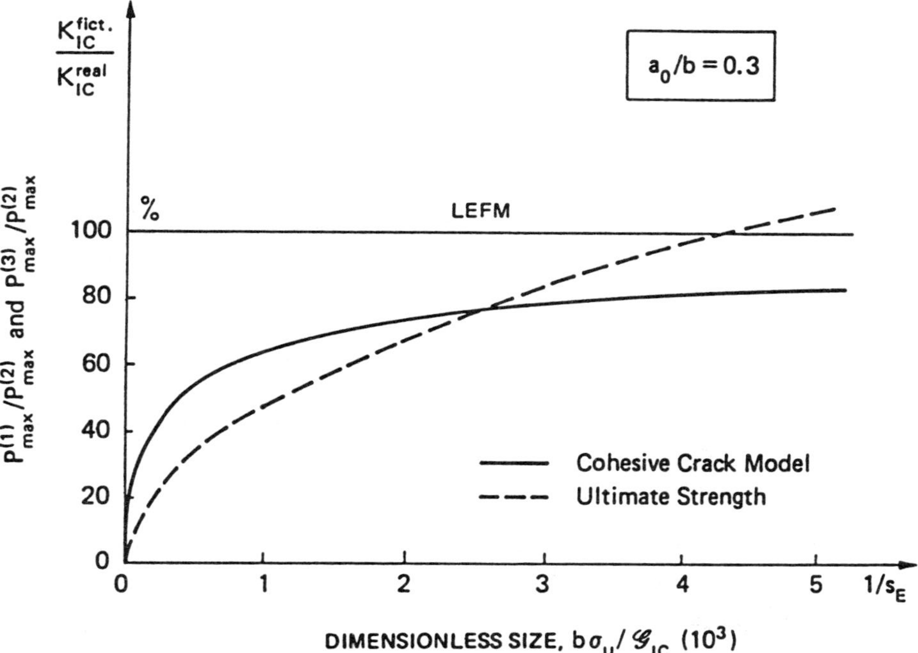

Figure 23: Increase in the fictitious fracture toughness $K^{fict.}_{IC}$ by increasing the specimen size ($a_0/b = 0.3$).

The values of the ratios $P^{(1)}_{max}/P^{(2)}_{max}$ and $P^{(3)}_{max}/P^{(2)}_{max}$ are reported as functions of the inverse of the brittleness number s_E in Fig. 23. The ratio $P^{(1)}_{max}/P^{(2)}_{max}$ may also be regarded as the ratio of the fictitious fracture toughness (given by the non-linear maximum load) to the true fracture toughness (considered as a material constant).

It is evident that, for high s_E numbers, the ultimate strength collapse results as being a more critical condition than that of LEFM ($P^{(3)}_{\max} < P^{(2)}_{\max}$), as well as the results of the cohesive crack model tend to those of LEFM for low s_{E} values

$$\lim_{s_{\mathrm{E}} \to 0} P^{(1)}_{\max} = P^{(2)}_{\max} \tag{62}$$

3 Loss of symmetry and bifurcations - mixed mode

3.1 Crack length control scheme

By neglecting tangential cohesive stresses and assuming, for normal cohesive stresses, a linear softening law, we can write

$$\sigma_c^+ = -\sigma_c^- = \sigma_c = \sigma_u(1 - w/w_c), \quad \text{for } 0 < w < w_c \text{ and } \dot{w} \geq 0 \tag{63a}$$

$$\sigma_c^+ = -\sigma_c^- = \sigma_c = 0, \qquad \text{for } w \geq w_c \tag{63b}$$

where the superscript + denotes the positive side of the crack (Fig. 24) and the superscript − the negative side, the dot denotes the derivation with respect to time (evolutionary problem).

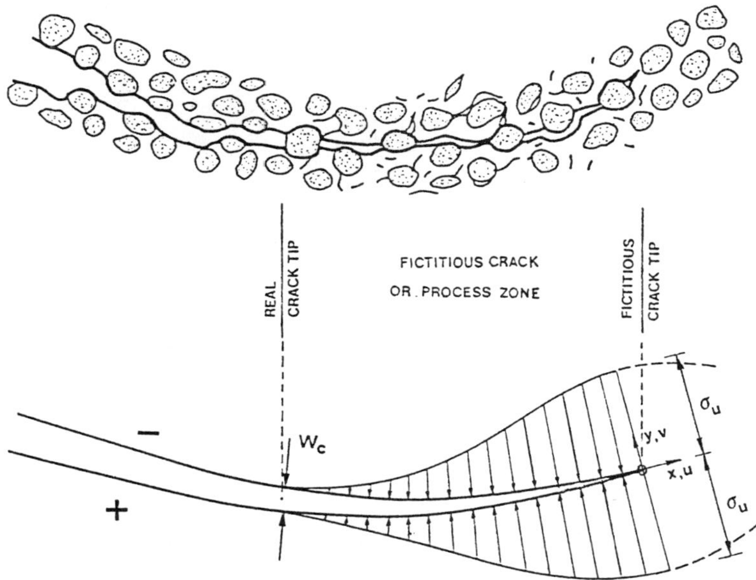

Figure 24: The cohesive model represents the process zone in the form of a fictitious crack.

The crack irreversibility condition $\dot{w} \geq 0$ entails that, if the crack tends to close, rigid unloading takes place which preserves the opening displacement ($\dot{w} = 0$), whilst eqn (63a) is no longer applicable.

In order to prevent new cracks from forming, at each point in the domain, the following condition must be fulfilled

$$\sigma_{\text{pt}} < \sigma_u \tag{64}$$

where σ_{pt} stands for the principal tensile stress.

By subdividing the domain into a finite number of elements and employing the Principle of Virtual Work, the following system of linear equations can be obtained[39–46]

$$(\mathbf{K} - \mathbf{C})\mathbf{u} = \mathbf{f}_1 + \lambda \mathbf{f}_2 \tag{65}$$

where

$$\begin{aligned}
\mathbf{K} &= \text{stiffness matrix } (n \times n) \\
\mathbf{C} &= \text{symmetrical strain-softening matrix} \\
\mathbf{K} - \mathbf{C} &= \text{effective stiffness matrix } (n \times n) \\
\mathbf{u} &= \text{vector of the } n \text{ unknown nodal displacements} \\
\mathbf{f}_1 &= \text{load vector depending on the crack length} \\
\mathbf{f}_2 &= \text{external load vector} \\
\lambda &= \text{external load multiplier}
\end{aligned}$$

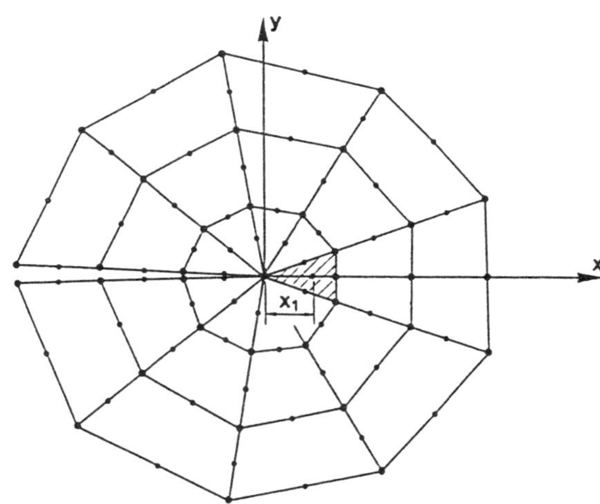

Figure 25: Finite element rosette at the fictitious crack tip[46]

The $(n+1)$th unknown λ is determined by setting that at the fictitious crack tip (the centroid of the dashed element in Fig. 25) the principal tensile stress reaches the value σ_u

$$\sigma_u^2 - (\sigma_x + \sigma_y)\sigma_u + \sigma_x\sigma_y - \tau^2 = 0 \tag{66}$$

Since σ_x, σ_y, τ can be expressed as functions of $\mathbf{u}$, eqns (65) and (66) make up a system of $(n+1)$ equations with $(n+1)$ unknowns, which must be solved at each step in the growth of the crack. Before the solution obtained can be regarded as acceptable, it is necessary to calculate the values of w in the process zone. If at any point $w > w_c$, then, based on the constitutive law (63b), the cohesive stresses vanish and the process zone shrinks. Similarly, if at any point $\dot{w} < 0$ then, based on the constitutive law (63a), the process zone is modified locally so as to enable rigid unloading ($\dot{w} = 0$) to take place.

If a modification occurs in the process zone, then it becomes necessary to reassemble the matrix (**K** − **C**) and to repeat the procedure starting from eqns (65). The iteration chain ends when the solution obtained does not bring about any variation in the process zone.

At this point it is possible to have the fictitious crack propagate, by a predetermined length, perpendicularly to the direction of the principal tensile stress, that is to say, in the direction given by

$$\vartheta = \arctan[2\tau/(\sigma_x - \sigma_y)]/2 \tag{67}$$

In order to follow the new position of the fictitious crack tip, the finite element rosette (Fig. 25) translates and rotates and a new mesh is automatically generated by the computer. Therefore, the solution process for the subsequent crack growth step can start. In other words, the numerical simulations are controlled by means of the length of the fictitious crack, whilst all the other parameters (**u**, λ, process zone) follow according to eqns (63) to (67). It should be noted that when $\sigma_x = \sigma_y$, Mohr's circle has degenerated to a point; hence it is not possible to determine from eqn (67) the subsequent direction of growth of the crack. This technique has been called 'Fictitious Crack Length Control Scheme'.[39–46]

3.2 Solution of the single crack growth step

During the evolution of the crack, the linear system (65) may go through different stages. Initially, the process zone is small and therefore matrix **C** (negative definite) is small compared with matrix **K** (positive definite). As a result, matrix (**K** − **C**) is positive definite and all its eigenvalues α_i are real and positive ($0 < \alpha_1 < \alpha_2 ... < \alpha_n$).

This initial behaviour is common to all problems, i.e. it does not depend on the geometrical and mechanical characteristics of the structure. As the crack grows, two different situations may occur

$$w_{max} = w_c \qquad \text{for } \alpha_1 > 0 \tag{68a}$$

$$w_{max} < w_c \qquad \text{for } \alpha_1 = 0 \text{ and } \alpha_2 > 0 \tag{68b}$$

where w_{max} is the maximum value of w in the process zone.

By way of exemplification, numerical simulations concerning the three-point bending test, performed by taking into account a wide range of geometrical and mechanical ratios, have shown that in this case situation (68a) invariably occurs. It involves a reduction in the process zone, with the ensuing increase of α_1, so that α_1 never becomes negative. Thus it is possible to simulate the test through the end, until structural collapse, by always using (**K** − **C**) positive definite matrices. System (65) always turns out to be well conditioned, even in the event of a catastrophic collapse (or snap-back) taking place.

If we now take into consideration Mixed Mode problems, for the geometrical and mechanical ratios analyzed in Refs 39-46, we find that situation (68b) invariably occurs.

Between situations (68a) and (68b) there is a significant difference: once the former has been reached, it tends to persist during the following steps. The latter instead represents a critical point which can be easily overcome in the course of the numerical simulation. Considering the importance of this situation, which may represent the start of bifurcated paths, it is advisable, as soon as α_1 becomes negative, to set back the fictitious crack tip, in order to approximate the condition $\alpha_1 = 0$ as closely as possible. Once this point has been analyzed, going on with the growth of the crack, we get $\alpha_1 < 0$. Matrix $(\mathbf{K} - \mathbf{C})$ is now invertible again, although it is no longer positive definite. Thus, it can be concluded that during crack growth the matrix $(\mathbf{K} - \mathbf{C})$ may, in some cases, become singular.

If the $(\mathbf{K} - \mathbf{C})$ matrix is not singular, it is possible to solve the system (65) for two right-hand sides

$$(\mathbf{K} - \mathbf{C})\mathbf{u}_1 = \mathbf{f}_1 \tag{69a}$$

$$(\mathbf{K} - \mathbf{C})\mathbf{u}_2 = \mathbf{f}_2 \tag{69b}$$

By denoting with σ_1 the stress at the fictitious crack tip (centroid of the element dashed in Fig. 25), corresponding to displacements $\mathbf{u}_1$, and with σ_2 the stress corresponding to displacements $\mathbf{u}_2$, we can write

$$\sigma = [\sigma_x, \sigma_y, \tau]^T = \sigma_1 + \lambda\sigma_2 \tag{70}$$

By substituting eqn (70) into eqn (66) we get a second order equation in λ which, if it has any solution, makes it possible to determine λ. Thus, the solution, in terms of displacements, can be written

$$\mathbf{u} = \mathbf{u}_1 + \lambda\mathbf{u}_2 \tag{71}$$

If the $(\mathbf{K} - \mathbf{C})$ matrix is singular and we assume that the eigenvalues α_i of $(\mathbf{K} - \mathbf{C})$ are distinct, the corresponding eigenvectors $\mathbf{q}_i$ are unique (within scalar multipliers). Then we have

$$(\mathbf{K} - \mathbf{C})\mathbf{q}_i = \alpha_i\mathbf{q}_i \tag{72}$$

Since $(\mathbf{K} - \mathbf{C})$ is symmetrical, the matrix $\mathbf{Q}$, defined as follows

$$\mathbf{Q} = [\mathbf{q}_1, \mathbf{q}_2, ..., \mathbf{q}_n] \tag{73}$$

turns out to be orthonormal. Hence it is possible to normalize the eigenvectors so as to obtain[46]

$$\mathbf{q}_i\mathbf{q}_j^T = 0 \quad \text{for } i \neq j, \text{ and } \mathbf{q}_i\mathbf{q}_i^T = 1 \tag{74}$$

Let us now consider new displacements $\mathbf{v}$ related to the nodal displacements by the following linear transformation (*singular value decomposition*)

$$\mathbf{u} = \mathbf{Q}\mathbf{v} \tag{75}$$

where $\mathbf{v}$ is called the *generalized displacements vector.*

By substituting eqn (75) into eqn (65) and pre-multiplying the latter by $\mathbf{Q}^T$ we get

$$\mathbf{Q}^T(\mathbf{K}-\mathbf{C})\mathbf{Q}\mathbf{v} = \mathbf{Q}^T(\mathbf{f}_1 + \lambda\mathbf{f}_2) \tag{76}$$

By pre-multiplying eqn (72) by $\mathbf{q}_j^T$, using eqns (74) and substituting it into eqn (76), we can write

$$\alpha_i v_i = \mathbf{q}_i^T(\mathbf{f}_1 + \lambda\mathbf{f}_2) \quad \text{for } i = 1, 2, \ldots, n \tag{77}$$

Equation (77) represents a system of n independent equations. Since we have assumed that all the eigenvalues are distinct, when $(\mathbf{K} - \mathbf{C})$ becomes singular, only one eigenvalue vanishes, i.e. $\alpha_k = 0$. Writing eqn (77) for $i = k$, noting that the first term vanishes and solving with respect to λ, we get

$$\lambda = -\left(\mathbf{q}_k^T\mathbf{f}_1\right) / \left(\mathbf{q}_k^T\mathbf{f}_2\right) \tag{78}$$

By solving eqn (77) in respect of v_i, we get

$$v_i = \mathbf{q}_i^T(\mathbf{f}_1 + \lambda\mathbf{f}_2)/\alpha_i \quad \text{for } i = 1, 2, \ldots, n; \quad i \neq k \tag{79}$$

The state of stress σ at the fictitious crack tip can be written as a function of the contributions σ_i, pertaining to the eigenvector $\mathbf{q}_i$, in the following form

$$\sigma = [\sigma_x, \sigma_y, \tau]^T = \sum_{i=1}^{n} \sigma_i v_i \tag{80}$$

By substituting eqn (79) into eqn (80) it is possible to make the state of stress σ depend solely on the unknown v_k. In order to impose the fictitious crack propagation conditions it is now sufficient to substitute eqn (80) into eqn (66) so as to obtain a second order equation in v_k which, if it has any real solution, makes it possible to determine v_k.

By substituting the generalized displacements $\mathbf{v}$ into eqn (75), we obtain the solution expressed in terms of nodal displacements $\mathbf{u}$.

If $\mathbf{q}_k^T\mathbf{f}_2 = 0$, it is necessary to check $\mathbf{q}_k^T\mathbf{f}_1$. If the latter is not zero, it becomes necessary to move back the fictitious crack tip and then make it grow at a slower rate. An infinite value of λ, in fact, is not compatible with eqn (64). If, instead, we find

$$\mathbf{q}_k^T\mathbf{f}_1 = \mathbf{q}_k^T\mathbf{f}_2 = 0 \tag{81}$$

it is useful to re-write eqn (77) for $i = k$

$$0 v_k = 0 \tag{82}$$

Regardless of the v_k value selected, by substituting v_i, given by eqns (79), and v_k into eqn (80), and then eqn (80) into eqn (66), we obtain a second order equation in λ which, if it has any real solution, makes it possible to determine λ.

Equation (82) represents a bifurcation point, as two solutions are possible, i.e. $v_k = 0$ and $v_k \neq 0$.

In order to reduce the computational effort in the calculation of the eigenvalues and eigenvectors, it is useful to reduce the order of the linear system by means of static condensation. This problem will be taken up again below, by providing a numerical example.

The case of multiple eigenvalues is not discussed in this paper.

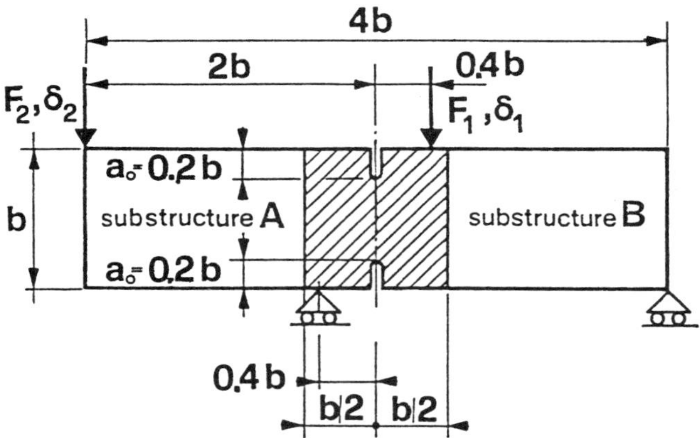

Figure 26: Layout of the four-point shear test[46]

As an example, let us consider a numerical simulation of the four-point shear test on a specimen with double notch (Figs 26 and 27). This is a structure with two axes of symmetry, in which the external loads and the support reactions make up a system of balanced external forces having a polar symmetry around the center of the specimen.

According to the polar symmetry, the application points of loads F_1 and F_2 undergo the same displacement $\delta_1 = \delta_2$ (Fig. 26). Numerical and experimental results, relating to different values of the geometrical and mechanical ratios of the specimen, are given in Refs 39-46.

In the example described below, the following geometrical ratios have been considered (Fig. 26)

$$\ell/b = 4, \quad c/b = 0.8, \quad a_0/b = 0.2, \quad t/b = 1 \tag{83}$$

where t stands for the thickness of the specimen, while the following dimensionless mechanical parameters have been taken into account

$$s_{\mathrm{E}} = \mathcal{G}_{\mathrm{IC}}/\sigma_{\mathrm{u}} b = 0.00025, \quad \nu = 0.1, \quad \epsilon_{\mathrm{u}} = \sigma_{\mathrm{u}}/E = 0.741 \times 10^{-4}$$

In the numerical simulation of this test, performed as described in the previous paragraphs, it is possible to identify the following stages.

(a) Initial stage (positive definite K–C)

This first stage involves the propagation of two cracks growing symmetrically from the two notches. Because of the symmetry, one equation only, and namely eqn (66), can impose the propagation condition of both fictitious cracks. The point, which describes the state of the structure, follows the portion O-A of the curves illustrated in Fig. 28.

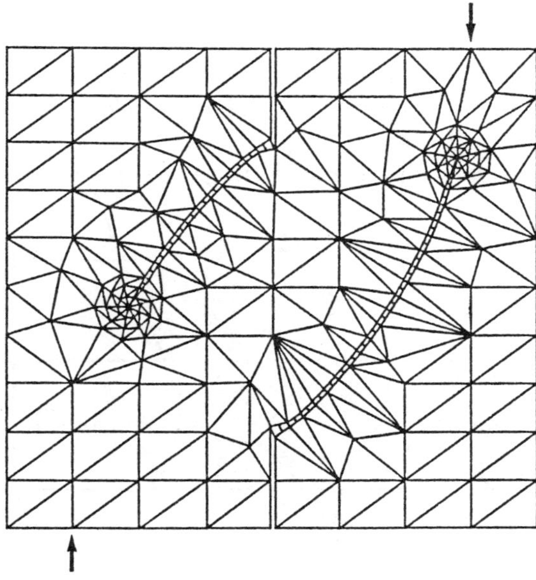

Figure 27: Example of a finite elements mesh [46]

(b) The K−C matrix is singular

In this case ($\alpha_k = 0$), the eigenvector $\mathbf{q}_k$, by neglecting a rigid motion, is seen to have a polar anti-symmetry (Fig. 29). This means that the eigen-displacements, in two polar-symmetrical points, are the same, in terms of both module and sign. Since the external loads, applied on two polar-symmetrical points, share the same module but have opposite signs, we find $\mathbf{q}_k^T\mathbf{f}_2 = 0$. Since this is the first eigenvalue to vanish, the cracking path up to this point is unique and polar-symmetrical, and hence $\mathbf{q}_k^T\mathbf{f}_1 = 0$. Equation (81) being satisfied, we have met a bifurcation point where two solutions are possible, i.e. $v_k = 0$ and $v_k \neq 0$.

(c) Symmetrical propagation

Solution $v_k = 0$ rules out any contribution on the part of eigenvalue $\mathbf{q}_k$ to the displacement field. For all subsequent crack growth steps, the polar-symmetry condition is imposed and therefore the (**K**−**C**) matrix immediately returns non-singular. After a small growth of the crack, the (**K**−**C**) matrix becomes singular again ($\alpha_k = 0$). The product $\mathbf{q}_k^T\mathbf{f}_2$ is other than zero, there are no further bifurcations and the analysis is carried on with (**K**−**C**) no longer positive definite without encountering any additional singularity. The point that describes the state of the structure follows the portion A-B of the curves plotted in Fig. 28.

(d) Non-symmetrical propagation

Without loss of generality, we can normalize the eigenvector $\mathbf{q}_k$ as illustrated in Fig. 29; i.e. the eigen-displacement field shows an opening discontinuity along the lower crack and an overlapping discontinuity, with the same absolute value, along the upper crack.

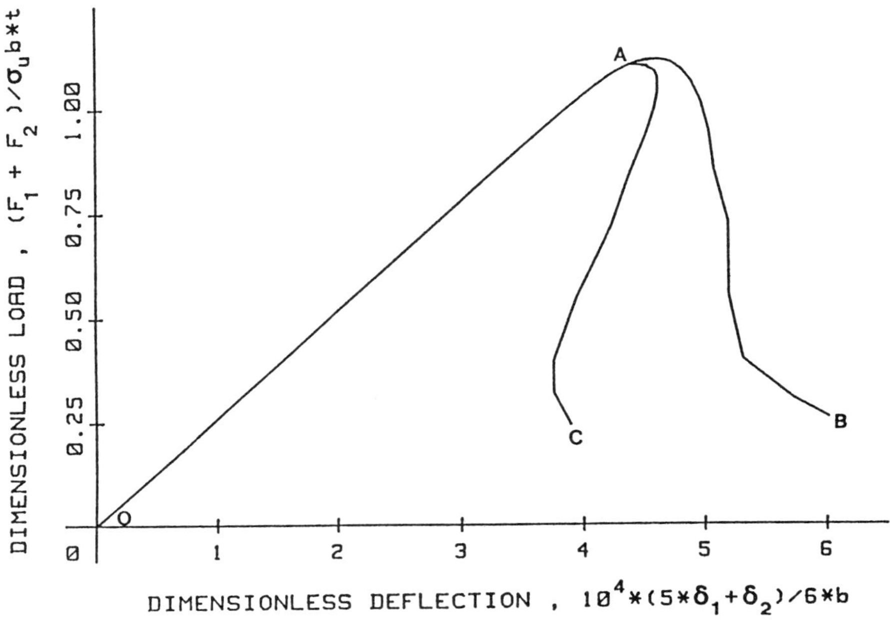

Figure 28: Total load vs. the displacement of its application point[46]

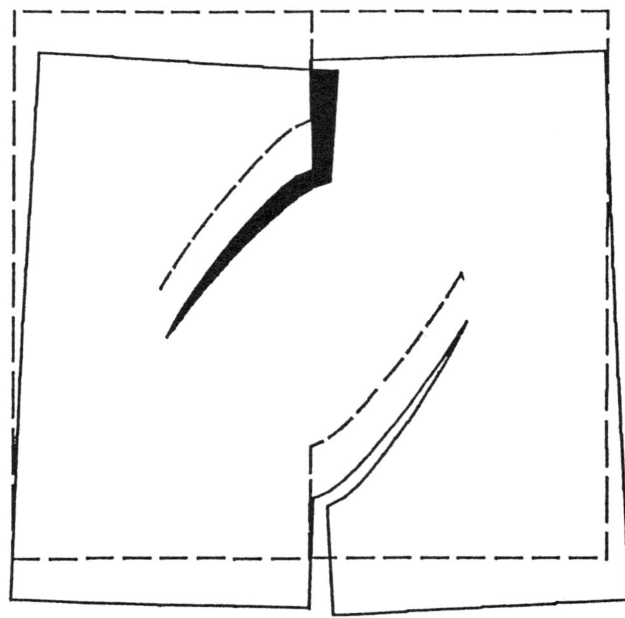

Figure 29: Eigenvector related to the vanishing eigenvalue[46]

In these hypotheses, regardless of the absolute value of $\mathbf{v}_k$, the assumption of $\mathbf{v}_k > 0$ results in $\dot{w} < 0$ along the upper crack. In order to follow the trend of the evolutionary process towards a loss of symmetry, it is therefore necessary to modify the constitutive law along the upper crack ($\dot{w} = 0$). This modification makes the $(\mathbf{K}-\mathbf{C})$ matrix invertible. But then the matrix becomes singular again ($\alpha_k = 0$) following a small growth of the lower crack. The product $\mathbf{q}_k^T\mathbf{f}_2$ is different from zero, no further bifurcation occurs and the analysis is carried on with $(\mathbf{K}-\mathbf{C})$ no longer positive definite without meeting any further singularity. During this stage, the tensile stress at the tip of the upper fictitious crack decreases monotonically. The point describing the state of the structure follows the portion A-C of the curves plotted in Fig. 28. It should be noted that in the symmetrical case δ_2 is always positive (downward); following the loss of symmetry it changes sign. The differences in the cracking trajectories observed in the two cases, on the other hand, are negligible.

As shown in Fig. 28, loss of symmetry in the crack propagation can cause snap-back instability. If we examine the evolution of the total load vs. the displacement of its application point (Fig. 28), we find that the area lying under the O-A-B curve (symmetrical case) is greater than the one under the O-A-C curve (non-symmetrical case).

4 Acknowledgements

The financial support of the Department of Scientific and Technological Research (M.U.R.S.T.) and of the National Research Council (C.N.R.) is gratefully acknowledged.

References

1. Bazant, Z.P. Instability, ductility and size effect in strain-softening concrete, *Journal of the Engineering Mechanics Division*, ASCE, 1976, **102**, 331-344.

2. Carpinteri, A. Interpretation of the Griffith instability as a bifurcation of the global equilibrium, *Proceedings of the NATO Advanced Research Workshop on Application of Fracture Mechanics to Cementitious Composites*, ed. S.P. Shah, pp. 287-316, Evanston, Illinois, September 4-7, 1984. Martinus Nijhoff,1985.

3. Carpinteri, A., Di Tommaso, A. & Fanelli, M. Influence of material parameters and geometry on cohesive crack propagation, in *Fracture Toughness and Fracture Energy of Concrete*, ed. F.H. Wittmann, pp. 117-135, Proceedings of the International Conference on Fracture Mechanics of Concrete, Lausanne, Switzerland, October 1-3, 1985. Elsevier, 1986.

4. Hillerborg, A., Modeer, M. & Petersson, P.E. Analysis of crack formation and crack growth in concrete by means of fracture mechanics and finite elements, *Cement and Concrete Research*, 1976, **6**, 773-782.

5. Wawrzynek, P.A. & Ingraffea, A.R. Interactive finite element analysis of fracture processes: an integrated approach, *Theoretical and Applied Fracture Mechanics*, 1987, **8**, 137-150.

6. Jenq, Y.S. & Shah, S.P. Two parameter fracture model for concrete, *Journal of Engineering Mechanics*, ASCE, 1985, **111**, 1227-1241.

7. Roelfstra, P.E. & Wittmann, F.H. Numerical method to link strain softening with failure of concrete, in *Fracture Toughness and Fracture Energy of Concrete*, ed. F.H. Wittman, pp. 163-175, Proceedings of the International Conference on Fracture Mechanics of Concrete, Lausanne, Switzerland, October 1-3, 1985. Elsevier, 1986.

8. Rots, J.G., Hordijk, D.A. & de Borst, R. Numerical simulation of concrete fracture in direct tension, pp. 457-471, *Proceedings of the Fourth International Conference on Numerical Methods in Fracture Mechanics*, San Antonio, Texas, March 23-27, 1987. Pineridge Press, 1987.

9. Bazant, Z.P. Size effect in blunt fracture: concrete, rock, metal, *Journal of Engineering Mechanics*, ASCE, 1984, **110**, 518-535.

10. Maier, G. On the unstable behaviour of elastic-plastic beams in flexure (in Italian), *Istituto Lombardo, Accademia di Scienze e Lettere, Rendiconti, Classe di Scienze (A)*, 1968, **102**, 648-677.

11. Carpinteri, A. Size effects on the brittleness of concrete structures, (in Italian), pp. 109-123, *A.I.T.E.C. Conference*, Parma,Italy, October 17-18, 1985.

12. Carpinteri, A. Limit analysis for elastic-softening structures: scale and slenderness influence on global brittleness in *Brittle Matrix Composites I*, ed., A.M. Brandt and I.H. Marshall, pp. 497-508, Euromech Colloquium 204, Structure and Crack Propagation in Brittle Matrix Composite Materials, Jablonna, Poland, November 12-15, 1985. Elsevier Applied Science, 1986.

13. Carpinteri, A. & Fanelli, M. Numerical analysis of the catastrophical softening behaviour in brittle structures, pp. 369-386, *Proceedings of the Fourth International Conference on Numerical Methods in Fracture Mechanics*, San Antonio, Texas, March 23-27, 1987. Pineridge Press, 1987.

14. Petersson, P.E. Crack growth and development of fracture zones in plain concrete and similar materials, Report TVBM 1006, Lund Institute of Technology, 1981.

15. Dougill, J.W. On stable progressively fracturing solids, *Zeitschrift für Angewandte Mathematik und Physik (ZAMP)*, 1976, **27**, 423-437.

16. Mazars, J. & Lemaitre, J. Application of continuous damage mechanics to strain and fracture behaviour of concrete, ed. S.P. Shah, pp. 507-520, *Proceedings of the NATO Advanced Research Workshop on Application of Fracture Mechanics to Cementitious Composites*, Evanston, Illinois, September 4-7, 1984. Martinus Nijhoff, 1985.

17. Carpinteri, A. Notch sensitivity in fracture testing of aggregative materials, Istituto di Scienza delle Costruzioni, Università di Bologna, Nota Tecnica n. 45, January 1980; *Engineering Fracture Mechanics*, 1982, **16**, pp. 467-481.

18. Carpinteri, A. Size effect in fracture toughness testing: a dimensional analysis approach, ed. G.C. Sih & M. Mirabile, pp. 785-797, *Proceedings of the International Conference on Analytical and Experimental Fracture Mechanics*, Rome, Italy, June 23-27, 1980. Sijthoff & Noordhoff, 1981.

19. Carpinteri, A. Static and energetic fracture parameters for rocks and concretes, *Materials & Structures*, RILEM, 1981, **14**, 151-162.

20. Carpinteri, A. Application of fracture mechanics to concrete structures, *Journal of the Structural Division*, ASCE, 1982, **108**, 833-848.

21. Carpinteri, A. & Sih, G.C. Damage accumulation and crack growth in bilinear materials with softening, *Theoretical and Applied Fracture Mechanics*, 1984, **1**, 145-159.

22. Carpinteri, A. Stability of fracturing process in R.C. beams, *Journal of Structural Engineering*, ASCE, 1984, **110**, 544-558.

23. Carpinteri, A., Colombo, G. & Giuseppetti, G. Accuracy of the numerical description of cohesive crack propagation, in *Fracture Toughness and Fracture Energy of Concrete*, ed. F.H. Wittman, pp. 189-195. Proceedings of the International Conference on Fracture Mechanics of Concrete, Lausanne, Switzerland, October 1-3, 1985. Elsevier,1986.

24. Carpinteri, A., Di Tommaso, A., Ferrara, G. & Melchiorri, G. Experimental evaluation of concrete fracture energy through a new identification method, in *Fracture Toughness and Fracture Energy of Concrete*, ed. F.H. Wittmann, pp. 423-436, Proceedings of the International Conference on Fracture Mechanics of Concrete, Lausanne, Switzerland, October 1-3, 1985. Elsevier, 1986.

25. Carpinteri, A. & Valente, S. Numerical modelling of mixed mode cohesive crack propagation, ed. S.N. Atluri & G. Yagawa, pp. 12-VI, *Proceedings of the International Conference on Computational Engineering Science*, Atlanta, Georgia, April 10-14, 1988, Springer-Verlag, 1988.

26. Carpinteri, A. Catastrophical softening behaviour and hyperstrength in low reinforced concrete beams, *25th CEB Plenary Session*, Treviso, Italy, May 11-13, 1987.

27. Carpinteri, A., Colombo, G., Ferrara, G. & Giuseppetti, G. Numerical simulation of concrete fracture through a bilinear softening stress-crack opening displacement law, ed. S.P. Shah & S.E. Swartz, pp. 178-191, *Proceedings of the SEM-RILEM International Conference on Fracture of Concrete and Rock*, Houston, Texas, June 17-19, 1987. Springer Verlag, 1989.

28. Barenblatt, G.I. The formation of equilibrium cracks during brittle fracture. General ideas and hypotheses. Axially-symmetric cracks, *J. Appl. Math. Mech.*, 1959, **23**, 622-636.

29. Dugdale, D.S. Yielding of steel sheets containing slits, *J. Mech. Phys. Solids*, 1960, **8**, 100-104.

30. Bilby, B.A., Cottrell, A.H. & Swinden, K.H. The spread of plastic yield from a notch, *Proc. R. Soc. A272*, pp. 304-314, 1963.

31. Rice, J.R. A path independent integral and the approximate analysis of strain concentration by notches and cracks, *J. Appl. Mech.*, 1968, **35**, 379-386.

32. Wnuk, M.P. Quasi-static extension of a tensile crack contained in a viscoelastic-plastic solid, *Journal of Applied Mechanics*, 1974, **41**, 234-242.

33. Schreyer, H. & Chen, Z. One-dimensional softening with localization, *Journal of Applied Mechanics*, 1986, **53**, 791-797.

34. Fairhurst, C., Hudson, J.A. & Brown, E.T. Optimizing the control of rock failure in servo-controlled laboratory tests, *Rock Mechanics*, 1971, **3**, 217-224.

35. Rokugo, K., Ohno, S. & Koyanagi, W. Automatical measuring system of load-displacement curves including post-failure region of concrete specimens, in *Fracture Toughness and Fracture Energy of Concrete*, ed. F.H. Wittmann, pp. 403-411, Proceedings of the International Conference on Fracture Mechanics of Concrete, Lausanne, Switzerland, October 1-3, 1985. Elsevier, 1986.

36. Biolzi, L., Cangiano, S., Tognon, G.P. & Carpinteri, A. Snap-back softening instability in high strength concrete beams, *Materials & Structures*, RILEM, 1989, **22**, 429-436.

37. Standard Method of Test for Plane Strain Fracture Toughness of Metallic Materials, E 399 - 74, ASTM.

38. Tada, H., Paris, P. & Irwin, G. *The stress analysis of cracks handbook*, Del Research Corporation, St. Louis, Missouri, pp. 2.16-17, 1963.

39. Carpinteri, A. & Valente, S. Size-scale transition from ductile to brittle failure: a dimensional analysis approach, in *Cracking and Damage*, ed. J. Mazars & Z.P. Bazant, Elsevier Applied Science, pp. 477-490, 1988.

40. Carpinteri, A. & Valente, S. Mixed mode crack propagation and ductile-brittle transition by varying structural size, *Atti del IX Congresso Nazionale AIMETA*, Bari, Italy, pp. 107-110, 1988.

41. Carpinteri, A., Valente S. & Bocca, P. Mixed mode cohesive crack propagation, *Proceedings of the Seventh International Conference on Fracure (ICF7)*, pp. 2243-2257, Houston, Texas, March 20-24, 1989, Pergamon Press, 1989.

42. Carpinteri, A. & Valente, S. Design and analysis of orthotropic composite materials through a mixed mode cohesive crack simulation, *Proceedings of 3rd Europ. Conf. on Composite Materials*, pp. 309-314. Elsevier Applied Science, 1989.

43. Bocca, P., Carpinteri, A. & Valente, S. Evaluation of concrete fracture energy through a pull-out testing procedure, *Proceedings Int. Conf. on Recent Developments on the Fracture of Concrete and Rock*, Cardiff, UK, pp. 347-356, 1989.

44. Bocca, P., Carpinteri, A. & Valente, S. Size effects in the mixed mode crack propagation: softening and snap-back analysis, *Engineering Fracture Mechanics*, 1990, **35**, 159-170.

45. Bocca, P., Carpinteri, A. & Valente, S. Mixed mode fracture of concrete, *International Journal of Solids and Structures*, 1991, **27**, 1139-1153.

46. Valente, S. Bifurcation phenomena in cohesive crack propagation, *Computer & Structures*, 1992, **44**, 55-62.